MuPAD User's Manual

MuPAD User's Manual

Multi-Processing Algebra Data Tool

MuPAD Version 1.2.2

The *MuPAD* Group

B. Fuchssteiner   K. Drescher   A. Kemper   O. Kluge   K. Morisse
H. Naundorf   G. Oevel   F. Postel   T. Schulze   G. Siek
A. Sorgatz   W. Wiwianka   P. Zimmermann

Translated by Julia McIntosh-Schneider

Springer Fachmedien Wiesbaden GmbH

MuPAD Copyright © Springer Fachmedien Wiesbaden
Originally published by Wiley-Teubner in 1996
Published in 1996 by John Wiley & Sons Ltd and B.G. Teubner

John Wiley & Sons Ltd.
Baffins Lane, Chichester
Chichester
West Sussex PO19 1UD
England

B.G. Teubner
Industriestraße 15
70565 Stuttgart (Vaihingen)
Postfach 80 10 69
70510 Stuttgart
Germany

National Chichester (01243) 779777
International +44 1243 779777

National Stuttgart (0711) 789 010
International +49 711 789 010

Designations used by companies to distinguish their products are often claimed
as trademarks. In all instances where John Wiley & Sons Ltd and B.G. Teubner
are aware of a claim, the product names appear in initial capital or all capital letters.
Readers, however, should contact the appropriate companies for more complete
information regarding trademarks and registration.

AXIOM is a trademark of the NAG, Ltd; HyTEX, © N. Köckler, University of Paderborn, Germany;
Maple is a trademark of Waterloo Maple Software; Macintosh is a trademark of Apple Computer, Inc.;
Macsyma is a trademark of Macsyma, Inc.; Mathematica is a trademark of Wolfram Research, Inc.;
Netpbm, © by J. Poskanzer; PARI, © by C. Batut, D. Bernardi, H. Cohen and M. Olivier; PostScript
is a trademark of Adobe Systems, Inc.; TEX is a trademark of the American Mathematical Society;
UNIX is a trademark of AT&T; X-Window is a trademark of MIT

Other Wiley Editorial Offices

John Wiley & Sons, Inc., 605 Third Avenue,
New York, NY 10158-0012, USA

Brisbane · Toronto · Singapore

Other Teubner Editorial Offices

B.G. Teubner, Verlagsgesellschaft mbH, Johannisgasse 16,
D-04103 Leipzig, Germany

The authors accept no responsibility for this manual or the programs described in it. The authors
of this manual and the developers of MuPAD assume no responsibility for direct or indirect damages
for losses, costs, claims for loss of profits or charges and expenses of any kind.

British Library Cataloguing in Publication Data
A catalogue record for this book is available from the British Library

ISBN 978-3-322-96650-6   ISBN 978-3-322-96649-0 (eBook)
DOI 10.1007/978-3-322-96649-0

Produced from camera-ready copy supplied by the authors

This book is printed on acid-free paper responsibly manufactured from sustainable forestation,
for which at least two trees are planted for each one used for paper production.

Contents

Chapter 1

Introduction

1.1 MuPAD: An Overview

MuPAD is a computer algebra system which, up to now, has been developed mainly at the University of Paderborn. Our aim was to design a program that can handle mathematical problems and computations of a new order of magnitude. Therefore, MuPAD was designed as a parallel system. However, MuPAD also runs on sequential computers, from small size machines to more powerful workstations.

Right from the start we were fascinated by the implications computer algebra systems can have on the way how mathematical formulae and techniques will be used in future applications of mathematics in science and engineering. We believe that the impact of modern technology is most beneficial for the scientific development if systems are created which facilitate the inclusion of the knowledge of the mathematical community. Therefore MuPAD was designed as a service to the community in the hope that the mathematical community reciprocates some of its knowledge and expertise to the system. As a consequence MuPAD is available to non-profit educational and scientific institutions free of charge.

1.1.1 What makes MuPAD different?

MuPAD differs from other computer algebra systems like Maple [4], Mathematica [10], Macsyma [9], Axiom [8] by four major points:

- MuPAD is a system that offers a window-based user-interface (hypertext on-line help, graphics and a source code debugger) on all platforms,

- MuPAD is the first system that provides native parallel instructions to the user,

- MuPAD offers tools for the dynamical linking of binary code objects and, last but not least,

- MuPAD is available free of charge for non-commercial institutions.

On the other hand, MuPAD is rather close to Maple regarding the syntax, but also close to Axiom in its object-oriented capabilities and its hypertext documentation.

1.1.2 History of MuPAD

The development on MuPAD began in 1989. The first official release was version 1.0 in August 1992. Then a new release appeared about every year (version 1.1 in March 1993, version 1.2 in March 1994, version 1.2.1 in September 1994). In November 1994, MuPAD was awarded the European Academic Software Award. This book corresponds to version 1.2.2, which was released in July 1995.

In comparison to other systems, MuPAD's time of development was rather short. One of the reasons for that is MuPAD's modular design and the incorporation of existing modules. As external well-defined subsystems we use the PARI system (from Cohen's group in Bordeaux) and, on some platforms, the HyTeX system (from Köckler's group in Paderborn), both contribute to MuPAD's efficiency. As modules that can be used in other applications we developed the memory management unit (MAMMUT, implemented in 1991) and the graphics tool VCam (1992). Apart from that the programming language, that was designed and implemented first in 1990, is particular to MuPAD.

1.1.3 What does this Book contain?

This book gives a detailed survey of MuPAD and contains the complete reference manual of the standard library. You will find in this book a complete description of the MuPAD programming language (chapter 2), including a description of the parallel statements, a description of the source-code debugger (chapter 3), how to produce graphics with MuPAD (chapter 4), an explanation of the user-interface on UNIX machines and on the Macintosh (chapter 5), and a survey of the functions of the MuPAD standard library (chapter 6). At the end of the book, you will find some interesting tables (appendix A), a description of the changes with respect to version 1.2.1 (appendix F) and the new functionalities of version 1.2.2 (appendix G).

1.1.4   Other Documentation about MuPAD

You will find a lot of information about MuPAD on the WWW server

$$\texttt{http://math-www.uni-paderborn.de/MuPAD/}$$

In particular several demonstration files, a detailed history of the system and a FAQ document for "Frequently Asked Questions".

In addition, the following hypertext documents are available on-line with your copy of MuPAD:

- Axioms, Categories and Domains, by Klaus Drescher.

- Combinat Package, by Karsten Morisse.

- Constructors of the domains Package, by Klaus Drescher.

- Gröbner Package, by Klaus Drescher.

- Linear Algebra Package, by Frank Postel.

- Network Package, by Karsten Morisse.

- Numlib Library, by Friedrich Schwarz and Paul Zimmermann.

- Orthpoly Package, by Karsten Morisse.

- Plot-Library, by Thorsten Schulze.

- Simple Lindenmayer Systems with MuPAD, by Klaus Drescher.

- Statistic Package, by Oliver Kluge and Paul Zimmermann.

- Tutorial and Demonstrations, by Benno Fuchssteiner and al.

- Type Package, by Klaus Drescher.

- Writing Library Packages for MuPAD, by Klaus Drescher and Paul Zimmermann.

You can also use the on-line index and a search mechanism, which enables you to search for a given word in the whole MuPAD manual.

1.1.5 Acknowledgements

The present development of MuPAD only was possible due to the generous support of many institutes. For continued support, we especially want to thank the Deutsche Forschungsgemeinschaft, the Mathematisches Forschungsinstitut Oberwolfach, the Universität-GH Paderborn (especially the university administration), the Minister für Wissenschaft und Forschung of Nordrhein-Westfalen and to INRIA (Institut National de Recherche en Informatique et Automatique). In addition we would like to thank the teams who allowed us to use their software especially the developers of PARI and of the HyTeX system. We could only gain our know-how because other teams have given us so much technical and scientific support. A special mention here for the Maple team.

MuPAD is the development of a large team of colleagues; some working marginally on the project and others working full time over a long period. In addition to the authors of this manual we would like to mention Ralf Hillebrand and Ralf Kraume. Furthermore, we thank Julia McIntosh-Schneider for the translation of this manual.

1.1.6 How to get MuPAD

You can get MuPAD via anonymous `ftp`. The home site of MuPAD is

```
math-ftp.uni-paderborn.de:/pub/MuPAD
```

There you can find additional documentation and binaries for several operating systems. This server is also reachable via WWW. Further ftp sites that mirror MuPAD periodically are

Germany
```
ftp.uni-paderborn.de:/pub/MuPAD
ftp.ask.uni-karlsruhe.de:/pub/education/mathematics/MuPAD
ftp.germany.eu.net:/pub/comp/applications/math/mupad
```
France
```
ftp.inria.fr:/lang/MuPAD
```
Switzerland
```
ftp.switch.ch:/mirror/MuPAD
```
USA
```
archives.math.utk.edu:/software/multi-platform/MuPAD
ftp.math.utah.edu:/pub/mupad
```
MuPAD for Linux is also available via

`sunsite.unc.edu:/pub/Linux/apps/math/MuPAD`

Before you try to get MuPAD, please refer to the mirror list to find the ftp site next to you.

For further information you can send either an electronic-mail to

`MuPAD-distribution@mathematik.uni-paderborn.de`

or a letter to

MuPAD Distribution
Automath (FB 17)
University of Paderborn
Warburger Straße 100
D-33098 Paderborn

Paderborn, October 1995

1.2   Getting started

Get a first impression of what MuPAD can do for you by trying some examples. First you have to start it, obviously enough. The following slightly depends on which operating system you have. On a UNIX box with graphical interface you may enter **xmupad** to start MuPAD with a graphical front-end, if you only have a tty terminal available you must enter **mupad** to start the terminal version. On a Macintosh you can double-click MuPAD's program icon as usual or use the finder's "Open" menu command to start it.

MuPAD is command-oriented, you textually enter commands and the result is printed. After some time, which is needed to initialize the system, a prompt appears and you may enter your first command. The prompt looks like **>>** on a UNIX box or like ● on a Macintosh.

1.2.1   A bit of Calculus

Do you want to obtain the antiderivative of $\frac{(x^2+2)(x-1)}{x^2(x^2+1)}$? Simply enter

```
>> int((x^2+2)*(x-1)/(x^2*(x^2+1)), x);


                                      2
    2                         ln(x  + 1)
  - + 2 ln(x) + atan(x) -  ----------
  x                               2
```

The function call `int(f(x), x)` computes the antiderivative $\int f(x)dx$. The result returned contains the functions `ln` and `atan` which represent the natural logarithm and the arcus tangens.

To start the execution of a command you must either hit the Return key (on a UNIX system) or the Enter key (on a Macintosh). If you hit the Return key on the Macintosh then only a line break is created and you may add further text to the command.

Each command must be followed by a semicolon or colon, they simply belong to a complete command. If you start execution without the command being complete (if you forgot to enter a semicolon for example) then a new prompt will appear which signals that the command is still incomplete. On a UNIX system this prompt looks like **&>**, on a Macintosh it is a *.

If the command is followed by a colon instead of a semicolon, then no output is printed. Let us check if the integral above is valid. First we compute it again and assign it to a new variable **f**

```
>> f:= int((x^2+2)*(x-1)/(x^2*(x^2+1)), x):
```

The result of the call of `int` is assigned to `f`. No output is printed because the command ended with a colon. Now let us compute the derivative $\frac{d}{dx}f$

```
>> diff(f, x);
```

```
  2   2      1         x
  - - -- + ------ - ------
  x   2    2         2
      x   x + 1   x + 1
```

This is quite similar to our original expression. The original factored form may then be obtained by the function `Factor`

```
>> Factor(%);
```

```
              2
  (x - 1) (x  + 2)
  ----------------
     2   2
    x  (x  + 1)
```

Note the percent sign `%`. It recalls the last result returned, which was the result of the differentiation before. Therefore the `%` is also called the "last-operator". It may also be conveniently used to work with the last 20 results. Enter `%2` to get the second-last result for example.

1.2.2   Getting Help

If you do not know how a command is called, a look at its help page will be useful. To get the help page for the command `diff` simply enter the command

```
>> ?diff
```

Ha — forgot the semicolon, you will say. But this is one of the two exceptions of the semicolon-rule. Help commands starting with ? need no semicolon.

If you simply enter ? without a command name you will get a help page about the help command. With the graphical front-end, you can navigate from there to the list of available functions, to the index or to the contents page of the on-line manual. The help system is organized as a hypertext system. Keywords are underlined or boxed, you may click on them to get their description. For each page there are also some "general" links, for example to the next section or the index. (Macintosh users: look for the entries in the Help menu.)

One may paste the examples from the help pages into the MuPAD session by clicking on the underlined or boxed prompts on the help pages.

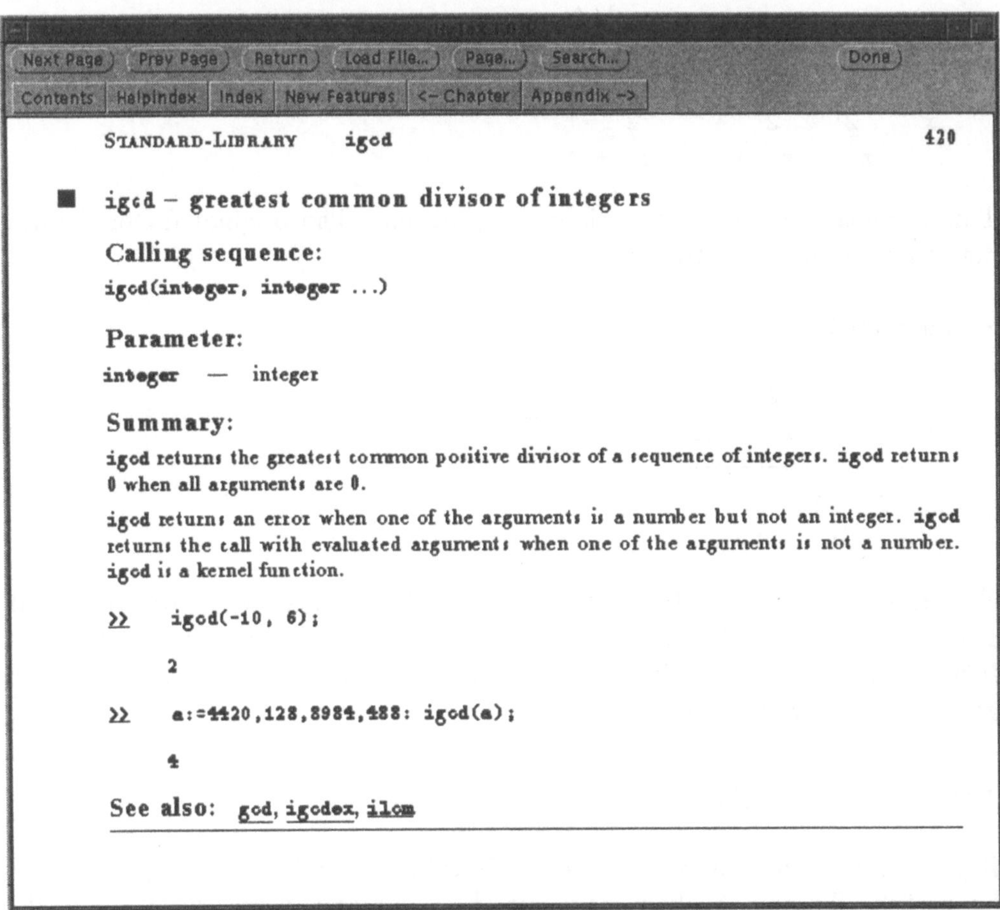

Figure 1.1: The Help Window

1.2.3 Nothing but Numbers

Expressions may not only be computed symbolically, one may also do numerical computations. Suppose we want to solve the equation $p = 0$ where p is the polynomial $x^8 + 24x^7 + 256x^6 + 1584x^5 + 6102x^4 + 14472x^3 + 19216x^2 + 10320x + 1836$. Simply type:

```
>> p:= x^8+24*x^7+256*x^6+1584*x^5+6102*x^4+14472*x^3+
        19216*x^2+10320*x+1836:
```

```
>> S:= solve(p=0, x);
```

$$\begin{array}{l}
[(-(13^{1/2} + 57)^{1/2} - 1)^{1/2} - 3,\; -(-(13^{1/2} + 57)^{1/2} - 1)^{1/2} - 3, \\[2ex]
-((13^{1/2} + 57)^{1/2} - 1)^{1/2} - 3,\; ((13^{1/2} + 57)^{1/2} - 1)^{1/2} - 3, \\[2ex]
-((-13^{1/2} + 57)^{1/2} - 1)^{1/2} - 3,\; ((-13^{1/2} + 57)^{1/2} - 1)^{1/2} - 3, \\[2ex]
(-(-13^{1/2} + 57)^{1/2} - 1)^{1/2} - 3,\; -(-(-13^{1/2} + 57)^{1/2} - 1)^{1/2} - 3]
\end{array}$$

The function `solve` returns a list containing all solutions of the equation —
symbolically of course. To get floating-point numbers for the solutions one has
to apply the function `float` to them. The following command applies `float`
to each solution:

```
>> map(S, float);
```

$$\begin{array}{l}
[-\,3.0 + 2.963942767\;I,\; -\,3.0 - 2.963942767\;I,\; -5.604794949, \\[1ex]
-0.3952050504,\; -5.511404093,\; -0.4885959066, \\[1ex]
-\,3.0 + 2.882212781\;I,\; -\,3.0 - 2.882212781\;I]
\end{array}$$

You need to have a higher precision? No problem:

```
>> DIGITS:= 40:
```

```
>> float(S[1]);
```

$$-\,3.0 + 2.963942767530546680097545463147697823331\;I$$

As you see MuPAD works with arbitrary precision. Integers and rational
numbers may also have as much digits as memory available.

1.2.4 Have a Look

If one has computed a function one may visualize it using MuPAD's graphics capabilities. One may plot a function in one or two variables with a simple command like

```
>> plot3d([Mode=Surface,
      [x, y, 5*besselJ(0, sqrt(x^2 + y^2))],
      x=[-20,20], y=[-20,20]]);
```

This command draws the Bessel function $J_0\left(\sqrt{x^2 + y^2}\right)$. Once the plot appears on the screen it may be manipulated via an intuitive user interface. Every aspect of the plot may be changed interactively. The result can be saved in different formats like PostScript.

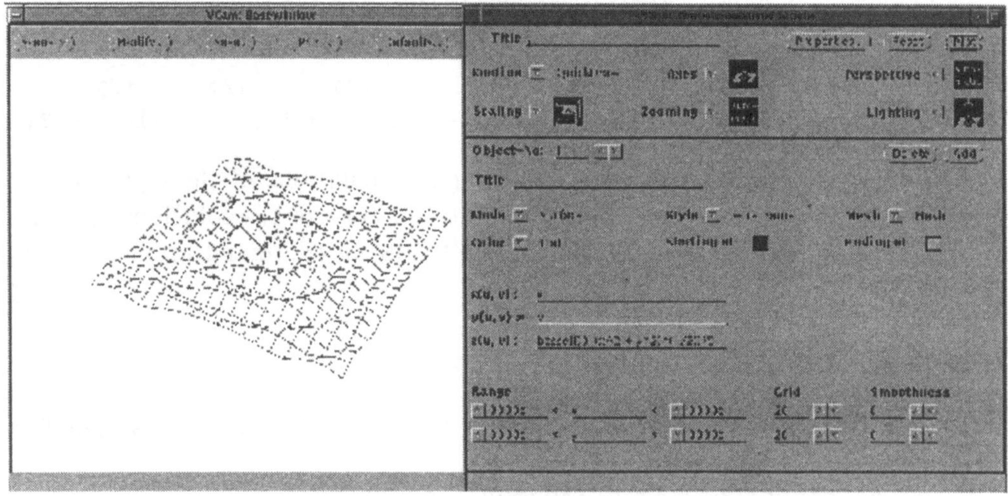

Figure 1.2: The Plot Window

Not only function graphs but also parametrized curves and surfaces and drawing primitives may be plotted.

1.2.5 Programming

One may not only enter simple commands like above but also procedures and control statements. The command language of MuPAD is a fully-fledged programming language. It has a Pascal-like syntax but is specially tailored towards symbolic manipulation.

The following is an implementation of the famous Euclidean algorithm

```
>> euclidean_alg:= proc(a, b)
       local t;
   begin
       if iszero(a) then return(b) end_if;
       if iszero(b) then return(a) end_if;
       a:= multcoeffs(a, 1/lcoeff(a));
       b:= multcoeffs(b, 1/lcoeff(b));
       while not iszero(b) do
           t:= divide(a, b, Rem); a:= b; b:= t
       end_while;
       multcoeffs(a, 1/lcoeff(a));
   end_proc;
```

You will have noticed that no types are given for the parameters a and b and the local variable t. MuPAD has an untyped language, a variable may be of any type.

There are many pre-defined structured types to work with, like expressions, lists, tables and arrays. The system automatically takes care for garbage collection. Procedures are first class data and may be manipulated or created at run-time.

MuPAD consists of a kernel, which is written in C, and a library, which contains most of the MuPAD functions, written in the MuPAD language. The library consists of readable text files which may be examined if one wants to study the algorithms used. Of course you may also implement your own commands in the MuPAD language and add them to your own, private library. Or you may offer them to the MuPAD community to be shared with others.

To aid programming a source-level debugger and several other tools like a profiler and a tool to create test coverages exist. The debugger (see figure 1.3) allows, among other features, to execute your code step by step, to set breakpoints and to view variables during execution.

1.2.6   Abstract Types

Besides the built-in types MuPAD has a notion for abstract types, which are called domains. There are domains to represent numbers, polynomials, fractions or matrices for example. Before you can do computations with matrices you must define the domain of their coefficients. Let us define the matrices over the rational functions and do some computations

```
>> loadlib("domains"):
```

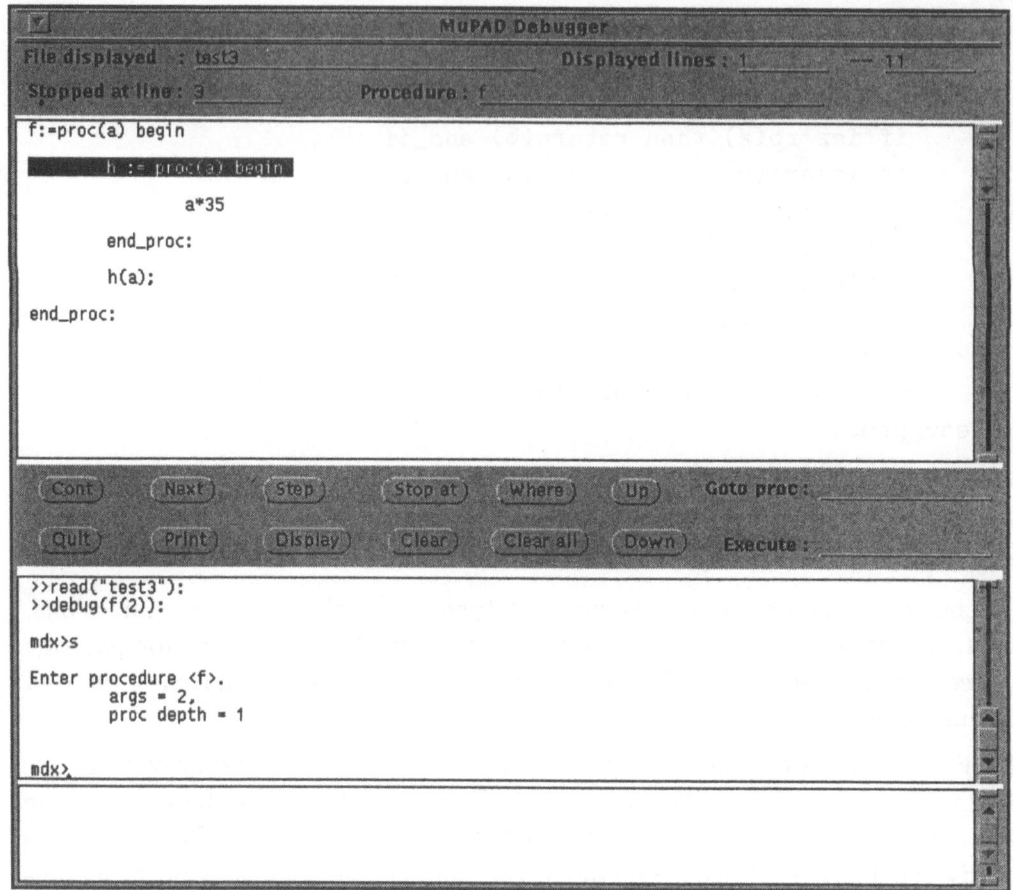

Figure 1.3: The MuPAD Debugger

```
>> MF:= Matrix(Fraction(Polynomial(Rational))):
   A:= MF([[a, 1], [0, 1/b]]);

       +-      -+
       | a, 1 |
       |       |
       |     1 |
       | 0, - |
       |     b |
       +-      -+


>> Id:= MF([[1, 0], [0, 1]]):
   (A - 1*Id)^(-1);
```

```
+-                                          -+
|    1                    b                  |
|  -----, ---------------------             |
|  a - 1                         2 |
|              - a + 1 + a b 1 - b 1  |
|                                            |
|                         b                  |
|    0,          - --------                  |
|                      b 1 - 1               |
+-                                          -+
```

The variable **MF** is the domain of matrices over the rational functions. A is
defined to be the 2×2 matrix $\left(\begin{smallmatrix} a & 1 \\ 0 & 1/b \end{smallmatrix}\right)$, **Id** is defined to be the 2×2 identity
matrix. Then the matrix $(A - \lambda\,\mathrm{Id})^{-1}$ is computed.

The elements of a domain generally may be manipulated with the usual operations like **+**, **-**, ***** and **/**, no special syntax is needed. You may define your
own domains together with their operations, which may then be used like the
built-in types.

You can define another 2×2 matrix B by

```
>> B := MF([[c, a], [a*d, b+c]]);
```

```
+-            -+
|  c ,    a   |
|             |
|  a d, b + c |
+-            -+
```

and now you are able to add or multiply A and B as usual.

```
>> A + B;
```

```
+-                      -+
|  a + c,       a + 1    |
|                        |
|                 2      |
|            b c + b  + 1 |
|  a d , ------------    |
|               b        |
+-                      -+
```

```
>> 3*A*B;
```

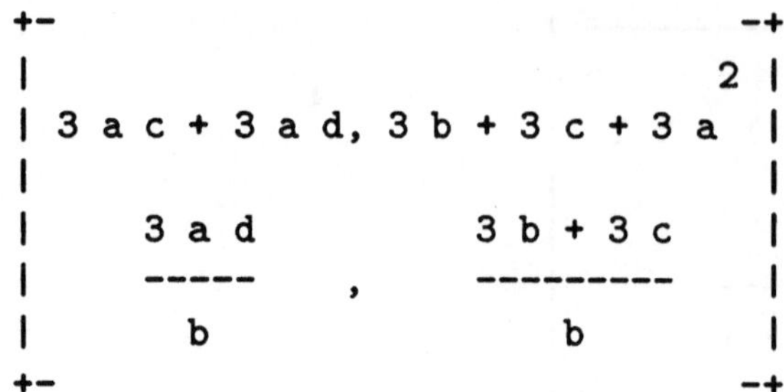

Chapter 2

The Language and its Functionality

2.1 Introduction

To perform the numerous tasks of a computer algebra system, a multitude of simple and complex data types, operators and statements are necessary. Important elementary data types are the various numerical data types like floating point numbers, complex numbers, rational numbers and integers of any length as well as identifiers and character strings. Additionally indexed identifiers are necessary to enable access to tables, arrays, lists and expression sequences. To control, for instance, the execution of statements, it must be possible to evaluate expressions to boolean values. In this context, the boolean constants TRUE and FALSE are important. Further available constants are E, EULER and PI. More complex data types allow objects to be combined to form a new object. These data types include lists, sets, tables and arrays and domains.

For combining different data types a series of operators are implemented. There are set operators, logical and relational operators, mathematical operators for the basic arithmetic operations and exponentiation, the sequence operator, the range operator, the concatenation operator as well as the mathematical operators mod and div.

The statement set includes, among others, loops and structures for the programming of conditional branches. For implementing parallel algorithms relevant programming constructs are available. Procedural programming in Mu-PAD is supported.

Input to the system consists of a statement in the MuPAD programming language or a valid expression and is concluded with a semicolon or colon. The colon prevents the result returned by the evaluator from being displayed on the monitor.

More than one statement can be entered in one line. These must be separated by a colon or semicolon. Furthermore input of more than one line is allowed. The line end must be masked by a backslash (\). The entry is complete when it has been concluded by a colon or semicolon.

The statements entered are then read-in and executed in the order they were entered. A single statement can consist either of a statement in the programming language or of a simple expression.

In the first case the statement is executed by the system and the result is displayed, in the second case the expression is evaluated and the result of the evaluation is returned.

All characters enclosed between two sharp symbols (#) are ignored by the system. In this way programs can be commented on. Comments can be more than one line long.

Executing can be interrupted by <Ctrl-C> on a UNIX system. The meaning of <Ctrl-C> is different in XMuPAD and MacMuPAD. Further information can be found in the description of the user interface.

2.1.1 Interactive Input

To simplify the input of commands to the operating system as well as calls of the on-line documentation on the interactive level, special linguistic constructs are available. These can only be used interactively and always begin with a special symbol followed by the calling sequence.

The input of a question mark followed by an identifier or a MuPAD language keyword is a short cut to call the on-line documentation. This identifier is not evaluated. If information is available about the identifier then this is returned by the information system.

Commands to the operating system begin with an exclamation mark. This is then followed by the command.

Example 1 *The commands*

```
>> ! ls -al
```

or

```
>> ? diff
```

list the contents of the current directory or displays the help page about the system function diff. *Commands to the operating system are only supported for UNIX.*

Additionally the system functions **system** (which is not supported in MacMu-PAD) and **help** are available to call operating system commands or the on-line documentation. These can also be used in more complex expressions and procedures. They expect the relevant operating-system command or the identifier to which the information is needed as an argument. This is given in form of a string. In the example above **system("ls -al")** or **help("diff")** yield the same results as the short cuts.

In order to communicate with the user from a MuPAD program the functions **input** and **textinput** can be used. These functions permit the output of text in the form of strings and expect one or more user entries. These entries are interpreted as expressions or strings according to the corresponding function.

2.1.2   The History Mechanism

MuPAD has an internal *history mechanism*, with which results are stored and can be re-called. Every result calculated by the evaluator is stored and can be addressed with the help of the system function **last**. **last(1)** returns the result calculated last and generally **last(i)** returns the last but i result calculated. The result is not re-evaluated.

A shorter notation is possible with help of the per cent sign. This must be followed by a positive integer. The penultimate result can be accessed with **%2**. Statements also enter values in the history mechanism. A detailed description of the history mechanism can be found in section 2.11.

2.1.3   System Initialization

Two files influence the behavior of the system after it has been started. The names of the files as well as their search paths are dependent on the operating system under which MuPAD is running. The following description applies to the MuPAD implementation under UNIX. On UNIX systems further details can be found on the man-pages concerning MuPAD. For the other MuPAD versions the relevant information can be found in the description of the corresponding interface (see section 5.3).

The file `.mupadsysinit` is read at the beginning of the system initialization. It contains default values valid for all users of the system. In addition the library functions are loaded or initialized with the help of `loadproc` or `loadlib`. The file is to be found in the directory of the MuPAD library.

Each user has the possibility of defining further presettings by using the `.mupadinit` file. This is read after the actual initialization of the system and if no further instructions are given is searched for in the user's home directory. The exact position of all initialization files can be specified by options (see the UNIX man-page) when MuPAD is started.

2.1.4 Language Elements

Valid expressions and statements of the language are built up by identifiers, numerical constants, keywords and special characters.

To create these basic elements MuPAD accepts the following characters:

- the 26 small letters `a` to `z`

- the 26 capital letters `A` to `Z`

- the 10 digits `0, 1, 2, 3, 4, 5, 6, 7, 8, 9`

- the special characters
  `% ; , / _ .^ + - * $ # ( ) { } [ ] < > = :  & " ! ? \ @`

The following words are keywords of the language and therefore cannot be used as identifiers:

and	begin	break	case	div	do
downto	elif	else	end_case	end_for	end_if
end_par	end_proc	end_repeat	end_seq	end_while	FAIL
FALSE	for	from	I	if	in
intersect	local	minus	mod	name	next
NIL	not	of	option	or	otherwise
parallel	parbegin	private	proc	quit	repeat
seqbegin	step	then	to	TRUE	union
until	while				

A list of all operators can be found at the end of section 2.4.12.

2.2 Details of Evaluation

In MuPAD, *evaluation* is understood as being the evaluation of an expression or the execution of a statement. This process is carried out by the system's evaluator or interpreter respectively. In this chapter some of the concepts and methods for influencing evaluation shall be introduced. For this it is first necessary to get an overview of the internal mechanisms of evaluation.

2.2.1 Expressions

An expression is an algebraic object that can be processed by the evaluator. In MuPAD expressions are represented with the aid of n-ary trees. The leaves of these trees are represented by elementary expressions. These can be identifiers, strings, boolean constants or numbers. Identifiers have the special characteristic that they can be assigned values. One important task of evaluation is to replace identifiers with their assigned values. This process is called *substitution*. The inner nodes of an expression tree are represented by function calls or more complex data structures like lists, sets or tables. They are used to combine (elementary) expressions. With their help new expressions can be formed.

2.2.2 Evaluation of Expressions

During the evaluation of a MuPAD expression it is transformed from one tree to another. This process is carried out recursively. Firstly, the children of a node are evaluated. Once these results are available they are used to evaluate the whole node. Evaluation is dependent on the actual form of the node. Identifiers are changed, if necessary, by the substitution described above. Except for some special objects like lists, arrays, tables or sets, the inner nodes are interpreted as function calls. The first child representing the operator is evaluated first. The result generally specifies the represented function and gives further information concerning its evaluation.

With the help of these data the complete evaluation of the node can be carried out. Normally, with expressions, this begins with the complete evaluation of all children. At the same time, tests are carried out to check if the operand types are valid for the function. Next the operands are processed and the internal *normal form* is built. This processing can, for instance, be of the addition of the numerical constants in the evaluation of a sum. The normal form is built by the operands being sorted and identical terms being combined. In many cases, during the evaluation of all children the flattening process described in section 2.3.9 is executed.

The sorting of operands is based on an internal order of expressions, which, in general, is not identical with the lexicographical order. During the sorting of sums and products, numbers are placed at the end of the expression and are either added or multiplied.

Example 2 *The expression*

```
>> a+1+2+c+4+b+3+a;
```

*is simplified to a*2+b+c+10. Firstly the 8 initial operands are sorted. Sorting enables a fast combination of the numerical values. Now, it is easy to see that the operand a is present twice. These two operands are then combined to one operand of the form a*2.*

The mechanism described here is used with many functions and data types. However, for some functions there is another method that can be used to ensure that either no or only some parameters are evaluated. Examples for these are **level** or **hold**.

If the evaluation of the operator returns an identifier then this is a formal function call of an unknown function. Now evaluation consists of the evaluation of the remaining children.

Example 3 *The function call*

```
>> f(a, 1+b, c+c);
```

*refers to the unknown function f. Therefore, after the evaluation of the parameters the expression f(a, b+1, c*2) is returned.*

If the evaluation of the operator results in a procedure definition then this procedure is executed using the further children as actual parameters (see section 2.6).

2.2.3 Evaluation of Statements

Statements are represented in the kernel like expressions: by system functions. This is described in detail in section 2.9. In contrast to standard operators not all operands of statements are evaluated before the statement itself is executed. For example it does not make any sense to evaluate the left hand side of an assignment before the assignment is executed. In this case only the second operand needs to be evaluated as this represents the value to be assigned.

During the execution of a statement only specific operands are evaluated, in loops these may be repeatedly evaluated. Other operands are completely ignored, like the `else` part of an `if` statement when the `then` part is run through.

2.2.4   Controlling Substitution

An important parameter for the evaluation of statements is the *substitution depth*. This is defined as follows: Before evaluation the substitution depth is 0. If an identifier is replaced by its value, the substitution depth is increased by 1 and the value of the identifier is further evaluated with this substitution depth. After the evaluation of this value has been completed the old substitution depth is reset.

The substitution depth also gives the recursion depth of the evaluation. Evaluation can be controlled by setting the *maximal substitution depth*. When the maximal substitution depth is reached the identifier is not evaluated further.

Interactively entered expressions are substituted completely. In this case, completely means that the user can set the maximal substitution depth with the help of the environment variable **LEVEL**.

Example 4 *After the execution of the statements*

```
>> a := b;   b := c;   c := 13;
```

the variable

```
>> a
```

is evaluated to 13 *in interactive input. The evaluation has reached the substitution depth* 3.

In procedures (see section 2.6) the maximal substitution depth is 1, in order to avoid any side-effects produced by evaluation of local variables and procedure parameters. This may lead to errors with a higher substitution depth.

Example 5 *After defining the following identifier and procedure*

```
>> c := d;
   f := proc(a, b)
        begin
           LEVEL := 2;
           a+b;
        end_proc;
```

the call

```
>> f(b,d);
```

leeds to the result d*2. *The reason for this unexpected behavior is the two-step evaluation of the formal parameter* a. *Similar effects can be produced, when local variables are used, which occur in subexpressions of the actual parameters. In this case a higher substitution depth leeds to an unwanted substitution of these identifiers.*

To change the maximal substitution depth for the evaluation of a single expression the function `level` is available. For this purpose the function should be called with two parameters. The first parameter specifies the expression to be evaluated and the second gives the substitution depth to be used. The use of `level` is logically equivalent to a temporary change of the environment variable **LEVEL**. If the second parameter is not given a maximal substitution depth $2^{31} - 1$ is assumed.

Example 6 *On the interactive level the input of*

```
>> aa := bb:  bb := cc:  cc := 3:
   level(aa, 1);
   level(aa, 2);
```

results in the values:

```
   bb
   cc
```

In connection with substitution the environment variable **MAXLEVEL** is also of importance. This is used to identify recursive definitions in the system. The system assumes that a recursive definition is present when a substitution depth of **MAXLEVEL** is reached. The default value of **MAXLEVEL** is 100. By changing this variable the user can generally control the recognition of a recursion. The value of **MAXLEVEL** should always be smaller than or equal to the maximal substitution depth **LEVEL** otherwise the given substitution depth will never be reached (except when using the function `level`).

Example 7 *The statement*

```
>> x := x+1:
```

is a recursive definition if **x** *has no value. The evaluation of* **x** *leads to a runtime error after 100 substitutions:*

```
>> x:
```

```
    Error: Recursive Definition
```

With the loop

```
>> for i from 1 to 100 do a.i := a.(i+1) end_for;
```

100 assignments are produced. These lead to a substitution depth of 100 for a total substitution of the identifier **a1***. Due to this, the input of*

```
>> MAXLEVEL:= 101: level(a1);
```

does not result in a runtime error, because the value of **MAXLEVEL** *is greater than 100.*

The substitution depth also has an effect on the evaluation of operators and statements. As already mentioned these are functionally represented. The name of such a (underline-) function must be evaluated to a function environment (see section 2.3.16) in order to be executed. This substitution is also affected by the user-specified substitution depth.

Example 8 *The following input*

```
>> LEVEL := 1:
   f := hold(_plus):
   a:=1:
   b:=2:
   f(a, b);
```

results in the value **1+2***. Due to the substitution depth 1 the operands of the function are fully evaluated, but the function name could only be substituted to* *the identifier _plus. So the execution of the addition was not possible. The system function* hold *prevents the evaluation of the identifier _plus, so that the identifier _plus and not its value is assigned to* f *(for further information see section 2.3.16).*

2.2.5   Influencing Evaluation

Further functions are available to the user for influencing the evaluation of an expression.

To prevent evaluation the function **hold** can be used. This function can have any expression as a parameter and returns it without any evaluation whatsoever, i.e. without substitution, execution or simplification. In a way the counterpart of the function **hold** is the function **context**. At first this evaluates its arguments as usual. The result of this evaluation is then re-evaluated in another context. This context consists of the environment from which the current procedure was called. The main application of the function **context** is the subsequent evaluation of procedure parameters that, for instance, have not been evaluated because of a **hold** option (for further information see section 2.6.5.1).

Example 9 *The function* hold *prevents the evaluation of both expressions and statements. Thus the expressions*

```
>> hold(1+2+a);
   hold((a := b));
```

return the values

```
   1+2+a
   a := b
```

without executing the assignment. Statements in expressions must be additionally set in parentheses (see section 2.5.10).

A further possibility of influencing the evaluation of an expression is the function **eval**. This is a system function which re-evaluates the results of certain special functions. These are functions which in one way or another return a non-evaluated result. Specifically these are the functions

```
   args     coeff     evalp     expr     hold     input     last
   lcoeff   nthcoeff  subs    subsex   subsop   tcoeff   text2expr
```

The additional evaluation is carried out using the substitution depth 0. The evaluation of all other data types are not influenced by the function `eval`. In particular `eval` has no effect on the results of calls of user-defined functions, because the result is calculated, i.e. evaluated, in the context of the procedure.

The mechanism of the function `eval` is described in the following, using the function `last` as a model:

Example 10 *The function* `last` *returns the result of a previous evaluation without re-evaluating it (see section 2.11). However, by using* `eval` *the result of the* `last` *call is evaluated. After the assignments*

```
>> a := b:  b := c:  c := d:
```

the expression

```
>> last(2), eval(last(2))
```

returns the sequence c, d. *Here the variable* c *is re-evaluated as the value of the* `last` *call.*

Notice that the function `eval` only has an effect on the functions described above; all other objects in the call are not affected.

Example 11 *Let us look at the procedure*

```
>> f := proc()
   local a;
   begin
       a*2;
   end_proc;
```

which returns a*2. *After the assignment*

```
>> a:=b;
```

the call

```
>> eval(f()+text2expr("a+1"));
```

returns the value a*2+b+1 *because the result of the call* f(), *in contrast to the result of the* `text2expr` *call is not subsequently evaluated although it is also in an argument of the* `eval` *call.*

The effect of the function `hold` is cancelled by a surrounding `eval` call. In the case of multiple nested `hold` calls, only one `hold` call is cancelled independent of the number of surrounding `eval` calls.

Example 12 *The input*

```
>> a:=b:
   eval(hold(hold(a)));
```

returns the result a. *But* `eval(eval(hold(hold(a))))` *does not evaluate to* b.

If the evaluation of statements in expressions is suppressed by setting the environment variable `EVAL_STMT` (see example 110) to `FALSE`, then the `eval` function can be used in this modus to evaluate such statements.

Example 13 *If the variable* `EVAL_STMT` *has the value* `FALSE`, *then the assignment* b:=c *is not executed in the sequence*

```
>> EVAL_STMT := FALSE:
   a := (b:=c): a;
```

If the call a *is replaced by*

```
>> eval(a);
```

then the assignment (b := c) *is evaluated.*

2.3 Basic Types

Although the declaration of variables and other objects in MuPAD is not necessary, each MuPAD datum has a type which is determined during runtime. Type determination takes place on two levels. Firstly, every MuPAD object is either an element of one of the *basic domains* listed below or an element of a user-defined domain. Basic domains and user-defined domains only differ in the possibility to overload miscellaneous operators and internally used methods. Whereas for user-defined methods every operator and internal method is able to be overloaded, this is not valid for basic domains. For some of theses only the functionality of the function call can be changed. In this case no operator or internal method can be influenced.

In this context we shall be talking about *basic types* or *user-defined types*. The system makes the following basic types available:

DOM_ARRAY	*DOM_BOOL*	*DOM_COMPLEX*
DOM_DOMAIN	*DOM_EXEC*	*DOM_EXPR*
DOM_FAIL	*DOM_FLOAT*	*DOM_FUNC_ENV*
DOM_IDENT	*DOM_INT*	*DOM_LIST*
DOM_NIL	*DOM_NULL*	*DOM_POINT*
DOM_POLY	*DOM_POLYGON*	*DOM_PROC*
DOM_RAT	*DOM_SET*	*DOM_STRING*
DOM_TABLE		

For example a non integer rational number is stored as an element of the domain *DOM_RAT*, and sets are represented by using the domain *DOM_SET*.

The domain *DOM_EXPR* includes all expressions that can be created by the operators of the system, all statements, procedure definitions, function calls as well as expressions created with the index operator. In order to examine these objects more closely and to manipulate them the classification above is far too rough. Due to this all MuPAD objects undergo a second classification. This second type is described as the *expression type* in the following. For all objects that do not belong to the domain *DOM_EXPR* the expression type and basic type are identical. The expression type is therefore only a finer structuring of the elements of the domain *DOM_EXPR*. For a detailed description of expression types see section 2.3.15.

At this point we would like to emphasize that a type declaration of identifiers is not necessary.

The system function **domtype** enables the user to determine to which basic or user-defined domain an object belongs. It returns the corresponding domain. For basic domains this is printed in form of an identifier starting with the prefix *DOM_*. For examining expression types the function **type** is available.

Example 14 *The expressions*

```
>> domtype(a+b+c);
   domtype(a*b*c);
```

yield the result:

```
DOM_EXPR
DOM_EXPR
```

Therefore, a distinction can only be made through the calls

```
>> type(a+b+c);
   type(a*b*c);
```

which return the expression types:

```
"_plus"
"_mult"
```

In the next sections the following types shall be described in detail:

- numerical types
- character strings
- FAIL
- boolean constants
- lists
- tables
- polynomials
- function environments

- identifiers or variables
- NIL
- empty objects
- numerical constants
- sets
- arrays
- graphical primitives
- expressions

2.3.1   Numerical Types

Numerical types are used to represent rational numbers (*DOM_RAT*) and integers (*DOM_INT*) as well as floating-point (*DOM_FLOAT*) and complex numbers (*DOM_COMPLEX*). The numbers can be of any length. With the system function `float` integers and rational numbers can be converted into floating-point numbers. The number of significant decimal places of floating-point numbers can be controlled by the environment variable `DIGITS`.

Syntactically the imaginary unit I is represented by the keyword I. Therefore I cannot be overwritten by an assignment.

Complex as well as rational numbers can be decomposed into real and imaginary parts or respectively in numerator and denominator with the function `op`. The operands of a complex number can be any real numbers. Their basic types need not be identical.

2.3.2   Identifiers

Identifiers are constructed by any combination of letters and digits and an underline, however, they may not begin with a digit. In MuPAD they are represented by elements of the domain *DOM_IDENT*.

In contrast to some other programming languages, for which identifiers are seen as identical as soon as they correspond in their first N positions (N = 8 is typical for old versions of Pascal, N = 31 is the minimum number in ANSI C), in MuPAD all positions are significant. In addition MuPAD distinguishes between capital and small letters.

With the concatenation operator, new identifiers and strings can be generated.

2.3.3   Character Strings

Character strings consist of a series of characters between quotation marks. The length of the character string is limited by $2^{32} - 1$ and therefore in standard applications can be seen as unlimited. This basic type is called *DOM_STRING*. The system makes a number of functions for manipulating character strings available. The user can, for instance, with the system functions **strmatch** and **strlen** compare character strings or respectively determine their length. To split texts into lists the system function **text2list** is available. With this, the texts, which must be in the form of a string, are split at the positions marked with a hyphen by the user. If the character string has to be converted into an expression then **text2expr** can be used.

The backslash (\) is of special importance in character strings. It is used to mask characters which are of a special meaning in character strings. Therefore, the backslash may only be followed by certain characters. If other characters are used here, a syntax error occurs.

A backslash may be followed by a multiplication sign or a question mark. These combinations are used as wild cards in the system function **strmatch** and, otherwise have no special meaning. They are stored as two characters in a string.

Furthermore the characters

$$\backslash, \text{ ", } t \text{ and } n$$

can be entered with a leading backslash. Here, the backslash is used to mask itself or respectively, to mask the quotation marks. The character strings \n and \t are used to enter a line break or a tabulator character. These four character strings are stored as one character and are therefore processed as one character by system functions like **strlen**.

In addition the output functions **print** and **fprint** enable the user to influence the output of these characters. By using the option **Unquoted** quotation marks at the beginning and end of a character string can be prevented. At the same time this option is responsible for the special characters being replaced as described above. Therefore, output of this form can not generally be re-read.

Example 15 *The character string*

```
>> "This is \n a \"special\" Text \t with a \\";
```

is returned unchanged by the system. In order to expand the special characters the function print *can be used. The expression*

```
>> print(Unquoted,"This is \n a \"special\" Text \t with a \\");
```

produces the output:

```
    This is
     a "special" Text          with a \
```

Without the option Unquoted *the character string is returned unchanged.*

2.3.4   NIL

The object **NIL** is the only element of the domain *DOM_NIL*. It is used to delete assignments so that, for instance, a variable once again stands only for itself. But also elements in lists, expression sequences, tables and arrays can be removed by assigning **NIL**.

In statements, empty statement parts are also labelled by the expression **NIL**. Arrays also use this expression to describe an unfilled entry and the missing elements of a procedure definition are marked identically.

Example 16 *(See sections 2.3.9, 2.3.11 and 2.3.12) Through*

```
>> L:= [x, y];
```

a list with the elements x *and* y *is created. With the assignment*

```
>> L[2] := NIL;
```

y *is then deleted from the list.*

2.3.5 FAIL

The object **FAIL** is the only element of the domain *DOM_FAIL*. It is usually used to signal that a computation could not be completed.

Example 17 *If a non-existent operand is to be accessed with the function* op *then the function returns the value* FAIL. *Therefore,*

```
>> op(a+b+c, 4);
```

returns the value FAIL.

2.3.6 Empty Objects

Some MuPAD functions do not return any results. For this purpose there is the basic type *DOM_NULL*. Expressions of this type create no visible character when displayed on the monitor. Nevertheless the user can still work as usual with this type, when, for instance the result of a calculation of the type *DOM_NULL* is requested. The system function **null** is used to explicitly create an object of this type. This system function is called without operands.

This basic type has a further meaning when evaluating expression sequences, sets, lists, function calls and indexed identifiers. If an expression of type *DOM_NULL* occurs as operand of an expression sequence, set or list, this operand is deleted. The occurrence of expressions of type *DOM_NULL* as one of the arguments of user-defined functions and most system functions leads also to a corresponding deletion.

Example 18 *The calls*

```
>> type(op([]));
   type(op(f()));
```

return the result

```
    DOM_NULL
    DOM_NULL
```

when f *has no value. The list*

```
>> [a, op([]), b];
```

evaluates to [a,b].

For information about the function `op`, with which the operands of an expression can be obtained, see section 2.10.3.

Some MuPAD functions are exclusively used for user information. These include, for instance, `history` or `help`. These functions return no result that can be processed further. Therefore, for instance, the result of the call

```
>> type(history());
```

is also `DOM_NULL`.

2.3.7 Boolean Constants

MuPAD provides the Boolean constants `TRUE` and `FALSE` of the basic type *DOM_BOOL*. They can be combined with the help of logical operators and other operands into new logical expressions. After a Boolean expression is entered no attempt is made to evaluate this to a Boolean constant (see section 2.4.4). Exceptions to this rule can be found in connection with the evaluation of certain statements. The comparison expression of an `if` statement for instance automatically undergoes Boolean evaluation in order to determine which branch of the statement is to be executed. The system function `bool` is available for the explicit evaluation of a Boolean expression.

2.3.8 Numerical Constants

The numerical constants include the values `E` ($= 2.7182...$), `PI` ($= 3.1415...$) and `EULER` ($= 0.5772...$). These are predefined identifiers which may not be overwritten. The assignment of a value to these identifiers — although syntactically correct — is not allowed. Therefore, strictly speaking `E`, `EULER` and `PI` do not form individual types. The user can use numerical constants to perform symbolic calculations. Therefore, these are not replaced by approximate numerical values. Their substitution take place only after an explicit call of the system function `float`.

2.3.9 Lists

A list is a basic type, formed by combining any number of expressions of different types. A list can be considered as an ordered series of expressions. In the notation of lists the individual expressions are separated by commas and the entire series is contained in square brackets. An empty list is also a valid expression. The name of this basic type is *DOM_LIST*. The evaluation of a list consists of the evaluation of all its elements. In addition empty expressions (of type *DOM_NULL*) are deleted and expression sequences are flattened.

Flattening is an internal mechanism which can be compared with using associativity when **a+(b+c)** is simplified to **a+b+c**. When an element of a list evaluates to an expression sequence this element is replaced by the individual elements of the expression sequence and, at the same time, the sequence is retained. The list is enlarged by the process of flattening. An analogue process takes place on evaluating sets, expression sequences, function calls and indexed identifiers.

Example 19 *The assignment*

```
>> x := a, b, c;
```

assigns the sequence **a, b, c** *to the identifier* **x** *(see section 2.4.5.1). If the list* **L** *is created by the assignment*

```
>> L := [x, d];
```

it now has 4 elements instead of 2: **a, b, c** *and* **d**.

The user can access individual elements of a list by using indices. The first element is given the index 1. Assignments to list elements also take place with indices. In this case the relevant index must already exist. Deletion of individual elements takes place by assigning **NIL** (see also example 16).

Example 20 *Let* **L** *be a list of the form*

```
>> L := [a, b, c, d];
```

Access to the third element of the list is achieved with the indexed identifier **L[3]**. *Therefore, with the assignment*

```
>> L[3] := 100;
```

the list

```
>> L
```

can be changed to **[a, b, 100, d]**. *Analogue an assignment of the form*

```
>> L[3] := NIL;
```

has the effect of deleting the third element of the list **L** *and its value is then the list* **[a, b, d]**.

If, by an assignment to an indexed identifier, an expression sequence is entered into a list, then the list is immediately flattened so that all operands of the entered sequence are taken up as new elements in the list.

Example 21 *A list is defined by the assignment*

```
>> L := [a, b, c, d];
```

With the assignment

```
>> L[2] := 1, 2, 3;
```

the value b *is removed from the list and* L *is extended by three new elements in this position so that the list*

```
>> L
```

finally looks like [a, 1, 2, 3, c, d].

With the operator . two lists can be concatenated. The function for appending new elements to already existing lists is **append**. To ascertain the existence of an element and its position in a list the function **contains** can be used.

Example 22 *The two lists*

```
>> L := [a, b]:
   M := [c, d]:
```

can be combined with the operation

```
>> L . M;
```

to the list [a, b, c, d]. *By calling*

```
>> append(L,c,d);
```

the same result is returned. In both cases the value of L *has not been changed, but for the result a new list has been created.*

2.3.10   Sets

Sets like lists consist of a finite series of expressions separated by commas. However the order of the expressions is irrelevant in sets and it is not possible that a given element is included more than once. The expressions in the set are enclosed in set brackets { and }. Empty sets are also valid objects. Analogue to lists, during the evaluation of a set all elements are completely evaluated, whereby the amount of elements may change. After evaluation the set contains only one copy of each element. Analogue to lists, elements that evaluate to objects of type *DOM_NULL* are not taken into account and expression sequences are flattened. For processing sets the operators **union**, **minus** and **intersect** are available. The system function **contains** checks for the existonce of an element in a set.

The order of the individual set elements is entirely random in the output (see example 23). Sets have the basic type *DOM_SET*.

Example 23 *The two commands*

```
>> M := {a, b, c};
```

```
>> N := {b, a, c, a, a, c, b, op([]), a};
```

both create a set with the elements a, b *and* c.

Because the output of sets depends on the order of the elements given in the internal data structure, two sets with the same elements can be represented and displayed differently.

2.3.11   Tables

A further important basic type is *DOM_TABLE*. The content of a table consists of a series of equations of the form

```
<index> = <value>
```

Both `<index>` and `<value>` represent any expression. The values stored in a table can be accessed with the relevant indices. There is the possibility of inserting, changing and deleting table entries.

Example 24 *If the table* T *contains the equation* a=5 *this means that for the index* a *the value* 5 *exists. The indexed identifier*

```
>> T[a];
```

gives the value 5.

To change the content of a table T or to enter new values an assignment of the form

```
T[<index>] := <value>
```

is necessary. `<index>` and `<value>` can once more be any expression. With this assignment the equation

```
<index> = <value>
```

is entered in the table. Any already existing entry with the same index is overwritten by the assignment above. If, before the assignment, the table T does not exist then the variable T is initialized with an empty table and finally the relevant entry is made.

Another possibility of creating a table is the system function `table`. It can be called with any number of parameters. The parameters have to be equations of the form `<index> = <value>`. With these the initial entries in the table are determined.

Example 25 *With the assignment*

```
>> T1 := table(a=c1, b=c2);
```

a table is set up that contains the value c1 *under the index* a *and* c2 *under the index* b. *The same table is created with the assignments*

```
>> T1[a] := c1;
   T1[b] := c2;
```

In contrast to lists and sets whose evaluation consists of the evaluation of each of their operands, tables evaluate to themselves, i.e. the stored indices as well as the values belonging to them remain unchanged.

Example 26 *The following table is not changed by the assignment of the variables* a *and* b:

```
>> T[a] := b:  T;
   a := 1:  b := 2:  T;
```

Both times the call of T *returns the result* `table(a=b)`.

When accessing with an indexed identifier firstly the index is evaluated and then the table is checked to see if a relevant entry exists. If this is the case then the relevant value is read from the table, evaluated and returned as the result of the call.

Example 27 *If* a *is not assigned the statements*

```
>> S[1] := b:  S[a];
   a := 1:  b := 2:  S[a];
```

return the output:

```
   S[a]
   2
```

If a check is to be carried out to see if an entry is stored under an index, then the function **contains** can be used for this purpose, because generally this information cannot be obtained by simple indexed accessing.

Example 28 *If the table* T *is initialized with*

```
>> T[a] := T[a]:
```

then the input

```
>> T[a];
```

returns the value T[a] *which can not be distinguished from the output we get when* T *is an uninitialized identifier.*

Using the function **contains** *this can be definitely determined. In the example above the call*

```
>> contains(T, a);
```

returns the value **TRUE**.

Copying entire tables takes place internally in MuPAD by creating a reference of the copy to the original. But when changes are made duplication occurs in memory. With this the *reference effect*, with which changes in the copied table have an effect on the original, is prevented.

Example 29 *To illustrate the avoidance of the reference effect, look at the following situation:*

```
>> T := table(a=1):
   S := T:
   S[b] := 2:
```

If S *would be only a reference to* T, T[b] *would also have had the value 2. However, in MuPAD* T[b] *remains undefined.*

Please note when comparing two tables, that neither the indices nor the relevant values are evaluated (see example 26).

Example 30 *Consider the tables* T1 *and* T2 *which are defined as follows*

```
>> T1 := table(a=c1, b=c2):
   T2 := table(a=1, b=1):
```

then after the assignments

```
>> c1 := 1:
   c2 := 1:
```

the calls T1[a] *and* T2[a] *as well as* T1[b] *and* T2[b] *each return the value 1. However the call*

```
>> bool(T1 = T2);
```

returns the result **FALSE**, *because the values* c1 *and* c2 *are still stored in table* T1, *and the table elements are not evaluated in the* **bool** *call.*

Evaluation only takes place when accessing with an indexed identifier. However, even in this case the table itself is not evaluated. The indexing expression is evaluated, before the relevant value is taken from the table, the result is subsequently also evaluated. Finally, with the assignment of **NIL** individual table entries or even the entire table can be deleted.

2.3.12   Arrays

Arrays (basic type *DOM_ARRAY*) are externally a special form of a table. The indices of an array must be integer expression sequences of a fixed length and lay within given limits. An array must be explicitly defined. For this purpose the system function **array** is available. In its simplest form **array** is called with a sequence of ranges.

Example 31 *With the statement*

```
>> A := array(1..3, 1..3);
```

the identifier A *is assigned a 3×3 array. The individual elements can be accessed with the expressions* A[1, 1], A[1, 2] *etc.*

The dimension of the array laid down in the call of **array** is subject to certain restrictions. Thus, only non-negative integers smaller than $2^{31} - 1$ are permitted. Access to indices that are outside these ranges lead to an error. If an indexed entry has no value then the indexed expression is returned.

Similarly to tables the values of arrays can be given in the call of **array**.

Example 32 *Through*

```
>> A := array(1..2):    A[1] := x:    A[2] := y:
```

an array with the components x *and* y *is defined. With the pretty-printer turned off the output of this array is:*

```
>> A;

    array(1..2, (1)=x, (2)=y)
```

This output is also a valid call of **array** *in which the complete array with its entries is defined in one expression.*

Another possibility to initialize an array is to give the complete data of all entries in form of embedded lists. In a two-dimensional array a list of lists is to be given whereby the i-th list contains the elements of the i-th row of the array. If an array element is to remain unoccupied then the value NIL must be given. If this list is given to the function **array** it has to be the last parameter. An additional initialization by giving equations is not allowed when using this initialization method. Higher dimensional arrays can be initialized with multiply embedded lists.

Example 33 *The call*

```
>> A := array(1..2, 1..3, [ [1, 2, 3], [4, 5, 6] ] );
```

creates a 2 × 3 *array with the two rows* [1, 2, 3] *and* [4, 5, 6].

Access to, and changing of elements of an array takes place analogue to that of tables. Also the evaluation behavior is the same as that of tables. This means, especially when copying entire arrays, that the reference effect does not occur and that the comparison of arrays is subject to the problem described in section 2.3.11 i.e. that analogue to tables, the array elements are not evaluated before comparison. The deletion of array elements is possible through assigning the value NIL.

While indexing arrays the evaluation of the actual array entry can be prevented by using the function index_val. index_val can also be used for this purpose while indexing tables.

Example 34 *Let an array be defined by the following assignment:*

```
>> A:=array( 1..4, (1) = b): b:=2:
```

Then the call

```
>> index_val(A,1);
```

returns the result b. *An evaluation of the array entry is not carried out.*

By replacing the function _index, which is called by MuPAD during indexing with square brackets, with the function index_val, the normal indexing can be adapted to this behavior (see also section 2.4.11).

For checking the existence of entries the function contains can also be used on arrays.

2.3.12.1   Subarrays

If one wants to work efficiently on arrays with parallel algorithms then one must be able to work on blocks of the array separately. For this a mechanism is necessary that allows access to subarrays. These subarrays are to be specified using the function array.

Example 35 *The call*

```
>> array(1..N, 1..M, [n,m]);
```

creates a N×M-array which as shown in figure 2.1 is split into blocks.

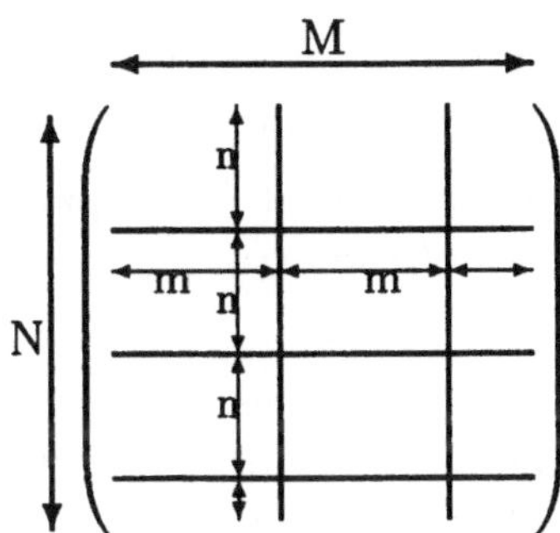

Figure 2.1: An array split into blocks

The matrix developed in this example consists of $\lceil \frac{N}{n} \rceil \cdot \lceil \frac{M}{m} \rceil$ blocks, which can all be individually addressed. This takes place through indexing, whereby the index runs through the blocks line by line. In its turn each block represents its own matrix.

Example 36 *The matrix*

```
>> A := array(1..4, 1..5, [2, 2]);
```

contains six subarrays. These are individually:

```
A[1] = array(1..2, 1..2);     A[2] = array(1..2, 3..4);
A[3] = array(1..2, 5..5);     A[4] = array(3..4, 1..2);
A[5] = array(3..4, 3..4);     A[6] = array(3..4, 5..5);
```

Because the subarrays are themselves returned as normal arrays these can be indexed individually.

Therefore the expression

```
>> A[2][1,3];
```

also returns the element `A[1,3]`*. Changing the entries of subarrays using assignments to identifiers indexed more then once is not possible.*

2.3.12.2   Operands of an Array

In contrast to all other basic types, apart from *DOM_EXPR*, arrays have a zeroth operand. This consists of an expression sequence in which the dimension, the ranges of the individual dimensions, as well as the partitioning into subarrays are stored. This partition is given in the form of a list of positive integers.

These informations, with exception of the dimension data, are equivalent to the first arguments of the **array** call.

The other operands in an array that is not partitioned further, consist of the row by row ordered entries of the array. If an entry is undefined then a NIL is entered as the operand.

The operands of an array partitioned into subarrays are defined somewhat differently. Here the operands are formed by the individual subarrays which are each given in form of a list. The elements of the subarray are ordered in the list row by row.

Example 37 *For the partitioned array*

```
>> A := array(1..4, 1..5, [2, 2]):
   A[1, 1] := 11:
   A[2, 2] := 22:
   A[4, 5] := 45:
```

the calls of the op *function*

```
>> op(A, 0);
   op(A, 1);
   op(A);
```

result in the values:

```
   2, 1..4, 1..5, [ 2, 2 ]

   [ 11, NIL, NIL, 22 ]

   [ 11, NIL, NIL, 22 ], [NIL, NIL, NIL, NIL],
   [ NIL, NIL],  [ NIL, NIL, NIL, NIL ],
   [ NIL, NIL, NIL, NIL ], [ NIL, 45 ]
```

The lists of the operands 3 and 6 are shorter, because the relevant subarrays contain fewer entries due to the total size of the array.

2.3.13 Polynomials

Special data structures and algorithms for polynomials exist in the MuPAD kernel. Thus the fundamental operations are carried out quickly on polynomials.

Arithmetical expressions (*DOM_EXPR*) are not used to represent polynomials in MuPAD because there the terms are sorted by the internal simplifier. This sorting hinders the canonical representation of polynomials which is however necessary for the efficient execution of special operations. Another reason that speaks against the use of expressions is the amount of memory space used by them. The polynomial data structure is far more compact.

Polynomials have the basic type *DOM_POLY*. Any arithmetical expression can be converted into an polynomial with the function `poly`. Conversely the function `expr` changes a polynomial back into an expression.

Example 38 *The call*

```
>> p:= poly(x^2 + x*y + 12, [x,y]);
```

creates the polynomial p *in the variables* x *and* y. *With the call*

```
>> expr(p);
```

the expression x^2 + x*y + 12 *is returned.*

Analogue to tables polynomials are returned in functional form by use of the function `poly`.

Example 39 *The polynomial* p *defined above is, for instance, returned by the system as* poly(x^2 + x*y + 12, [x,y]).

As a special feature the coefficient ring can be given for polynomials. Apart from polynomials with arbitrary expressions as coefficients, polynomials with coefficients from the residue class rings $\mathbb{Z}/n\mathbb{Z}$ (for positive integers n) as well as with domain elements as coefficients are permitted.

Internally a sparse distributed representation of the polynomials is used, where the terms are ordered lexicographically. The terms are stored in a special internal type. However, this representation is not directly accessible to the user.

The exponents of the terms are stored as machine integers. Therefore, the sum of the exponents of a term may not be greater than $2^{31} - 1$. If the exponents get too large a run-time error will occur. 

Although internally the terms of a polynomial are ordered lexicographically, one may also select the leading term with respect to a couple of other orderings. The available orderings, which may be used with functions like `lcoeff` or `lterm`, are the following:

- **LexOrder** The lexicographical ordering, which is generally used as default ordering.

- **DegreeOrder** Ordering by total degree, where terms with equal degree are ordered lexicographically.

- **DegInvLexOrder** Also total degree ordering, but terms with equal degree are ordered inverse lexicographically.

Polynomials are not implicitly evaluated. Even functions with polynomials as arguments and expressions as results do not evaluate the result. If the result of such a function is to be evaluated then the call must be embedded in **eval**.

Example 40 *Let the polynomial*

```
>> p := poly(a*x^2 + y, [x]);
```

be defined in the variable x. *The value of* p *remains the same* poly(a*x^2 + y, [x]), *even when the identifiers* a, x *or* y *are assigned values. If, for instance, subsequently*

```
>> x := 2; y := 3;
```

is entered then the value of

```
>> p
```

remains the expression poly(a*x^2 + y, [x]). *The call*

```
>> expr(p);
```

also returns the result a*x^2 + y *after the assignments to* x *and* y. *Only with the call*

```
>> eval(expr(p));
```

the result a*4 + 3 *is returned.*

2.3.13.1 Operands of a Polynomial

From the user's point of view, a polynomial always has three operands:

- The 1st operand is the polynomial expression (*DOM_EXPR*): a sum of monomials. The order of the monomials in the sum is given by the lexicographical order.

- The 2nd operand of the polynomial is a list (*DOM_LIST*) with the variables. The order of the elements in the list gives the order of the variables: the first element is the main variable, then the secondary variables follow.

- The 3rd operand of the polynomial gives the coefficient ring of the polynomial. This can be either the identifier `Expr`, an expression of the form `IntMod(n)` or a domain (*DOM_DOMAIN*).

The residue class ring $\mathbb{Z}/7\mathbb{Z}$ is written as `IntMod(7)` („integer modulo 7") in MuPAD. Like `Expr`, `IntMod` simply is an identifier and has no value. To represent a polynomial with the residue class ring `IntMod(n)` the symmetrical modulus is used. The coefficients of a polynomial over `IntMod(7)` are therefore integers between -3 and 3.

Polynomials can be manipulated as usual with the functions `op`, `subsop`, `subs` and `subsex` (see section 2.10). Please note that when using the substitution functions the result must be a valid polynomial.

Furthermore, these functions do not manipulate the 1st operand (the polynomial expression) very efficiently. Instead one of the special polynomial operations, e.g. `coeff` or `nthmonomial`, should be used whenever possible.

2.3.13.2 Polynomials over Domains

Apart from the special coefficient rings `IntMod(n)` and `Expr` other, user-defined structures, called domains,2.3.18 can also be given as the coefficient ring when calling `poly`. To work with a polynomial defined over a domain, certain methods for the fundamental arithmetical operations must be available for the domain. These are in particular:

- `Dom::_plus(e1,e2)`: returns the sum of two domain elements `e1` and `e2`.

- `Dom::_mult(e1,e2)`: calculates the product of `e1` and `e2`.

- `Dom::negate(e)`: returns the negative of the domain element `e`.

- `Dom::_power(e,i)`: raises the domain element `e` to the power of a positive integer `i`.

Furthermore the domain must contain the constant entries `Dom::zero` (the 0 in the coefficient ring) and `Dom::one` (the 1). Of course, these constants must be elements of `Dom`.

Depending on the application the methods `divide` and `norm` can also be of use. Both these methods are required by the polynomial functions with the same names:

- `Dom::divide(e1,e2)`: divides the domain elements `e1` and `e2`. If division is not possible the method returns `FAIL`.

- `Dom::norm(e)`: returns the norm of the domain element `e` as a number.

If a polynomial over the domain `Dom` is to be differentiated with the function `diff` then, in addition, the methods `diff` and `intmult` are necessary:

- `Dom::diff(e,x)`: returns the derivative of the domain element `e` with respect to `x`.

- `Dom::intmult(e,i)`: calculates the multiple $i \times e$ of the domain element `e` for the positive integer `i` (For example for `diff(poly(e*x^3), x)` = `poly(3*e*x^2)`. the multiplication $3 \times e$ must be calculated in the domain `Dom`.)

Finally, to convert polynomials over domains with the function `poly` the methods `convert` and `expr` must be given:

- `Dom::convert(e)`: changes the expression `e` into a domain element. If this is not possible then the method returns `FAIL`.

- `Dom::expr(e)`: conversely, converts the domain element `e` into an expression.

The method `convert` is used to newly create polynomials over domains from expressions. The method `expr` is used to convert polynomials over domains into expressions.

2.3.14 Graphical Primitives

Special data structures are made available in the MuPAD kernel in order to generate lists of polygons. They can be created by use of the system functions `point` and `polygon`.

The simplest graphical primitive *DOM_POINT* can be created by use of the function `point`, which serves for generating two- or three-dimensional points.

Example 41 *The commands*

```
>> point2d := point(1, 2, Color = [1, 0, 0]):
   point3d := point(1, 2, 3):
   op(point2d);
   op(point3d);
```

return the result:

```
   1, 2, [1.0, 0.0, 0.0]
   1, 2, 3, NIL
```

As can be seen in the example above the data structure for a two- or three-dimensional point consists of the co-ordinates of the point as well as the color of the point. The color is internally stored in a list, containing three real values between 0.0 and 1.0, which are used to describe the amount of red, green and blue of the specified color. If the color is not specified, a NIL is inserted for the last operand of the data structure *DOM_POINT*.

Higher order primitives can be created by use of the system function `polygon`. Polygons themselves are composed out of an arbitrary number of either two- or three-dimensional points. Furthermore there exist additional options in order to influence the graphical representation of the polygon. These options are used to define the color of the polygon and to determine whether the polygon is to be closed and/or filled. A call of `polygon` results in creating an entry of the data structure *DOM_POLYGON*. The first operands of this data structure are the different points, i.e. entries of the type *DOM_POINT* followed by the specified options.

Example 42 *The commands*

```
>> point1 := point(0, 0):
   point2 := point(1, 0):
   point3 := point(0, 1):

   poly2d := polygon(point1, point2, point3,
                     Color = [0, 1, 0],
                     Closed = TRUE,
                     Filled = FALSE):
```

serve for creating a two-dimensional polygon consisting of three points, which is to be closed but not filled. Furthermore the color of the polygon is specified. The different operands of this polygon can be obtained by the command

```
>> op(poly2d);
```

which results in

```
    point(0, 0), point(1, 0), point(0, 1),
          [1.0, 1.0, 1.0], TRUE, FALSE
```

As can be seen, the first operands of the data structure *DOM_POLYGON* are the different points, followed by the specified values for the color and the options `Closed` and `Filled`. If the color is not specified, the corresponding operand will be filled with a `NIL`. If the options `Closed` and `Filled` are not specified, the default values (`FALSE`) for these options will be entered into the structure *DOM_POLYGON*.

2.3.15 Expressions

The most important basic type in MuPAD is an expression. This type has the name *DOM_EXPR* and contains all forms of expressions formed by the operators described in the next sections. Also function calls, indexed expressions, procedure definitions and statements, such as assignments, loops, control structures are of type *DOM_EXPR* reflecting the fact that statements and procedure definitions are usual MuPAD objects.

With all the basic types described so far the results of the functions `type` and `domtype` are identical. This is not the case with expressions. The function `type` returns more detailed information about the individual expression types than `domtype`.

The type of an expression is usually determined by its lowest priority operator or the function representing this expression. In general this operator is a function environment (see section 2.3.16) or can be evaluated to one. In this case the string stored in the attribute table under the index `"type"` is used as expression type and therefore also returned by the function `type`. If the evaluation of the operator does not result in a function environment or if this function environment has no `"type"` entry then the string `"function"` is used as type.

The expression types formed by the operators and statements of the system are summarized in the following table:

`"_and"`	`"_assign"`	`"_break"`	`"_case"`
`"_concat"`	`"_div"`	`"_equal"`	`"_exprseq"`
`"_fconcat"`	`"_for"`	`"_for_down"`	`"_for_in"`

```
"_for_in_par"   "_for_par"    "_if"        "_index"
"_intersect"    "_leequal"    "_less"      "_minus"
"_mod"          "_mult"       "_next"      "_not"
"_or"           "_parbegin"   "_plus"      "_power"
"_procdef"      "_quit"       "_range"     "_repeat"
"_seqbegin"     "_seqgen"     "_stmtseq"   "_unequal"
"_union"        "_while"      "function"
```

As a rule all other predefined functions having the property, that through the evaluation of a corresponding function call the operator can remain unchanged, define their own type. This type is implemented as described above. The type stored in the corresponding function environment is the function name converted to a string. All remaining functions do not have a special type and therefore the function **type** returns the string "function".

Example 43 *The call*

```
>> type(sin(2));
```

returns the result "sin".

Example 44 *Let the identifiers* f *and* T *be undefined. Then the expressions*

```
>> a:=b:
   f(a);
   T[a];
```

evaluate to:

```
   f(b)
   T[b]
```

The expressions have the expression type "function" *respectively* "index".

2.3.15.1 Evaluation of Expressions

All objects of the basic type *DOM_EXPR* have in common that their operators determine the way the expression is evaluated. This operator represents the function to be called and all other operands are considered as arguments of this function call. Generally the operator can be of any type. The aim of this section is to describe the functionality of the different operator types.

The first step in the evaluation of an expression is the evaluation of its operator, taking into account the current substitution depth. The result of this evaluation determines the further evaluation.

- In the case of a function environment (see section 2.3.16) the first operand is regarded as operator and the expression is evaluated as follows.

- If the operator is evaluated to a procedure definition, then all other operands of the expression are evaluated and used as actual parameters of the function call.

- In the case of a directly executable object, the expression is evaluated as described in section 2.3.17.1.

- If the operator is evaluated to a polynomial, then the variables are substituted by the operands.

- If the operator is a list or set, then each element of the operator is viewed as a function and applied to the operands. The elements are then replaced by the evaluated function calls and the resulting list or set is returned.

- If the operator is an array or table, then each entry again is viewed as a function and applied to the operands. The evaluated function calls replace the original entries.

- If the operator is an identifier the operands are evaluated and used as arguments of a new formal function call.

- If the operator evaluates to a domain D, then the method D::new() is called. (see section 2.3.18)

- In all other cases first it is checked if the method "func_call" is defined for the domain of the operator. If a definition is present then the associated method is called. Otherwise the operands are not evaluated and the operator itself is regarded as a constant function and returned as result of the expression.

A special situation occurs when the evaluated operator is again an expression of type *DOM_EXPR*. See section 2.4.10 for a detailed description.

Example 45 *Let* s *be assigned the character string* "Hello" *through*

```
>> s := "Hello":
```

Then the function call

```
>> s(a);
```

returns the value "Hello". *The parameter* a *is not evaluated. If* s *is assigned the set* {sin, 7} *through*

```
>> s := {sin, 7};
```

the function call

```
>> s(a);
```

returns the value {sin(a), 7}. *In order to use a list or set directly as operator one has to inclose it in brackets:*

```
>> ([ f, 2/3, {g, "Hello"} ])(x, y);

   [ f(x, y), 2/3, {"Hello", g(x, y)}  ]
```

2.3.15.2   Output of Expressions

Describing the output of expressions two cases are to be distinguished. If the operator of an expression is a function environment or an identifier to which a function environment is assigned, then the output is controlled by the second operand of this function environment. To do so a function call is constructed whose operator is this second operand of the function environment and whose only argument is the original expression. This function call is evaluated and the result is displayed through a renewed call of the output mechanism. The output is displayed unquoted, to allow any desired output.

The exact meaning of *DOM_EXEC* objects as second operand of a function environment is described in detail in section 2.3.17.2. If the second operand of the function environment consists of an expression of type *DOM_NIL*, then this expression is displayed as a function call using the operands of the expression as arguments. If the considered function environment was obtained by substitution of the operator of the expression to be displayed, then the original operator (which has to be an identifier in this case) is used as the function name. Otherwise the output of the function environment itself is used as operator.

In all other cases the expression is regarded as a formal function call and therefore the output of the operator is used as function name and the output of the other operands as arguments.

Example 46 *The output of sums shall be converted into functional notation with the identifier* new_plus *as function name. For this purpose the function environment assigned to* _plus *has to be modified. The second operand of this is replaced by the following procedure:*

```
>> f := proc(x)
   begin
       subsop(x, 0 = hold(new_plus));
   end_proc:

   _plus := subsop(_plus, 2 = f):
```

If the expression

```
>> a+b+c;
```

is to be displayed then firstly the operator of the expression, i.e. the identifier _plus *is evaluated to the modified function environment. Then a function call with this function environment as operator and the call* _plus(a, b, c) *as only operand is created.*

The evaluation of this call returns the value _plus(a,b,c) *(a call of an unknown function because the identifier* new_plus *has no value). Then this is used as the output of the original expression.*

At this point please note that by returning an expression, that contains the original expression to be displayed, an infinite recursion may be generated.

2.3.16   Function Environments

The following description of function environments is primarily directed to the advanced user, who has already gained experience in using the system and now seeks new possibilities to change the functionality of existing operators and system functions or who wants to construct new functions with them.

Function environments are objects of the basic type *DOM_FUNC_ENV*. All information necessary for evaluation, output and type determination of expressions is stored in a function environment. In general predefined MuPAD functions, regardless if they are system functionsor library functions are represented as a function environment that is assigned to the corresponding identifier. Only some library functions and user-defined procedures, which have no further attributes (see next section), consist of a procedure definition which is assigned directly to the corresponding identifier. Function environments have to have three operands. As mentioned above their first operand is responsible for the evaluation of function calls. To perform this evaluation the function environment is substituted by its first operand. The second operand controls the output of such an expression. Both operands can be of any type. The third operand can be `NIL` or a table, where special attributes of the function to be represented can be stored.

The function `func_env` can be used to create new function environments. It is called with three arguments which correspond to the three operands of the function environment to be created. Independent of its operands the evaluation of a function environment leads to the function environment itself.

2.3.16.1   Attributes of Function Environments

As mentioned above the third operand of a function environment can be used to store special information about the corresponding function. For example under the index `"diff"` a procedure can be stored which computes the derivative of the represented function. The exact meaning of this entry is described in connection with the function `diff`. Other examples are the entries `"D"` and `"expand"` which determine how the corresponding operator is to be differentiated respectively how the expression is to be expanded. There are many of such predefined attributes and the user is able to add its own ones.

To simplify the entry of values into the attribute table the function `funcattr` is available. It is called with three arguments, the function environment to modify, the index and the value to be entered, and returns the changed function environment. To read-out a value from the attribute table the function `funcattr` is called with two arguments — the function environment and the index. If no value is stored under the given index then `funcattr` returns the value `FAIL`.

A special attribute is stored under the index `"type"`. This attribute defines the expression type of a function call of type *DOM_EXPR* whose operator this function environment is respectively whose operator can be evaluated to this in a one-step substitution.

Example 47 *The call*

```
>> type(a+b+c);
```

evaluates to the character string "_plus" because the operator of the sum **a+b+c** *consists of the identifier* **_plus** *which can be evaluated in a one-step substitution to a function environment containing the character string "_plus" under the index* **"type"**.

Another important attribute can be stored under the index **"print"**. This attribute determines the output of the function environment itself as described in the following.

2.3.16.2   Output of Function Environments

For the output of a function environment, firstly the attribute table is searched for an entry under the index **"print"**. If such an entry exists a function call is created with the associated value as its operator and the function environment as its only argument. The result of evaluating this call is then used to display the function environment. The output takes place unquoted i.e. if the return value is a character string then this is displayed without quotation marks and with expanded special characters. By this the definition of any desired output is possible.

At this point it should be noticed again that function calls, whose operator is of a basic type like character strings or numerical values return this operator as their result. Therefore an expression of this kind entered into the attribute table under the index **"print"** leads to the use of this expression as output of the function environment.

If no entry can be found under the index **"print"** in the attribute table then the first operand of the function environment is printed instead.

Example 48 *The identifier*

```
>> time;
```

leads to the substitution of the identifier by the assigned function environment. This is then returned as **time** *because no entry is stored in the attribute table under the index "print" and therefore the output of the first operand of the function environment, an expression of type* **DOM_EXEC**, *is used instead.*

2.3.17 Directly Executable Objects

Directly executable objects are of the basic type *DOM_EXEC* and serve as an interface of the C functions of the MuPAD kernel to the user's level. The main idea concerning directly executable objects is to have a kernel function, stored in the first operand, which is to be executed. This possible generic function needs further arguments for its evaluation and these arguments are stored in the other operands of the object. These arguments are usual MuPAD objects which can be manipulated by the user.

Directly executable objects have at least four operands. The types of the first four operands of a *DOM_EXEC* object are subject to certain limitations, because *DOM_EXEC* objects primarily occur as the first or second operand of function environments and in this context they have a special meaning.

DOM_EXEC objects can be distinguished between those objects which are used to evaluate function calls (occur as first operand of a function environment) and objects which influence the output of expressions (second operand of a function environment). In the following these two types are referred to as objects of the first respectively second type. Each of these types of executable objects determine the types of their first four operands. The types of further operands underlie no limitations.

The system function `built_in` is used to create directly executable objects. The arguments of the call are used as operands of the object to be created. Some of the operands contain a direct link to C-functions located in the MuPAD kernel. For that the accessible functions are numbered by positive integers. These numbers are used as arguments of the function `built_in` to specify a certain C-function. They are also returned as operands of the directly executable object when asking for them using the function `op`.

The output of *DOM_EXEC* objects is determined by the third operand (first type) respectively the fourth operand (second type). The corresponding operands have to be character strings and are displayed unquoted.

The exact restrictions for the operands of a directly executable object are presented in detail together with their special functionality in the next section.

Example 49 *The first operand of the function environment assigned to the identifier* _plus *is of type DOM_EXEC. Therefore the call*

```
>> type(op(_plus, 1));
```

returns `DOM_EXEC` *and the call*

```
>> op(op(_plus, 1));
```

returns the operands of this object:

```
817, NIL, "_plus", NIL
```

2.3.17.1   Directly Executable Objects used for Evaluation

In this section directly executable objects which occur as the first operand of a function environment or directly as an operator of an expression shall be described.

In this case the first operand of the directly executable object has to consist of a positive integer representing an internal C-routine. This function carries out the actual evaluation of the expression. For this purpose the object to be evaluated and the three remaining operands of the *DOM_EXEC* object are necessary. The second operand is either empty, i.e. it contains an object of the type *DOM_NIL*, or it again consists of a positive integer which refers to an additional evaluation routine in the MuPAD kernel.

A series of mathematical functions are realized in this manner. The first operand consists of a common function which carries out standard operations such as the checking of the arguments. The second operand contains the individual evaluation routine.

The third operand of a *DOM_EXEC* object consists of a character string. The corresponding identifier generally forms the operator of function calls whose evaluation is performed through this directly executable object. If a function call evaluates to itself then in many cases the operator of the result is replaced by this identifier.

The fourth operand consists of an optional remember table, in which, in the case of assignments to a corresponding function call, entries are made. If a remember table is present, before each evaluation it is checked if an appropriate entry already exists. If no entry is found then the evaluation of the function call continues unaffected. A missing remember table is represented by NIL. Apart from a few exceptions all functions react to remember tables. Through substitution of the remember table or through an assignment to a function call, the functionality of the function can be influenced.

Example 50 *Through the assignment*

```
>> _plus(a, b, c) := 100;
```

the value 100 is entered under the index (a, b, c) in the remember table of the first DOM_EXEC object of the function environment assigned to the identifier _plus. Subsequently the call a+b+c returns the value 100, while the call a+b+c+1 just as a+1+b+c-1 is not taken into consideration because the remember table is searched for an expression sequence corresponding to the arguments of the complete call. Only when this search yields no result the actual evaluation and simplification does take place. Evaluation of the individual operands occurs before the remember table is searched. The same functionality can be achieved with the call

```
>> _plus := subsop(_plus, [1, 4] = table((a, b, c) = 100)):
```

whereby the remember table is replaced directly.

2.3.17.2   Directly Executable Objects used for Output

In this section directly executable objects which occur as the second operand of a function environment shall be described. As described above these objects are executed when an expression of type *DOM_EXPR* has to be displayed. If the operator of such a function call is a function environment or if the operator can be substituted to a function environment by one-step substitution, then the second operand of the directly executable object determines the output.

The fundamental idea is that the first operand of the *DOM_EXEC* object contains a standard routine for output which reads-out further information like function names, operator symbols and operator priorities from the other operands. Thus, there are standard routines for the output of operators and functions as well as a series of different functions for the output of various statements. As already mentioned, the operands of *DOM_EXEC* objects of this type have to conform to certain conditions:

The first operand of the DOM_EXEC object has to contain a positive integer representing a C-routine. This function computes the actual output of the expression to be displayed. For this purpose the object to be displayed and the three remaining operands of the *DOM_EXEC* object are necessary.

The second operand has to be a positive integer which by the C-routine in the first operand is interpreted as the operator priority of the considered operator, function or statement. The higher this value is the greater is the priority of the operator to be displyed. This value has no effect on the input.

The third operand is a character string in which the operator symbol is stored. This is the symbol that appears between the individual operands of the expression if the expression is printed in operator notation. Otherwise a **NIL** is stored here. In this case output is generally functional. Exceptions are statements which are output neither functionally nor in operator notation because generally they have their own syntactical form. In this case the C-routine stored in the first operand is implemented in such a way that the other operands of the *DOM_EXEC* object are not needed.

In the fourth operand there is a further character string which, in the case of functional output, contains the identifier to be used as function name. Generally this character string is identical to the name of the system function: in the case of operators it is the name of the relevant underline function. This operand always exists because, for instance, system functions which describe an operator and therefore are, as a rule, returned in operator notation, are functionally represented when they only have one operand. Furthermore, as already described, this fourth operand is needed for the output of directly executable objects themselves.

Example 51 *The expression*

```
>> hold(_plus(a));
```

is returned as `_plus(a)`. *In this case, because only one operand is present, functional output is chosen. The fourth operand of the DOM_EXEC object is used as function name.*

Changing the operands of a *DOM_EXEC* object should be carried out with great caution because the C-functions stored in the first operand presuppose that certain types are present during their execution. When operands are changed in arbitrary manner a system crash can result.

Example 52 *The system function* `_plus` *represents the addition operator of the system. The second operand of the corresponding function environment*

```
>> op(op(_plus, 2));
```

has the following operands:

```
    1100, 12, " + ", "_plus"
```

The value 1100 represents the C-routine which is generally responsible for the output of any system operator. The value 12 represents the operator priority, which is needed for a correct parethis. The third value is a character string containing the operator symbol +. *The name of the corresponding underline function converted into a character string forms the last operand.*

Example 53 *With the exception of operators and statements all other system functions are displayed using a unique kernel routine. Thus, the second operand of the function environment of* sin

```
>> op(op(sin, 2));

    1101, 0, NIL, "sin"
```

has the same first operand as cos, ln *and many other functions. In this case the operator priority is of no importance and no operator symbol is defined. The last operand gives the function name.*

Example 54 *The system function* _if *represents an* if-*statement. The second operand of its function environment*

```
>> op(op(_if, 2));
```

has the operands:
```
    1111, 17, NIL, "_if"
```

The number 1111 *describes a kernel routine whose only task is the output of* if *statements. All statements have the priority* 17 *and have no operator symbol. Functional output can never take place with statements thus the last operand is only valid for the output of DOM_EXEC objects.*

By changing the operands of *DOM_EXEC* objects the output of various functions can be influenced. The following example explains this in more detail.

Example 55 *By changing the second operand of a DOM_EXEC object the bracketing of expressions can be changed. For instance the operator* and *initially has a stronger priority than the* or *operator. This is mirrored in the priority values* 4 *and* 3. *If these values are exchanged then the bracketing of the output is changed. This has no effect on the input priority. Before changing the priority values the input*

```
>> d and ( c or a and b );
```

returns the expression unchanged. After exchanging the priorities with

```
>> _and := subsop(_and, [2,2] = 3):
   _or := subsop(_or,  [2,2] = 4):
```

the expression

```
>> d and ( c or a and b );
```

is incorrectly returned in the form d and c or (a and b).

Example 56 *As already described, the third operand is the operator symbol to be used. By changing this operand the output of the operator can be changed.*

If one wants the elements of an expression sequence to be separated by a pipe character instead of a comma, then the following substitution must be carried out:

```
>> _exprseq := subsop(_exprseq, [2,3] = " | "):
```

The output of the expression sequence

```
>> a, b, c, d;
```

then has the form a | b | c | d.

Using the functions `func_env` and `built_in` new function environments can be created and existing definitions can be modified. Both functions should be used only with great caution because the MuPAD system always assumes that function environments are correct and therefore does not perform a check.

2.3.18 Domains

In MuPAD types are ordinary objects: they are of type *DOM_DOMAIN* and are called domains. Naturally there are some domains predefined by the system (basic domains or basic types), but the user also has the possibility of defining new domains (user-defined domains or user-defined types). Most system functions determine their return value with the help of functions defined within the domains of their arguments – called methods – when called with arguments of user-defined type. Thus, object-oriented programming is possible with domains.

The contents of a domain consists of a set of equations of the form

$$\texttt{<index> = <value>}$$

Both `<index>` and `<value>` stand for any expression. The entries stored in a domain can be accessed by using the corresponding index and the function **domattr**. Entries, which should be interpreted as functions, are called methods. There is the possibility of adding, changing and deleting domain entries.

Example 57 *If the domain* D *contains the equation* "a"=5 *this means that the entry 5 exists under the index* "a". *Then*

```
>> domattr(D,"a");
```

returns 5.

If the index is a character string which represents a valid identifier the parser offers an alternative notation for this function call, namely `D::a`.

Example 58 *If, as above, the domain* D *contains the equation* "a"=5, *then*

```
>> D::a;
```

also returns the value 5.

To change the contents of a domain or enter new values, an assignment of the form

```
domattr(D, <index>) := <value>;
```

is used where `<index>` and `<value>` can be any expression. Through this assignment, the equation

```
<index> = <value>
```

is added to the domain. If `<index>` is a character string that represents a valid identifier then the assignment can also be written as

```
D::<index> := <value>;
```

When modifying domains, please note that, in contrast to all other data structures, domains are subject to the so-called reference effect, i.e. a domain that is to be changed is not copied beforehand. Hence the change in the domain can also affect other data.

Example 59 *If* D1 *is a domain then the following entry*

```
>> D2 := D1:
   D1::a := 2:
   D2::a;
```

returns the value 2, i.e. through the change in D1, D2 *has also been changed.*

With this reference effect of domains it is possible to create recursive data structures. Thus a domain can contain itself which is useful in many cases because in this way the domain knows itself.

Example 60 *If* D *is a domain then the following entry*

```
>> D::a := D:
   bool(D = D::a);
```

returns the value TRUE.

To create a new domain the system function **domain** is available. In its simplest form **domain** is called without arguments.

Example 61 *With the assignment*

```
>> D := domain();
```

a new domain, with no entries, is assigned to the identifier D.

domain() can be given a parameter which describes the contents of the domain to be created in more detail. This parameter can either be a table or a domain.

Example 62 *If* T *is a table and* D *a domain then with the assignments*

```
>> D1 := domain(T):
   D2 := domain(D):
```

two new domains are created. The contents of the domain D1 *are the same as the contents of* T *and* D2 *is a copy of the domain* D.

Exactly like tables, domains evaluate to themselves, i.e. the entries and indices are not evaluated. Only when reading-out of a domain the index is evaluated and then the domain is searched for the corresponding entry. If this entry exists it is returned. If no entry is stored under the corresponding index then

domattr searches the domain for a value with the index "domattr". If again no value is found **domattr** returns the value FAIL. Otherwise the found value is called with the domain and the original index as parameters and the result is returned by **domattr**. That value is stored in the domain under the original index for finding the value quickly the next time.

Example 63 >> D::domattr:=proc(dom, index)

```
    begin
        "Not yet implemented [".expr2text(index)."]"
    end_proc:
  domattr(d, 2);
```

returns

```
        "Not yet implemented [2]"
```

With the help of this mechanism an infinite number of methods and all kind of inheritance can be implemented.

Because a domain may be a recursive data structure and therefore no meaningful output exists, the name of the domain is searched for under the index **"name"** and this is used for the output. If this is a character string then this is returned unquoted.

Example 64 *With the following assignments*

```
>> D := domain():
  D::name := "Name";
```

a domain that is printed as **Name** *is created.*

2.3.19 Domain Elements

Each datum in MuPAD has a type which is represented by a domain, i.e. is a domain element. Apart from the elements of the basic domain users can create elements of their own domains and therefore adapt the system to suit their own wishes and ideas. Elements of a domain are created with the system function **new**. This function expects a domain as its first argument and any number of further arguments of any type. It creates a domain element with the type of its first parameter and the other arguments as its entries.

Example 65 *With the assignments*

```
>> D := domain():
   e := new(D, 1, 2);
```

a domain element, with type D and entries 1 and 2, is created.

How a domain element is processed in the various operations is determined entirely by its type. If MuPAD carries out a system operation with an element of user-defined domain it is checked if a method with an index associated with the operation is in the domain of the element. If this is the case then this method is applied and the result is returned. Failing this, dependent on the operation, an error is returned or a standard operation is carried out.

Example 66 *Let e be the domain element defined in the example above. Evaluating this domain element MuPAD checks if there is an entry with the index "evaluate" in the domain. Because this is not the case the domain element evaluates to itself.*

Conversely the domain element f is created as follows,

```
>> D := domain():
   D::evaluate :=
      proc(n)
      begin
         _mult(op(n))
      end_proc:
   f := new(D, 4, 3):
   f;
```

evaluates to 12.

2.3.19.1　Internally Used Methods

The domain indices used by MuPAD are:

convert: (see 2.3.13.2).

divide: (see 2.3.13.2).

domattr: If no value is found under an index in a domain then a method is searched for under the index "domattr". If one is found then it is called with the index as argument. The result of the method is entered under the relevant index in the domain. The entry is made explicitly in the domain so that the complicated search need not be carried out again at a later time. If no entry is found under the index "domattr" then a FAIL is entered under the index and returned as the result.

elemattr: If the function domattr is applied to a domain element with further arguments, then a method is searched for under the index "elemattr" in the domain of the element. If a method is found it is applied to the arguments of the function call and the result is returned.

Otherwise domattr works as if it was called with the domain of the element as parameter.

evaluate: If a domain element e is to be evaluated then MuPAD searches for a value f associated with the index "evaluate" in the domain of e. If this is the case then the domain element evaluates to f(e), otherwise it evaluates to e.

posteval: If a variable is evaluated with a remaining substitution depth of 1 (i.e. LEVEL−1 substitutions have already been carried out) then generally the value of this variable is only read-out. However, in some cases it is necessary to re-evaluate this value. For this reason a variable is evaluated as follows with a remaining substitution depth of 1:

1. The value w stored in the variable is read-out.

2. If the domain of w has a method f under the index "posteval" then f(w) is returned as the result, otherwise w is returned.

expr: (see 2.3.13.2).

func_call: If a function call h(a1, ..., an) (an object of type DOM_EXPR) is evaluated then firstly h is evaluated to the value e. If e is a domain element of a user defined domain then MuPAD searches for an entry f under the index "func_call" in the domain of e. If an entry f is found the result of f(e, a1, ..., an) is returned as the result of the function call, otherwise e is returned as result. It is important to notice that f does not evaluate its arguments. If some of the arguments are needed in an evaluated form the function **context** can be employed.

set_func_call: In MuPAD statements of the form h(a1, ..., an):=value are allowed when h represents a procedure or function environment. Similarly it is also possible to enter this statement when in h represents a domain element e. In this case a method f with the index "set_func_call" is searched for in the domain of e. If such a method is found then the result of the function call f(e, a1, ..., an, value) is assigned to h, otherwise an error is returned.

This procedure also works automatically on multiply indexed assignments when under the index `"func_call"` an appropriate method is stored.

_index: If an expression of the form `h[a1, ..., an]` is evaluated, whereby in `h` a domain element `e` is stored then a method `f` with the index `"_index"` is searched for in the domain of `e`. If an entry is found the result of the call `f(e, a1, ..., an)` is returned as the result of the expression, otherwise an error is returned.

set_index: In MuPAD statements of the form `h[a1, ..., an] := value` are allowed when a table or array is stored in `h`. Similarly it is also possible to enter this statement when a domain element `e` is stored in `h`. In this case a method `f` with the index `"set_index"` is searched for in the domain of `e`. If such a method is found then the result of the function call `f(e, a1, ..., an, value)` is assigned to `h`, otherwise an error is returned.

This procedure also works automatically on multiply indexed assignments when an appropriate method is stored under the index `"_index"`.

intmult: (see 2.3.13.2).

invert: If MuPAD calculates the multiplicative inverse of a domain element `e` then the system searches for an entry `f` with the index `"invert"` in the domain of `e`. If an entry is found the result of the function call `f(e)` is taken as the inverse, otherwise the standard result `1/e` is returned.

name: If an entry under the index `"name"` is stored in a domain, this entry is used to display this domain. Otherwise a standard output is provided (see page 63).

negate: If MuPAD calculates the additive inverse of a domain element `e` then the system searches for an entry `f` with the index `"negate"` in the domain of `e`. If an entry is found the result of the function call `f(e)` is taken as the inverse, otherwise the standard result `-1*e` is returned.

new: When creating a new domain element the user should never use the function **new** directly but should use a constructor provided under the index `"new"` (see 2.3.19.4).

norm: (see 2.3.13.2).

not: If MuPAD calculates the logical complement of a domain element `e` then the system searches for an entry `f` with the index `"not"` in the domain of **e**. If an entry is found the result of the function call **f(e)** is taken as the complement, otherwise the standard result **not e** is returned.

one: (see 2.3.19.4).

print: If a domain element **e** is to be output then MuPAD searches for an entry **f** with the index **"print"** in the domain of **e**. If an entry is found the result of the call **f(e)** is displayed, otherwise a standard output is displayed.

zero: (see 2.3.19.4).

In most cases the definition of methods in basic domains does not have any effect.

2.3.19.2 Overloading of Functions

A domain does not only determine how these above described internal MuPAD operations work on its domain elements but determines how most system functions work on them, i.e. most system functions are overloadable. The various system functions are overloadable in different ways. The exact overloading mechanism is described on the help page of the relevant function. It can be roughly understood as follows:

If a function g discovers that it does not know how it should react to its arguments (at least one of the function arguments is an element of a user-defined domain) then it searches in the domains of certain arguments (which arguments are chosen depends on the function) for an entry f with the name **"g"** of the function as index. If the function finds such an entry it applies f to its arguments and returns the result. If it finds no entries it returns an error or returns a datum which has been processed in a standard manner.

Example 67 *Let* D *be a domain representing numbers in their prime number factorization. Each domain element has the prime factors in their relevant multiplicities as its entries. Now the system function* isprime *may also be applied to the domain elements of the type* D, *whereby a domain element represents a prime number exactly when it has one entry. This can be achieved with the following statement:*

```
>> D::isprime :=
   proc(n)
   begin
```

```
    if extnops(n)=1 then
       TRUE;
     else
       FALSE;
     end_if;
  end_proc;
```

2.3.19.3 Manipulation of Domain Elements

To work efficiently with domain elements functions are needed to manipulate
already existing domain elements. For this purpose the normal manipulation
functions like **op**, **subsop** etc. are available. In order to allow the programmer to offer the user of domain elements a logical representation of them that
differs from the physical representation these functions are overloadable. However if the logical representation differs from the physical representation the
normal manipulation functions cannot be used to implement the logical representation. For this reason there are some manipulation functions that only
work on domain elements and are therefore not overloadable. The following
functions exist:

extop: This function is used to access the entries of a domain element. The
function must always have a domain element as its first operand.

1. If the function is called with only one argument then it returns an
   expression sequence with the operands of the domain element.

2. If the function is called with two arguments then the second argument must be a positive integer i. The function returns the i-th
   operand of the domain element.

   If i is equal to 0 then the function returns the domain of the element.

3. If the function is called with more than two parameters then it returns an expression sequence of the operands of the domain element
   whose indices correspond to the arguments of the function.

extsubsop: This function gets the domain element to be manipulated as its
first parameter. As its other arguments the function gets equations which
describe the manipulation to be carried out. The left hand side of each
equation is a nonnegative integer that gives the index of the operand
of the domain element to be changed. The right hand side gives the
replacement for the relevant operand. The indices can also be greater
than the number of operands of the domain element. In this case the

appropriate number of operands with the value NIL are entered in the domain element.

extnops: This function has a domain element as its single argument. It returns the number of operands of this domain element.

2.3.19.4   Logical Interfaces

When designing a new type you usually have an exact idea to which operations this type should react. However the optimal physical implementation may change at a later time. For this reason the developer of a domain must have the possibility to hide the inside of a domain from the user. This possibility is given by the ability of individual functions to be overloaded. In order to give the developer the possibility to program a logical interface the functions **new** and the functions introduced in section 2.3.19.3 are not overloadable. For this reason these functions should not belong to the interface of this domain.

To guarantee correct working with domains the logical interface of a domain should contain certain methods and entries as a standard:

1. Each domain should have a method "**new**" which creates a new element of this domain. This function should check for incorrect parameters.

2. Each domain that represents an algebraic structure with a neutral element regarding addition should store this under the index "**zero**".

3. Each domain that represents an algebraic structure with a neutral element regarding multiplication should have this entered under the index "**one**".

Domains representing coefficients of polynomials and requirements for these domains are described in section 2.3.13.2.

2.3.19.5   The Efficient Application of Domains

With the possibility of overloading functions it is very easy for the user to calculate, for instance, the sum of several domain elements because he can work with the usual + operator. However it is more efficient when programming not to use the overloadability of functions but to directly read-out the correct method from the domain and to apply it to the arguments. If one method is often used in a function then it is worthwhile reading this out of the domain once and then temporarily store it in a local variable. This can save much runtime because although extracting data from domains is fast it does take some time.

Example 68 *In the following a domain is defined that represents $\mathbb{Z}_7$ as a group regarding addition. It is simple to create a domain that represents $\mathbb{Z}_7$ as a field but for reasons of clarity only the group is implemented.*

```
>> Z7 := domain():
   # The method new is defined #
   Z7::new :=
      proc(x)
      begin
         if type(x) <> DOM_INT then
            error("Wrong type in Z7::new");
         end_if;
         new(Z7, x mod 7);
      end_proc:
   # The output is defined #
   Z7::print :=
      proc(x)
      begin
         "Z7(".op(x,1).")";
      end_proc:
   # The n-ary addition is defined #
   Z7::_plus :=
      proc()
         local x, i;
      begin
         x:=extop(args(1),1);
         for i from 2 to args(0) do
            x:=(x + extop(args(i),1)) mod 7;
         end_for;
         new(Z7, x);
      end_proc:
   # The unary minus is defined #
   Z7::negate :=
      proc(x)
      begin
         new(Z7, (7-extop(x,1)) mod 7);
      end_proc:
   # The binary minus is defined #
```

```
Z7::substract :=
   proc(x,y)
   begin
      new(Z7, (extop(x,1) - extop(y,1)) mod 7);
   end_proc:
# The 0 is defined #
Z7::zero := Z7::new(0):
# The name of the domain is defined #
Z7::name := "Z7":
```

The so created domain still has some unpleasant characteristics. For example, this domain must be stored under the name Z7 otherwise the methods of the domain do not work properly. However this can be changed by, for instance, the command

```
>> subs(Z7, hold(Z7)=Z7):
```

After this everywhere in the domain where the identifier Z7 was before there is now the domain that represents Z7, i.e. Z7 contains itself as datum. Notice that the variable Z7 now contains the new value of the domain due to the reference effect. Since the methods of this domain do not need the variable Z7 to store the domain any longer, the new domain can be assigned to another variable and Z7 can be deleted.

2.4   Operators

In this section the unary and binary operators of the system are described. With their aid operands can be combined to create a new element belonging to the domain *DOM_EXPR*. The following operators or groups of operators are available:

- mathematical operators,

- the operators **mod** and **div**,

- relational operators,

- logical operators,

- range and sequence operators,

- self-defined operators,

- set operators,

- concatenation operators,

- function calls,

- the index operator,

- any system functions.

The evaluation of the expressions in this section is subject to some rules. If an operator cannot be directly evaluated due to the type of its operands two different reactions are possible:

- The type of an operand is non-valid for the corresponding operator, e.g. a set is a non-valid operand of a sum. In this case a runtime error is produced and evaluation is terminated with the corresponding error message.

- The type of an operand can become a valid operand after a re-evaluation. This is possible, for example, with identifiers, unknown function calls or indexed identifiers. In this case the expression stays temporarily unevaluated, i.e. without being further executed, but with already evaluated operands. No run-time error is generated.

2.4.1 Mathematical Operators

The mathematical operators $+$, $-$, $*$, $/$ and $\hat{}$ are available for addition, subtraction, multiplication, division and exponentiation. Except for exponentiation, which can only be used as a binary operator, all of these can be used as n-ary operators. In contrast to addition and multiplication, subtraction and division have to contain at least two operands. An unevaluated addition or subtraction has the expression type "_plus", an unevaluated multiplication or division the expression type "_mult", an unevaluated exponentiation has the expression type "_power".

Subtraction and division play a special role because these are not implemented as separate operators. Subtraction of an expression is represented as addition of the corresponding opposite element. Similar, division is the multiplication with the inverse. The inverse in this case is obtained by raising to the power of -1 (see example 158).

MuPAD also contains the two operators `div` and `mod`. These will be described separately in the next section.

Example 69 *The subtraction* `a - b` *is understood by the system as*

    a + (-b)

The quotient `a / b` *is only a shorter version of the expression*

    a * b ^ (-1).

2.4.2 The Operators `mod` and `div`

The operators `mod` and `div` are binary operators. For integers $m \neq 0$ and a, a `div` m and a `mod` m are the integers with

$$a = (a \text{ div } m) \cdot m + (a \text{ mod } m)$$

and

$$0 \leq a \text{ mod } m \leq |\, m \,| - 1.$$

If a and m have a numerical type but do not evaluate to integers a runtime error occurs. In case of a non-numerical argument the function call with the evaluated operands is returned. The result is of expression type `"_div"` or `"_mod"`.

Example 70 *The expression*

```
>> (-3*7) div 4;
```

returns the result -6.

Besides this classical definition the operator `mod` has a further function in MuPAD. If a, b and m are integers, $m \neq 0$ and b and m are both prime, then

$$\frac{a}{b} \text{ mod } m$$

calculates the solution of the equation

$$a = b\,x$$

in $\mathbb{Z}/m\mathbb{Z}$.

Example 71 *The command*

```
>> (2/3) mod 5;
```

calculates the result 4 *of the equation* $3x = 2$ *in* $\mathbb{Z}/5\mathbb{Z}$.

With the aid of the functions **modp** and **mods** the user can choose between a positive and a symmetrical representation of the integers modulo m.

Example 72 *The function* mod *works initially with a positive representation. This can be changed with the assignment*

```
>> _mod := mods;
```

to a symmetrical representation. The input

```
>> 3 mod 4;
```

then returns the value −1. *The initial behavior is recovered by assigning the value* modp *to the identifier* _mod.

2.4.3 Relational Operators

Relational expressions (in short relations) are formed by combining two expressions with one of the *relational operators*

$$< \qquad <= \qquad > \qquad >= \qquad = \qquad <>$$

The relational operators are used to describe equations and inequalities. Inequalities of the form **x < y** or **x > y** have the expression type "**_less**", inequalities of the form **x <= y** or **x >= y** have the expression type "**_leequal**". These inequalities are always converted into one of the forms **x < y** or **x <= y**. Equations have the expression type "**_equal**", and inequalities of the form **x <> y** have the expression type "**_unequal**".

In a Boolean context, for instance in the evaluation of the condition of an **if** statement or the explicit Boolean evaluation of a relation by the system function **bool**, an attempt is made to evaluate the relation to one of the Boolean constants **TRUE** or **FALSE**. With inequalities this is only possible when both operands of the relation have a real value. Any object can be tested for equality.

Example 73 *With*

```
>> equ := a < 4*5 + 3;
```

equ *is assigned the inequality* a < 23.

```
>> type(equ);
```

results in "_less". *According to this, the call of*

```
>> a := 14:
   bool(equ);
```

results in the value TRUE.

2.4.4   Logical Operators

The logical operators **and**, **or** and **not** are available. They are used to create a new logical expression by combining several logical expressions. The operators **and** and **or** are binary, the operator **not** is unary.

MuPAD simplifies expressions using the rules of DE MORGAN even when an explicit evaluation is not possible. Unevaluated expressions have the expression types "_or", "_and" or "_not".

Example 74 *The expression*

```
>> TRUE or 1>2;
```

returns the value TRUE. *The expressions*

```
>> not(aa>bb);
   not(cc=dd);
```

are simplified to

```
   aa <= bb
   cc <> dd
```

Thus

```
>> not( (aa<=bb) or (cc<>dd) );
```

returns bb>aa and cc=dd.

Please note that in case of a Boolean evaluation at first all operands of the logical expression undergo Boolean evaluation. The result is then calculated.  The process, common in other programming languages, whereby the operands of the operator **and** are evaluated one after the other from left to right until the first expression is found that returns **FALSE**, is not implemented here. Therefore no argument of a logical expression may lead to a runtime error, even when a previous operand already determines an unambiguous result.

2.4.5 Ranges and Sequences

2.4.5.1 Expression Sequences

Expression sequences consist of a series of expressions separated by commas. The comma is a binary operator. During the evaluation of an expression sequence each single operand is evaluated and then returned at its position in the sequence. In addition flattening of the operands, as described in section 2.3.9, as well as deletion of objects of the basic type *DOM_NULL* takes place. The individual expressions in the sequence can be accessed with the help of indexing. Similar to lists, individual elements of expression sequences can be changed by assignments to the corresponding indexed identifiers. This expression type is called "**_exprseq**".

Example 75 *The command*

```
>> a := 3: b := 4: c := 5:
   S := a+b, c, c*5, 10, d;
```

gives the sequence 7, 5, 25, 10, d. *The identifier* d *can be accessed as the fifth element of the sequence with*

```
>> S[5];
```

Expression sequences can also be used as parameters of a function call or of an indexed identifier. In this case flattening also takes place.

Example 76 *After the assignment*

```
>> a := 1, 2, 3;
```

the call

```
>> max(a);
```

returns the value 3.

2.4.5.2   The Range Operator

The range operator .. is another binary operator. It is used for the specification of series of integers. Apart from other uses ranges are, for instance, useful for the definition of arrays with the help of the function **array**. When combined with sequence operators, ranges make the explicit construction of expression sequences possible (see example 77). Ranges have the expression type "**_range**".

2.4.5.3   The Sequence Operator

The sequence operator \$ can be used both as a binary and unary operator. An evaluation as a unary operator can only take place when the operand represents a range. During the evaluation of such an expression the range is converted into the corresponding expression sequence.

Example 77 *The expression*

```
>> $ 1..5;
```

evaluates to the expression sequence 1, 2, 3, 4, 5.

A unary sequence always remains unevaluated when the operators of the range are not integers. If the first operand of the range is larger than the second then an element of the domain *DOM_NULL* is returned as the result. The unevaluated sequence operator represents an object of the expression type "**_seqgen**".

There are two ways of using the sequence operator as a binary operator. In the first case it has to be an expression of the form

```
<expr1> $ <ident> = <expr2> .. <expr3>
```

where the range `<expr2> .. <expr3>` has to evaluate to an integer expression sequence. `<ident>` has to be a non-assigned identifier. In this case the evaluation leads to a sequence. Its elements are formed by all elements of the form `<ident>` in the expression `<expr1>` being replaced by the values `<expr2>` to `<expr3>`.

In the second kind of using the binary sequence operator the expression has to be of the form

```
<expr1> $ <expr2>
```

The result of the evaluation is an expression sequence which contains `<expr1>` exactly `<expr2>` times. Of course this only makes sense when `<expr2>` evaluates to a non-negative integer. With this form of the sequence generator `<expr1>` is evaluated only once and the result is copied.

Example 78 *The expression*

```
>> 1/i $ i=1..3;
```

returns the expression sequence 1, 1/2, 1/3. *Through*

```
>> abc $ 10;
```

the sequence

```
    abc, abc, abc, abc, abc, abc, abc, abc, abc, abc
```

is created.

The first form of the binary sequence operator leads to an error when the index variable already has a value before execution. Because the second argument of the binary sequence opererator, in contrast to the first argument, is fully evaluated before execution the index variable is also replaced by its value and a reasonable execution of the sequence operator is prevented. An evaluation of the index variable can be prevented by using the system function `hold`.

Example 79 *If the identifier* i *already has a value the problem of premature evaluation of the identifier can be prevented with the call*

```
>> 1/i $ hold(i)=1..3;
```

As expected this returns the sequence 1, 1/2, 1/3.

2.4.6 Self-defined Operators

With the aid of the **&** symbol the user has the possibility to define a new binary associative operator. An expression of the form

```
<expr1>  &<ident>  <expr2>
```

is represented internally as a function call

```
<ident>(<expr1>,<expr2>)
```

Therefore unevaluated expressions have this form i.e. the expression type `function`. By assigning a procedure definition to the identifier `<ident>` the corresponding binary operator can be defined.

Example 80 *An expression of the form*

```
>> (a+b) &f (c*d);
```

is equivalent to the function call `f(a+b, c*d)`.

If, with the help of the **&** symbol, the operator notation is used in an n-ary manner, then the operator is supposed to be associative and the operands are flattened. This does not occur in the functional notation. This process can be prevented by the explicit use of brackets.

Example 81 *The call*

```
>> a &f b &f c;
```

is automatically flattened, i.e. the expression `f(a, b, c)` *is created. This can be prevented by the explicit use of brackets. Thus*

```
>> a &f (b &f c);
```

is equivalent to `f(a, f(b, c))`.

Naturally, with the aid of the **&** operator, binary system functions can also be addressed.

Example 82 *The assignment*

```
>> a := b;
```

can be written by using the underline function **_assign** *(see section 2.9):*

```
>> _assign(a, b);
```

or

```
>> a &_assign b;
```

2.4.7 Set Operators

For the processing of sets the system provides the binary set operators

`union`, `minus` and `intersect`.

With their help the user can determine the union, difference and intersection of sets. If the operators cannot be fully evaluated, objects of the expression types "`_union`", "`_minus`" or "`_intersect`" are created.

Example 83 *Let the following sets be defined:*

```
>> M := {a, b, c, d}:
   N := {c, d, e, f}:
   O := {g, h}:
```

Then the three expressions

```
>> M union N;   M intersect O;   N minus M;
```

have the values:

```
   {a, b, c, d, e, f}
   {}
   {e, f}
```

2.4.8 The Concatenation Operator for Identifiers and Character Strings

With the aid of the concatenation operator . various operations can be carried out according to the operand type. The operator is binary, i.e. it affects two operands. There are five different type combinations possible. The mechanism of each of the five combinations is different. Expressions of this form that are not fully evaluated have the expression type "`_concat`".

The concatenation

- `<list1>.<list2>` creates a new list, in which all elements of the initial lists `<list1>` and `<list2>` are contained (see example 22).

- `<string1>.<string2>` creates a new character string.

Example 84 *Through*

```
>> "hello "."world";
```

both character strings are concatenated to "hello world".

- **<string>.<ident>** also creates a string. The same result is obtained as if the concatenation of the character strings **<string>** and "**<ident>**" had been carried out.

- **<ident>.<string>** creates a new identifier. Because this identifier, in contrast to character strings, can only consist of a combination of letters, numbers and the underline, **<string>** has also to satisfy the same conditions that makes this expression result in a valid MuPAD identifier.

Example 85 *The concatenation*

```
>> a."bc";
```

creates the new identifier abc. *However, a call of the form*

```
>> a."b c";
```

leads to an error because ab c *does not represent a valid MuPAD identifier.*

- **<ident>.<expr>** also creates an identifier, as long as the expression **<expr>** evaluates to a non-negative integer or an identifier.

Example 86 *The input*

```
>> b := 123:
   a.b;
```

evaluates to the identifier a123. *The expression*

```
>> a.i $ i = 0..3;
```

returns the expression sequence: a0, a1, a2, a3.

Please note that the concatenation operator, exactly like all other operators, firstly evaluates its arguments. This can lead to errors when concatenating identifiers that have been previously assigned a value. This can be corrected by embedding the corresponding identifier in a hold call. 

To enable the use of the hold function on the left-hand side of the assignment it is for syntactical reasons necessary to place the entire concatenation expression in brackets.

Example 87 *If the variable* a *already has the value* 10 *then the identifier* a.1 *can be created by the concatenation:*

```
>> hold(a).1;
```

If the so-created identifier is to be assigned a value this can be accomplished by

```
>> (hold(a).1) := value;
```

An identifier created by a concatenation is not re-evaluated independently of the substitution depth.

Example 88 *Let the following identifiers be defined:*

```
>> a:=b: c:=d: bd:=e: e:=100:
```

Independent of the substitution depth set — the environment variable LEVEL *should have at least the value* 1 *so that the concatenation operator is evaluated — the call*

```
>> a.c;
```

returns the result bd. *The identifier is not re-evaluated even with a higher substitution depth.*

2.4.9   The Concatenation Operator for Functions

The concatenation operator for functions @ is used for the execution of functions one after the other and widens the possibilities of working with functions. Generally concatenated operators are used in the operator of an expression (see section 2.4.10) and result in the execution of the operators one after the other. It can be used as an n-ary operator. Expressions created in this manner have the expression type "_fconcat".

Example 89 *Let* f, g *and* h *be unassigned identifiers. Then the call*

```
>> (f @ g @ h)(100);
```

returns f(g(h(100))). *The command*

```
>> f @ g @ h;
```

is returned unchanged.

2.4.10 Operations on Functions

With the aid of function arithmetic, functions can be combined by operators and can be used immediately in the form of a new function call on given arguments. The operator of such a new function call generally consists of a bracketed expression which contains the function operator or operators as well as the corresponding operands. For the operations addition, subtraction, multiplication and division the arguments are used on each operand of the operator and the result is calculated with the help of the function operators.

Example 90 *Let* f *be the following procedure. The identifier* g *is unassigned.*

```
>> g:=NIL:
   f:=proc(n)
      name f;
      begin
        n^2;
      end_proc:
```

The call

```
>> (f + g)(2);
```

returns the result g(2)+4.

Further pre-defined operators of this sort are the exponentation as well as the concatenation of functions (see section 2.4.9). If the operator of a function call consists of an expression created with the aid of the power operator then the arguments of the function call are used on the first operand of the power. Finally the result is raised to the power of the second operand of the power.

Example 91 *The call*

```
>> (f^2)(x);
```

then returns the result f(x)^2.

All other operators of the system have no special effect in this context. Thus a corresponding function call only evaluates its arguments. With the aid of the system function **funcattr** function arithmetic can be defined for further system operators. For this purpose a corresponding routine has to be added to the attributes of the function environment of the respective operator or the relative system function. This has to be stored under the index "operator". There are two different methods of defining such an extension. On the one hand a function already in the system kernel can be used. These are already present in the function environments of the operators described above and have the names `op_plusmult` for the four fundamental arithmetic operators and `op_power` or `op_fconcat` for the exponentiation and concatenation respectively. Which function to be used is dependent on the functionality of the operator to be extended.

Example 92 *If the* union *operator is also to be used as a function operator this can be carried out with the aid of the function* `op_plusmult`. *Firstly the function environment of the operator* union *has to be extended by the corresponding attribute:*

```
>> _union := funcattr(_union, "operator", op_plusmult);
```

The already present routine `op_plusmult` *can be used because, like for addition and multiplication, the argument of the function call is applied to all the operands of the function operator and after this the evaluation of the function operator takes place. Let* f *and* g *be functions, defined as follows. The call*

```
>> f(1) := {1}:
   g(1) := {2}:
   (f union g)(1);
```

then returns the value {1, 2}.

If the arguments should not be used on all operands of the function operators, it is also possible to realize the functionality with the help of a user-defined procedure. (If the arguments are to be used on the first operand and then this result is to be calculated with the other operands the already present routine `op_power` can be used.) This procedure can also be added with the help of the function `funcattr`. This procedure is called with two parameters: the unevaluated function call as well as the already evaluated operator. The result of the procedure is used as the result of the entire function call.

Example 93 *The range operator shall now be used to realize a n-fold nesting. This can be accomplished with the following procedure definition. For this purpose the function* `op_range` *is entered into the attribute table of the system function* `_range` *under the index* "operator". *With this it controls the functionality of the range operator in the operator of an expression.*

```
>> op_range := proc(func_call, operator)
   local num, func, args;
   begin
      num  := op(operator, 2);
      func := op(operator, 1);
      args := context(op(func_call));
      result := func(args);
      for i from 2 to num do
         result := subsop(func("dummy"), 1 = result);
      end_for;
      result;
   end_proc:

   _range := funcattr(_range, "operator", op_range);

   f := proc(n)
   begin
      n + 1;
   end_proc:
```

The call of

```
>> (f..3)(a);
```

then returns the value a+3.

Here the function op_range *is called with the as yet not evaluated function call
and the expression* f..3*. In the second argument the identifier* f *has already
been replaced by the procedure definition above. The procedure* op_range *now
extracts both operands of the second argument and evaluates the arguments of
the call in the correct context. Finally, in accordance with the number given
by the second operand of the expression* f..3*, a nested call of this procedure
is created and the result of this call is returned.*

2.4.11 The Index Operator

As already mentioned in the description of the basic types, expression sequences, lists, tables and arrays can be accessed with the help of the index
operator [ ].

Identifiers and formal function calls can also be indexed. Indexed expressions
have the expression type "_index". Assignments to indexed expressions are
also possible. This is described in section 2.5.2.2. The exact meaning and

the usage of the index operator on the different basic types can be found in the description of the corresponding types. For user-defined types there is the possibility of defining the functionality of the index operator using "_index" (see section 2.3.19.1).

With indexing of tables and arrays a subsequent evaluation of the value read-out of the table or array is carried out. This subsequent evaluation is carried out because tables and arrays — independent of their content — evaluate to themselves and therefore this evaluation has to be made up for on accessing. This subsequent evaluation can be prevented by the use of the function index_val. The usage of index_val and _index — the system functions with which the indexed expressions are represented — are identical, the only difference is the missing evaluation after reading from tables and arrays.

Example 94 *Let* T *be a table, which has been generated by the assignment*

```
>> T[a] := b;
   b := 2;
```

The calls

```
>> T[a];
```

and

```
>> _index(T, a);
```

are equivalent and both return the value 2. To read the unevaluated value b *from the table the function* index_val *can be used. The call*

```
>> index_val(T, a);
```

returns the value b. *Through the assignments*

```
>> index_val := subsop(index_val, 2 = op(_index, 2)):
   _index := index_val:
```

the entire indexing can be adapted to this method. The first assignment ensures that the output of the function index_val *corresponds to the output of the function* _index. *For this purpose the second operand of the corresponding function environment of the function* _index *is copied into the function environment of the function* index_val.

If an indexed object does not refer to a table, list, expression sequence or array, it can be returned with an evaluated index. Before indexing an object the indices are flattened.

Example 95 *The identifier* G *has no value. Let the identifier* a *be assigned an expression sequence of the form* 1, 2, 3. *Then the call*

```
>> a:=1,2,3;
   G[a, 4];
```

returns the value G[1, 2, 3, 4]. *The indexed identifier* G *is returned with an evaluated and flattened index.*

2.4.12 Priority of the Unary and Binary Operators

The priority of the unary and binary operators is given in the following table. The operators are noted from top to bottom with decreasing priority.

```
::
.
@
not
&<ident>
^
*  /
+  -
mod  div
intersect
minus
union
..
<  <=  >  >=  =  <>
$
and
or
,
;  :
```

2.5 Statements

MuPAD makes a variety of different *statements* available. These can be rough-ly subdivided into

- assignments,

- control structures for the conditional execution of statement sequences,

- loops,

- control structures for controlling parallel and sequential executions.

In this section the statements of the programming language will be individually introduced. These are objects of the basic type *DOM_EXPR*. As with other objects of this type, with the function **type** a finer distinction can be made. Statements can be used interactively as well as in the body of a procedure.

Just like usual expressions statements also display their result when they are entered interactively. The result of an assignment is the evaluated right-hand side, the result of a statement sequence is the result of the last evaluated statement. If the last element of a statement sequence is an expression then the result of its evaluation is considered as the value of the statement sequence.

In all other sequential statements the result is formed by evaluating the body of the statement. This body is generally a statement sequence, that under certain conditions may have to be multiply executed. The result of the final execution of this statement sequence then determines the result of the whole statement. If the body is empty then an element of the domain *DOM_NIL* is returned as the result. The commands **next** and **break** influence the result of a statement as follows: The result of the sequence is the entry preceeding the **next** or **break**. The result of a parallel statement is determined by a different mechanism (see section 2.8.1.1).

With the environment variable **PRINTLEVEL** it is possible to display interme-diate computations during the evaluation of a statement or procedure.

2.5.1 Statement Sequences

The statements described in more detail in the following sections can be used singly or combined into *statement sequences*. The latter has the expression type "_stmtseq".

Syntax:

```
<stmtseq>    ::=    <stmt>
                |  <stmtseq> <sep> <stmt> .
<sep>        ::=    ; | : .
```

When statement sequences are not used in a parallel context processing takes place in the order of input.

Example 96 *By assigning a statement sequence to an identifier it is possible to write a form of primitive program. Through*

```
>> c := hold((a:=1; b:=2));
```

the variable c *is assigned a statement sequence without this being evaluated. Only a new evaluation of* c *leads to both assignments being executed.*

2.5.2   Assignments

The statement that occurs most often is the *assignment*.

Syntax:

```
<stmt>   ::=   <name> := <expr> .
```

2.5.2.1   Assignments to Identifiers

In the simplest form of an assignment <name> is an identifier. <expr> is completely evaluated and then assigned to <name>.

The assignment of NIL to an identifier is a special case. With this the identifier is reset i.e. it evaluates to itself. Exceptions to this are the environment variables which, when NIL is assigned to, are reset to their default value (see section 2.6.4). The expression type of an assignment is "_assign".

Example 97 *Through the assignment*

```
>> a := 100;
```

the identifier a *with the value* 100 *is entered in the internal list of variables. After*

```
>> a := NIL;
```

a *is deleted from this list and the identifier* a *stands again for itself. The assignments*

```
>> a := 50;    a := b;    b := c;    c := d;    d := 80;
```

form a statement sequence. After they are executed a evaluation of

```
>> a;
```

returns the value 80. *However this is not the value entered for* a *in the variable list. There the value for* a *is the identifier* b *because* b *was the last value assigned to* a. *The output* 80 *is the result of the complete evaluation. This mechanism guarantees that a new assignment to the identifier* b *also has an effect on* a.

Only the right-hand side is evaluated during an assignment to an identifier. The left-hand side of the assignment remains unevaluated. This changes when a more complex expression forms the left-hand side.

2.5.2.2   Assignment to Expressions

Apart from a simple identifier on the left-hand side of an assignment an identifier which is created with the help of the concatenation operator is also allowed. In this case the left-hand side is firstly evaluated to the identifier. If this is not possible a runtime error occurs. Subsequently the assignment mechanism is the same as described in the previous section. In this context please note the special cases described in section 2.4.8 for the use of the concatenation operator on the left-hand side of an assignment.

It is also possible to change the elements of tables, arrays, expression sequences and lists by an assignment. For this purpose indexed identifiers are used on the left-hand side of an assignment. This indexed identifier describes the element in which a new value is to be inserted as well as the exact place of the insertion.

Hereby the left-hand side of the assignment is evaluated in a special manner. The indices are completely evaluated and the place of insertion is determined. If a correct indexing is not possible because, for instance, if an array is indexed with values outside its ranges, then an error is returned. Otherwise the right-hand side of the assignment is inserted into the corresponding object. If the operator does not evaluate to one of the types given above, then an empty table is created and the values are inserted into this.

Example 98 *When executing the assignment*

```
>> T[2] := e*f;
```

firstly it is checked if the variable T *evaluates to an array, a list, an expression sequence, a table or an element of a user-defined domain.*

If T *is an array, a list or an expression sequence then it is checked if the value 2 is a valid index. Dependent on this either an error message is given or the value* e*f *is entered under the index 2 in the array, list or expression sequence. If* T *is a table no further check is necessary because any expression is valid as index. The value* e*f *can be entered directly. Any previous value stored under the index 2 is overwritten.*

If T *is not of one of the types mentioned above then its value is deleted and it is assigned a new table. In this the value* e*f *is entered under the index 2.*

The case where the left-hand side of the assignment is a function call is explained in detail in section 2.6.8.

Furthermore, a combination of the concatenation operator, indexed identifiers and function calls is possible on the left-hand side of the assignment.

In the case of user-defined domain elements an entry is searched for under the index "_index" and the method found there is evaluated. A more detailed description can be found in section 2.3.19.1.

2.5.3 The if Statement

The if statement enables the user to conditionally execute expression sequences.

Syntax:

```
<stmt>        ::=   if <expr> then <stmtseq> end_if
                |   if <expr> then <stmtseq> <elsepart> end_if .


<elsepart>    ::=   else <stmtseq>
                |   elif <expr> then <stmtseq>
                |   elif <expr> then <stmtseq> <elsepart> .
```

Evaluating the `if` statement at first the complete expression following the `if` is evaluated in a Boolean context, even if it is a compounded expression. If this evaluation results in a Boolean constant then dependent on the result, the first statement sequence, or respectively the second sequence, if present, is executed.

If the Boolean evaluation is not possible then an error message is given. The expression type of an `if` statement is `"_if"`.

To simplify nested `if` statements the `elif` construct is available. This is equivalent to `else` `if` but does not require a new `end_if`.

Example 99 *The execution of the statement*

```
>> a := 1: b := 2: c := 3:
   if a>b then
       c := 1;
   elif b<c then
       c := 2;
   else
       c := 3;
   end_if;
```

results in the evaluation of the statement `c := 2`. *The statement above is equivalent to*

```
>> a := 1: b := 2: c := 3:
   if a>b then
      c := 1;
   else
      if b<c then
          c := 2;
```

```
    else
        c := 3;
    end_if
  end_if;
```

2.5.4 The case Statement

The **case** statement is a control structure that extends the concept of the **if** statement. Here an expression is compared with a number of given values and one or more statement sequences are executed. The **case** statement has the expression type `"_case"`.

Syntax:

```
<stmt>        ::=  case <expr> <ofpart> <otherpart> end_case .

<ofpart>      ::=  of <expr> do <stmtseq>
               |  of <expr> do <stmtseq> <ofpart> .

<otherpart>   ::=  otherwise <stmtseq>
               |
```

During the evaluation of these statements firstly the expression following the keyword **case** is evaluated and then compared with the values of the expressions in the **of** items. If these expressions are identical then the corresponding statement sequence is executed. If such a statement sequence is not ended by a **break** statement then all statement sequences of the following items are also executed. This process is only interrupted by the execution of a **break** statement or the end of the **case** statement. This behavior enables the user to re-use the same statement sequence for different comparison expressions.

Optionally the **case** statement can contain an **otherwise** part. This is the alternative to the checked cases and is always executed at the end of the statement unless one statement has not previously been ended with execution of **break**.

When the first identical expression has been discovered during the evaluation of a case statement all following statements are executed without the corresponding conditions being checked, unless a **break** occurs. This distinguishes the **case** statement in MuPAD from the functionality of the case statement

in Pascal where only those statements that correspond to the first identical expression are executed. However, the Pascal functionality is easily emulated by ending every of-do sequence with a **break**.

The functionality of the **case** statement in MuPAD is equivalent to the switch statement in the C programming language. In comparison to nested **if** statements the **case** statement is to be preferred for its higher processing speed and its compact notation.

Example 100 *The* case *statement*

```
>> a := 1: b := 2: c := 3: d := 4:
   case d
      of a   do s := 1;
      of a+c do s := 2;
      of b   do s := 3; break;
      of d   do s := 4;
   end_case;
```

leads to the evaluation of the assignments s := 2 *and* s := 3.

MuPAD makes further possibilities for influencing the execution of the **case** statement available. By using the command **next**, which like **break** interrupts the execution of the current statement, the next matching of item is executed.

Example 101 *If the following* case *statement is entered:*

```
>> i := 4:
   case i
      of 1 do s := 1;
      of 2 do s := 2;
      of 4 do s := 4; next;
      of 5 do s := 5;
      of 4 do s := s*2;
      of 5 do s := s^2; break;
      of 4 do s := s-2;
   end_case;
```

the assignments s := 4, s := s*2 *and* s := s^2 *are carried out.*

2.5.5 The for Loop

The **for** loop is used to execute a statement sequence several times. For this purpose MuPAD offers two different kinds of loop.

2.5.5.1 Running Through a Range of Values

In the first form of the **for** loop, execution is controlled by a counter, which runs through a range with a fixed step width. This form of the **for** loop has one of the expression types "_for" or "_for_down".

Syntax:

```
<stmt>        ::=   for <ident> from <expr> to <expr>
                       <steppart> do <stmtseq> end_for
                 |   for <ident> from <expr> downto <expr>
                       <steppart> do <stmtseq> end_for .


<steppart>    ::=   step <expr>
                 |
```

When evaluating the **for** loop at first the range of the index variable **<ident>** is determined. For this, both expressions **<expr>** that specify this range, as well as the expression that determines the step width **<steppart>**, if present, are evaluated. If the step width is not given then the default value 1 is used. The step width has to be positive. After the assignment of the starting value to the index variable the statement sequence is executed.

In the next step the index variable is newly calculated by adding or subtracting the step width depending on if it is the **to** or **downto** form of the loop. Then the statement sequence is once more executed. If the index variable exceeds the range the loop is terminated. After the execution the index variable retains this last value. If the loop body is never executed, the index variable is set to the starting value.

Example 102 *The loop*

```
>> for i from 1 to 5 step 2 do A[i] := i^2 end_for;
```

enters 1=1, 3=9, 5=25 *into the table* A.

If the expressions that specify the range of the index variable cannot be evaluated to a real value, or if a non-real value is assigned to the index variable in the loop body, then an error occurs during the evaluation of the **for** loop.

The range of the index variable as well as the step width are only evaluated once, at the beginning of the loop execution. Therefore later changes of these values have no effect on the evaluation of the loop.

In contrast to this the index variable itself can be changed in the loop body and correspondingly influences the evaluation of the loop.

Example 103 *The following statement*

```
>> for i from 1 to 5 do
     A[i] := i^2;
     i := i+1;
   end_for;
```

enters the same values into the table as the loop in example 102. However, as in example 102 here the index variable i has the final value 7.

2.5.5.2   Running Through the Operands of an Expression

Often a given statement is to be executed for each operand of an expression. Using the functions **op** and **nops** and the **for** loop introduced in the previous section this is possible.

Example 104 *After the assignment*

```
>> L := [1, 3, 5]:
```

the loop

```
>> for i from 1 to nops(L) do
     A[op(L, i)] := op(L, i)^2
   end_for;
```

leads to the same result as in example 102 except that here the loop variable

```
>> i
```

has the value 4.

However, the execution of this loop is very expensive because the extraction of each individual operand by the system function op has to be carried out more than once in each run through the loop. The following is a more elegant solution: Firstly all operands of the expression are determined, these are stored in a list and for each element of the list the loop body is executed. The elements of the list can be directly accessed so that the explicit determination of individual operands is not necessary.

In practice running through the operands of an object takes place quite often and for this purpose MuPAD makes a further loop construct available to the user. This is the second form of the **for** loop and it has the expression type "**_for_in**".

Syntax:

```
<stmt>   ::=   for <ident> in <expr> do <stmtseq> end_for .
```

In this form of the **for** loop the variable **<ident>** runs through the operands of the expression **<expr>**. The operands are determined at the beginning of execution by a single internal call of the **op** function. This method has a speed advantage over the use of the standard **for** loop.

Example 105 *The result of example 102 can also be obtained with the loop*

```
>> for i in L do A[i] := i^2 end_for;
```

2.5.6   The while Loop

A **while** loop in MuPAD has the following syntax.

Syntax:

```
<stmt>   ::=   while <expr> do <stmtseq> end_while .
```

It is used to repeatedly execute a statement sequence until the condition given by the expression is no longer true.

During the execution of the **while** loop firstly the expression to be tested is evaluated. If a Boolean evaluation is not possible then a runtime error occurs. If the result has the value **TRUE** then the loop body is executed until the evaluation of the condition results in the Boolean value **FALSE**. The expression type of the **while** loop is "**_while**".

Example 106 *The evaluation of the loop*

```
>> i := 2:
   while i<5 do i := i+1 end_while;
```

results in the loop body i := i+1 *being run through three times. After the execution of the loop the variable*

```
>> i;
```

has the value 5.

2.5.7   The repeat Loop

The **repeat** loop is a further possibility of repeatedly executing a statement sequence in dependence on a condition. It has the expression type "**_repeat**".

Syntax:

$$\langle stmt\rangle \quad ::= \quad \text{repeat } \langle stmtseq\rangle \text{ until } \langle expr\rangle \text{ end\_repeat} \,.$$

Here the statement sequence is executed until the termination condition is evaluated to **TRUE**. In contrast to the **while** loop the body of a **repeat** loop is run through at least once because the condition is checked at the end of the loop. If a Boolean evaluation of the condition is not possible the evaluation is terminated with an error.

Example 107 *The evaluation of the loop*

```
>> i := 2:
   repeat i:=i+1 until i=5 end_repeat;
```

also results in the loop body i := i+1 *being run through three times. This example also returns the value 5 for the variable*

```
>> i;
```

after the loop has been executed.

2.5.8   The Commands next and break

Two further commands are available for influencing the execution of a loop or a **case** statement. While executing a loop the command **next** leads to the current iteration being interrupted and if applicable re-starting with the next iteration. With the command **break** the entire evaluation of the innermost loop or **case** statement can be interrupted.

If the commands **next** or **break** are used in a procedure then the procedure is terminated (see section 2.6). These statements have the expression types "**_break**" and "**_next**".

Example 108 *The execution of the statements*

```
>> i:=1;
   while i<5 do
      if i=3 then
         break;
      else
         i := i+1;
      end_if;
   end_while;
```

results in the identifier

```
>> i;
```

having the value 3.

2.5.9   The quit Statement

The statement **quit** is used to exit MuPAD, i.e. the current session is terminated. There is has not to be followed by a semicolon or colon. If the **quit** command is given in a procedure it results in a jump back to the interactive level without MuPAD being quit. The **quit** statement has the expression type "**_quit**".

2.5.10   Statements as Expressions

Because statements can be evaluated to any kind of expressions (see page 88) their usage is allowed in every context. Consequently statements can therefore also be used as expressions. In this case the statement has to be inclosed by additional brackets. Dependent on the value of the environment variable **EVAL_STMT** the statements are evaluated or not. The default value is **TRUE**. If **EVAL_STMT** is set to **FALSE** the statements are only executed when one of the following conditions is fulfilled:

- The statement is entered interactively as a complete expression and not as a subexpression of another statement.

- The statement is part of a statement sequence that forms the body of another statement or a procedure definition.

By preventing the automatic evaluation of statements the user is able to access parts of the statement and manipulate them.

Example 109 *After setting the environment variable* EVAL_STMT *to* FALSE *the if statement remains unevaluated.*

```
>> EVAL_STMT := FALSE;
   stmt := (if a then b else c end_if);
```

With

```
>> op(%);
```

the operands a,b,c *of the* if *statement are obtained.*

Example 110 *The following statements return the given results:*

```
>> EVAL_STMT := TRUE:
```

```
>> c   :=  (a := b);

   b
```

```
>> f((a := b));

   f(b)
```

```
>> xx := (if TRUE then
            1; a := 2;
         else
            100;
         end_if);
```

2

```
>> (a := b) + (n := m);

    b+m
```

This demonstrates that statements can be evaluated in any context and the values returned by them can be used for further processing. In the case of **EVAL_STMT = FALSE** *none of the inner statements would have been evaluated because they all occur as part of a further expression.*

Apart from the possibility to control the evaluation of statements with the global variable **EVAL_STMT** there are two other functions in MuPAD, **eval** and **hold**, that control the evaluation of statements. With **eval** the user can explicitly force the evaluation of a statement when **EVAL_STMT** is set to FALSE. With **hold** the evaluation of a statement can be prevented.

Example 111 *Although* **EVAL_STMT** *is set to* TRUE, *the variable* **a** *is assigned the assignment* **b:=c** *through*

```
>> EVAL_STMT := TRUE;
   a := hold((b := c));
```

This is only executed when evaluating **a**, *as depicted in the following example:*

```
>> S := 1 + hold((x := y));

    1 + (x := y)
```

```
>> x;

    x
```

```
>> S;

    1 + y
```

```
>> x;

    y
```

Similarly the user can force the evaluation of a statement despite the assignment **EVAL_STMT := FALSE**:

```
>> EVAL_STMT := FALSE:

>> a := (b := c):

>> a;

     b := c

>> b;

     b

>> eval(a);

     c

>> b;

     c
```

This is not the usual usage of **hold** and **eval** (see section 2.2.5).

2.6   Procedures

A procedure definition is a further possible language expression and is a basic element of structured programming in MuPAD. Procedure definitions are elements of the domain *DOM_PROC*. A procedure element contains the following parts:

- formal parameters,

- local variables,

- options,

- procedure name,

- procedure body,

- remember table.

Apart from the procedure body, all of these parts are optional i.e. they need not
to be specified in a procedure definition. By assigning a procedure definition to
an identifier the procedure is indirectly given a name. This is later accessible
in the procedure through the variable **procname**.

Procedures satisfy the following

Syntax:

```
<stmt>           ::=   proc(<identseq>) <decl> <procbody>

<decl>           ::=   <decl> <decl_part>
                 |     <decl_part> .

<decl_part>      ::=   <localpart>
                 |     <optionpart>
                 |     <namepart>
                 |     .

<localpart>      ::=   local <identseq> ; .

<optionpart>     ::=   option <identseq> ; .

<namepart>       ::=   name <exprseq> ;
                 |     name ; .

<procbody>       ::=   begin <stmtseq> end_proc .

<ident_seq>      ::=   <id_seq>
                 |     .

<id_seq>         ::=   <ident>, <id_seq>
                 |     <ident> .
```

Example 112 *Through the assignment*

```
>> f := proc(n, m)
   local s;
   begin
```

```
      s  := n+m;
      s/2;
   end_proc:
```

the identifier f *is assigned a procedure which calculates the arithmetical mean of its two arguments. The variable* s *is local in the procedure (see section 2.6.3) and does not influence the value of* s *outside the procedure definition. With the call*

```
>> f(a*b, c);
```

the procedure is evaluated with the actual parameters a*b *and* c.

MuPAD does not distinguish between the assignment of a procedure definition or of any other object to an identifier. Generally, a procedure is available when the corresponding assignment to an identifier has been explicitly executed (see example 114). Procedures can also be called directly. For this, instead of the procedure name, the definition itself is used.

Example 113 *The assignment*

```
>> T[a] := proc(b) begin b^2 end_proc:
```

enters a procedure definition directly in a table. With the call

```
>> T[a](5);
```

the procedure defined above can be directly called with the argument 5. *A procedure name in form of an identifier is not necessary.*

The evaluation of the actual procedure definition does not change the definition itself, i.e. no simplification of expressions takes place in the procedure definition. Only when the procedure is called the evaluation of the procedure body takes place.

Example 114 *Let the procedure f be defined as follows:*

```
>> f := proc(x)
   begin
      g := proc(x) begin x^2 end_proc;
      x*2
   end_proc:
```

Then the procedure calls

```
>> g(5);     f(3);     g(5);
```

return the results:

```
g(5)
6
25
```

Only by evaluating f *the global variable* g *is assigned a procedure definition.*

2.6.1 The Procedure Body

The body of a procedure consists of a statement sequence, which is embedded between the keywords **begin** and **end_proc**. This statement sequence is sequentially processed during the procedure execution (see section 2.8) and can be empty.

The result of the final evaluated expression is also the return value of the procedure. To exit the procedure, apart from the possibilities described in section 2.5.8, the function **return** is available. Calling **return** in a procedure causes the procedure to be immediately exited. The argument of **return** is evaluated and then is used as the result of the procedure call.

Example 115 *The function* sqrt *is used to calculate the square root of its argument. Because this calculation should only be used for numerical arguments the following procedure firstly checks the type of the actual parameter. In case of a non-numerical argument the procedure returns the result* FAIL.

```
>> g := proc(x)
   begin
     if not(testtype(x, NUMERIC)) then
        return(FAIL);
     else
        float(sqrt(x));
     end_if;
   end_proc:


>> g(12);
   g(symb);

     3.464101615
     FAIL
```

The following property should be especially noted when evaluating the procedure body: In the procedure body complete substitution does not take place unless this is specifically requested by the call of **level** or when the substitution depth (see section 2.2.4) has been changed by using the environment variable **LEVEL**. The depth of substitution is restricted to 1 as can be seen in the following example.

Example 116 *Let the assignments*

```
>> a := b:      b := c:      c := d:
```

as well as

```
>> f := proc() begin a end_proc:
   g := proc() begin level(a) end_proc:
```

be carried out. Then the entries

```
>> a;      f();      g();
```

return the results:

```
    d
    b
    d
```

The identifier **a** *is fully substituted only on the interactive level or by explicit use of the function* **level***. In the procedure* **f** *the already mentioned one-step substitution takes place.*

Example 117 *Let the function* **f** *be defined as follows:*

```
>> f := proc(x)
   begin
      sqrt(x);
   end_proc:
```

The function **f** *is called with the value* x^2*. Because of the missing of name spaces a complete substitution of* **x** *would lead to a recursive definition. During parameter passing, as described in the following section, the value* x^2 *is assigned to the identifier* **x***. A multi-step substitution leads to the value* **x** *being multiply substituted and this leads to the recursive definition described. This effect is avoided in the single step substitution.*

2.6.2 Parameter Passing

Formal parameters are those variables that are specified as parameters in the definition of a procedure. The *actual parameters* are those values that are actually inserted during later calls of the procedure.

In example 112 the identifiers n and m are formal parameters. The actual parameters are the expressions a*b and c.

If an expression sequence is passed to a procedure the sequence is flattened before the actual parameters are assigned to the formal parameters (see section 2.3.9), as can be seen in the following example: 

Example 118 *The following simple procedure* f *returns the number of passed arguments.*

```
>> f := proc()
   begin
       args(0);
   end_proc:
```

If f *is called by*

```
>> f((a, b, c), d);
```

the result is 4 and not 2 (see section 2.6.2.2).

2.6.2.1 Evaluation of Actual Parameters

Parameter passing in MuPAD takes place through *call by value*. Firstly all the actual parameters are completely evaluated. Then a local variable is created for each of the formal parameters specified in the procedure definition and is then assigned the value of the corresponding formal parameter. This local copy of the actual parameter is used for the evaluation of the procedure. This does not change the actual parameters, i.e. assignments to a formal parameter in the procedure body have no effect on the value of this variable outside the procedure. In concrete terms this means that the value of the actual parameter outside the procedure cannot be changed by the execution of this procedure. If NIL is given as the actual parameter then the corresponding formal parameter behaves like a non-initialized local variable.

2.6.2.2 Access to Actual Parameters

Many programming languages only allow an equal number of actual and formal parameters calling a procedure. With this a fixed definite assignment between the actual and formal parameters is created in the order of their specification. Therefore, in the procedure body every parameter can be addressed through the formal name used in the definition.

It is possible in MuPAD to call a procedure with different numbers of actual and formal parameters. If the number of actual parameters is smaller than the number of formal parameters the additional formal parameters are treated as non-initialized local variables. If there are more actual parameters than formal ones a mechanism is necessary to access these parameters. In a procedure — and only there — the system function **args** is available for this purpose. The call **args()** returns an expression sequence that contains the actual parameters in the order they were entered. The call **args(0)** returns the number of the actual parameters. Furthermore the n-th parameter can be accessed with **args(n)**, whereby n lies between 1 and **args(0)**. Naturally, **args(n)** can be used instead of the formal parameter.

Example 119 *The following procedure calculates the sum of its arguments independent of their number.*

```
>> sum := proc()
   local i, S;
   begin
     if args(0) = 0 then
       return(procname())
     else
       S := 0;
       for i from 1 to args(0) do
         S := S + args(i);
       end_for;
     end_if;
   end_proc:
```

When calling the procedure without actual parameters the call is returned by using procname.

```
>> sum();

   sum()
```

2.6.3 Local Variables – Scoping

The declaration of identifiers is not necessary in MuPAD. However it is possible to declare identifiers as local in a procedure by using the keyword `local`. This limits the validity of the identifier to this procedure or any procedures called from it (dynamic scoping). In this manner local procedures can also be defined.

An identifier is visible in the procedure in which it is defined as local as well as in all procedures called directly or indirectly from this procedure. This is valid as long as none of these procedures contains a new `local` definition of the identifier or uses this identifier as formal parameter. In these cases the identifier is assigned a new value. The scope of this identifier then applies to all procedure calls resulting from this procedure.

Please note that the scope of a variable, contrary to programming languages like Pascal or C, is not statically determined but dynamically.

If the definition of an identifier is missing then it is considered as global. Consequently global identifiers can be understood as local on the uppermost level of the system.

The following example should demonstrate the scoping.

Example 120 *If the procedure* f *is defined as follows,*

```
>> a := 2: b := 3:
   f := proc(c)
   local a;
   begin
       a := 7+c;
       b := a
   end_proc:
```

then the call

```
>> f(4);
```

returns the result 11. *The assignments to the variables* a *and* b *in* f *have only an effect on* b *outside the procedure* f, *because* a *is locally defined in the procedure. This means that* a *still has the value* 2 *after* f *has been called whereas* b *is assigned the value* 11.

Example 121 *If the procedure* f *defined by*

```
>> a := 2: b := 2:
   f := proc(a)
   begin
      g := proc()
      local b;
      begin
         h := proc() begin a := 5*b end_proc;
         b := a;                                    #(1)#
         h() + a;                                   #(2)#
      end_proc;
      a*h()*g()*h()*a*b;                            #(3)#
   end_proc:
```

is now called with the parameter a

```
>> f(a);
```

then the evaluation of f(a) *takes place in step (3) of the evaluation of the
expression. This is evaluated as follows: The identifier* a *is equal to* 2, *because
the procedure is called with the actual parameter* a. *Because at this point
the procedure* g *has not yet been executed the procedure* h *is not yet known
and therefore the function call* h() *remains unevaluated. Evaluation of the
procedure* g *leads to the local* b *in* g *being set to* 2 *in step (1). Subsequently,
with the call of the now known procedure* h, a *receives the value* 10. *Therefore,
the procedure* g *returns the value* 20 *in step (2). The following call of* h *returns
the value* 10. *The variables* a *and* b *now have the values* 10 *and* 2 *and therefore
in step (3)* 2*h()*20*10*10*2 *is evaluated to* 8000*h().

Please note that the procedures g *and* h *defined in* f *are global variables after*
f *has been called i.e. they can be addressed outside the procedure* f.

2.6.4 Environment Variables

Apart from local and global variables there is a further kind of variable in
MuPAD. These are the so-called *environment variables*. These are system
variables that can be used as local variables after being defined as local in
a procedure. With this the scope of the variable is limited to the actual
procedure as well as the procedures called from this one. In contrast to the
usual local variables, environment variables when used as local variables are
initialized with their actual value outside the procedure.

Because the scope of a local variable is limited to the actual procedure, in this case assignments to the environment variables also have no effect on the values of the variables outside the procedure. Furthermore, if an assignment takes place, with most of the environment variables a special check of the assigned value is carried out because in many cases the set of possible values of these variables is limited.

Example 122 *In the procedure* f *the environment variable* DIGITS *is defined as a local variable:*

```
>> f  :=  proc()
   local DIGITS;
   begin
      print(float(PI));
      DIGITS := 20;
      print(float(PI));
      return();
   end_proc:
```

If the value of DIGITS *is set to* 5 *before* f *is called then the procedure returns:*

```
>> DIGITS := 5:  f();

   3.1415
   3.14159265358979323384
```

The subsequent input of

```
>> float(PI);
```

also returns the value 3.1415.

The following system variables are environment variables. They are given with their default values in the following table:

Environment Variable	Default Value
DIGITS	10
LEVEL	100
PRINTLEVEL	0
MAXLEVEL	100
HISTORY	[20, 3]
TEXTWIDTH	75
LIB_PATH	
READ_PATH	
WRITE_PATH	
EVAL_STMT	TRUE
PRETTY_PRINT	TRUE

The initial values of the path variables LIB_PATH, READ_PATH and WRITE_PATH are not given here because these are dependent on the local MuPAD installation.

The environment variable LEVEL plays a special role, because it is set up as a new local variable with the initial value 1 in each procedure in order to guarantee a substitution depth of 1 in procedures.

2.6.5   Options

By giving predefined identifiers after the keyword option further characteristics of a procedure can be stored. Until now MuPAD provides two options.

2.6.5.1   The Option hold

Before the evaluation of the procedure body usually all actual parameters of the procedure call are completely evaluated. The procedure then is executed with these evaluated parameters. This is usually the type of parameter passing that is wanted. However, in some cases this early evaluation can be inconvenient. By setting the option hold the automatic evaluation of the parameters can be prevented.

If necessary, a subsequent evaluation in the procedure can then be started by using the system function context. The function context has to be called with one argument. This argument is evaluated as usual. This evaluation does not take place in the actual context but in the context outside the procedure. Therefore, an evaluation prevented by using the option hold can be carried out later by using the function context. Problems created by new local identifiers hiding their values outside can be bypassed in this way. If global variables are changed by the procedure then these variables naturally influence the later evaluation with **context**. In the same way by using **context** the status outside the procedure can be changed, e.g. when assignments are executed in **context**.

The use of the function **context** is subject to some restrictions. The function has not to be embedded, i.e. a **context** call has not to be used as an argument of the function itself. Furthermore, the use of **context** in any parallel statements is forbidden.

Example 123 *The following procedure* **newassign** *enables the user to define an assignment by which the left-hand side is evaluated with the substitution depth 1 before the actual assignment is carried out. For this purpose the procedure uses the option* hold. *This is necessary in order to prevent a premature evaluation of the left-hand side.*

The left and right-hand side of the assignment are passed to the procedure **newassign**. *Then an assignment is created and embedded in* hold *in order to prevent immediate execution. In the following* subsop *call the arguments of the assignment are replaced by the actual parameters. The value of the left-hand side is substituted with level 2. This is equivalent to a usual one-step substitution when the option* hold *is not used. The value of the right-hand side is subsequently completely evaluated. In* subsop *no evaluation of the assignment takes place because the evaluation of a* last *call does not include a re-evaluation of the corresponding value. A subsequent evaluation can be achieved by using the function* eval *and is executed in the third instruction.*

```
>> newassign := proc(left, right)
   option hold;
   begin
     hold((a:=b));
     subsop(%, 1=level(left, 2), 2=level(right));
     eval(%);
   end_proc:
```

With the assignments

```
>> b := NIL: a := b:
```

and the procedure call

```
>> newassign(a, 2);
```

the variable b *is assigned the value 2. The value of* a *remains unchanged by the call of* newassign. *But* a *evaluates also to 2 when a multilevel substitution takes place.*

```
>> val(a); a;

      b
      2
```

The implementation above leads to an error, when the actual parameters and the formal parameters are correlated. Let, for instance, the identifier left *have the value* b *and the identifier* b *be unassigned. Then a call of the form*

```
>> newassign(left, 2);
```

leads to an unmeant result, because the formal parameter left *contains the value* left *and so a subsequent evaluation to* b *is no longer possible. The function* context *offers one possibility of avoiding this problem. For that the procedure* newassign *should be modified as follows:*

```
>> newassign := proc(left, right)
   option hold;
   local  value, stmt;
   begin
      value := context(right);
      hold((a:=a));
      stmt := subsop(%, 1=context(left), 2=value);
      context(stmt);
   end_proc;
```

With this implementation any name conflicts between actual and formal parameters of the procedure no longer produce errors. This is ensured by a consequent usage of the function context.

2.6.5.2 The Option remember

In applications it often happens that a function has to be evaluated again with the same arguments. Therefore it is a good policy to store results in order to avoid re-calculation. In MuPAD this can be achieved by the so-called remember mechanism.

A *remember table* is bound to each procedure in MuPAD. It is a table, in which the arguments of the procedure represent the index and the result of the function call represent the corresponding calculated values. When a procedure is called, irrespective of the option **remember** being set or not, the remember table is checked to see if an entry exists for the actual argument. If this is the case then the result is taken directly from the table. Otherwise the result is calculated and if the option **remember** is set then this result is entered in the remember table. With this, time-consuming function evaluations can often be replaced by a fast search in a table.

If the option **remember** is not set an entry in the remember table can be made with a direct assignment to the corresponding function call. This shall be described in the next section. Using the option remember the number of function evaluations can be reduced drastically, especially in recursive procedures.

Example 124 *The calculation of the n-th Fibonacci number* fib(n) *requires an exponential amount of calls when the calculation is made recursively with the following procedure:*

```
>> fib:= proc(n)
   begin
      if n<2 then
         n;
      else
         fib(n-1)+fib(n-2);
      end_if;
   end_proc:
```

However with the option **remember** *the amount of function evaluations becomes linear. At best this is also true for the runtime behavior. However, this is dependent on the access time of the remember table. In the example of the calculation of* fib(n) *in addition to the n+1 function evaluations there are n+1 writing and n-2 successful reading accesses to the table.*

The remember table is represented by a MuPAD table in the procedure definition. This table is the fifth operand in the data structure of a procedure and can be accessed with op. Caution is advised while using the option **remember** when the procedure has side effects on global variables or when the value of a global variable influences the procedure result. Both cases are not recognized by MuPAD and can therefore lead to errors.

Example 125 *The procedure*

```
>> f:= proc(n)
     option remember;
     begin
        w*n;
     end_proc:
```

multiplies its argument by the global variable **w**. *If the value of* **w** *changes, the procedure values taken from the remember table may become wrong:*

```
>> w := 10:
   f(5);

     50
```

```
>> w := 20:
   f(5);

     50
```

2.6.6 The Procedure Name

Every MuPAD procedure can explicitly be assigned a name. Generally such a name consists of an identifier but any expression is possible. The name is laid down in the procedure definition and can be used during execution through the variable **procname**.

There are many possible ways of assigning a name to a procedure. Generally, this is done during the input or the reading of a procedure definition as the right-hand side of an assignment. In this case the unevaluated left-hand side of the assignment is used as the procedure name. It is laid down in the data structure of the procedure if no explicit definition is given inside the procedure definition. A direct definition occurs when the keyword **name** is used. This can be followed by any expression. This expression is not evaluated and is stored in the data structure of the procedure. A directly defined name cannot be overwritten by an assignment of the procedure definition. Before execution of the procedure body the local variable **procname** is initialized with the name of the procedure. This name is either taken from the data structure of the procedure or, if this contains no name the unevaluated operator of the actual procedure call is used.

Example 126 *The procedure definition*

```
>> proc(a, b)
   name f;
   begin
      procname(a+b);
   end_proc:
```

contains the name **f** *through the definition* **name f**. *Therefore, a subsequent direct call with the parameters 3 and 4 returns the result* **f(7)**. *The same result will be obtained if the procedure is first assigned to the identifier* **f** *and then called using* **f**.

With an assignment a name can be stored when no other definition exists. Therefore the assignment

```
>> f := proc(a, b)
   begin
      procname(a+b);
   end_proc;
```

creates a procedure of the following equivalent form:

```
   proc(a, b)
   name f;
   begin
      procname(a+b);
   end_proc
```

Naming through an assignment or an explicit definition both lead to identical procedure definitions. The only difference is the assignment of the identifier **f**. *By defining the name in the procedure body no assignment is made and therefore a call of the procedure using this name is not possible.*

Giving only the keyword **name** without any argument the entry of a name by an assignment of the procedure can be prevented. In this case the procedure definition remains unchanged by the assignment.

Example 127 *Through the assignment*

```
>> f := proc() name ;
   begin procname
   end_proc;
```

*a procedure is assigned to the identifier **f**. However, the identifier **f** is not entered as the name of the procedure. Therefore, after the following assignment and the call*

```
>> g := f:
   g();
```

*the value **g** is returned. Without the definition* **name** *; this call would have returned the value* **f**.

2.6.7   The Operands of a Procedure Definition

A procedure definition contains six operands: the operator is an integer and represents the internal function which interprets the procedure. The operands are the formal parameters, the local variables, the options as well as the procedure body, the remember table and the procedure name. If an operand is not specified, then a **NIL** is returned as operand.

The procedure name belongs to the definition part of the procedure and is to be given before the procedure body it creates the sixth operand of the procedure.

Example 128 *The procedure is defined by*

```
>> f := proc(a, b)
   local c;
   option remember;
   begin
      c := a+b;
   end_proc:
```

and then called with the parameters 2 and 3.

```
>> f(2,3):
```

A call of **op** *returns the operands of the procedure in the form of an expression sequence:*

```
>> op(f);

   (a, b), c, remember, (c := a + b), table((2, 3) = 5), f
```

The last parameter forms a special case. If the assignment of a name is prevented by the entry **name** ; then the procedure definition has an empty expression sequence as its last operand.

Example 129 *The procedure*

```
>> f := proc(a,b)
   name ;
   local c;
   begin
      c := a+b;
   end_proc;
```

has the operands:

```
>> op(f);

   (a, b), c, NIL, (c := a + b), NIL, _exprseq()
```

2.6.8 Assignments to Function Calls

Like assignments to indexed identifiers as described in section 2.5.2.2, assignments to function calls are also handled as special cases. If a function call is assigned a value a check is carried out to see if the function name already contains a procedure definition. If this is the case then a remember table is stored in the procedure definition if one does not already exist. The corresponding entries are made there. If the function name does not contain a procedure definition then an empty procedure definition in the form of

```
proc() option remember; begin procname(args()) end_proc;
```

is assigned to it.

Example 130 *During the execution of the assignment*

```
>> f(1) :=  {c, d};
```

firstly it is checked if the identifier f *contains a procedure definition. If this is the case then a remember table for* f *is created if one does not already exist and the set* {c,d} *is entered in it under the index 1. Any entries already stored under this index are overwritten. If the identifier* f *does not contain a procedure definition then* f *is assigned an empty procedure definition with the option* remember *and the set* {c,d} *is also entered in the remember table of* f *under the index 1.*

Hence, similar to the creation of tables, the user can also create a function by assigning a value.

Example 131 *In order to show another, more complex example that combines a function call and indexing on the left-hand side of an assignment, we shall look at the assignment:*

```
>> f(a)[b] := z;
```

The identifier f *has no value before the assignment. Then with this statement* f *is assigned a formal procedure definition and a remember table is created for* f. *Because* f(a)[b] *is an indexed expression a table is created as described in section 2.5.2.2 in which the value* z *is entered under the index* b. *This table is the value which is entered in the remember table of* f *under the index* a.

After the assignment above

```
>> op(f);
```

results in the expression sequence:

```
    NIL, NIL, remember, procname(args()), table(a=table(b=z)), f
```

2.7   Pure Functions

A pure function is an expression of the type *DOM_EXEC*. It can be created using the functions **func** and **fun**.

Pure functions do not support local variables and therefore a functional style of programming is required.

Example 132 *Through the assignment*

```
>> f := func(
      (if x mod 2 = 0 then
          x/2
       else
          (x+1)/2
       end_if), x):
```

the identifier f *is assigned a pure function, which halves its arguments and then rounds to the next greater or equal integer.*

Local variables are not supported by pure functions. In this way the time necessary for the invocation of a function can be considerably reduced. However, this is only noticeable when the execution time of the function body is relatively short.

2.7.1 The Function Body

The function body is a MuPAD expression, which is executed when no corresponding entry is contained in the remember table for this parameter. The result of the execution then is the result of the function invocation.

The function body is evaluated with the substitution depth defined in the context of the function call.

Example 133 *The command series*

```
>> f:=fun(a*args(1)):
   a:=b: b:=c: c:=2:
   LEVEL:=1:
   f(1);
   LEVEL:=100:
   f(1);
```

returns:

```
   b
   2
```

This is due to the function body being evaluated in the first call with the substitution depth of 1 and in the second call with a substitution depth of 100.

The only change in the context concerns the history mechanism. Evaluating the body of a pure function, there is neither access to the history of the calculation before the evaluation nor does the evaluation of the body effect the history of the calculation after the evaluation. On the history of the evaluation of the body the history mechanism works as usual.

2.7.2 Parameters of a Pure Function

As in procedures *call by value* is used for parameter passing in pure functions.

At first the parameters are completely evaluated. During the evaluation of the function body the evaluated parameters can be accessed with the function

args. Furthermore by using this function the number of parameters and a sequence of all arguments can be obtained.

Formal Parameters for the specification of a pure function can be used with **func**. However, **func** substitutes these formal parameters Errors can be created because these formal paramters are sometimes substituted in positions where this is not desirable.

Example 134 *If the user defines the function*

```
>> f:=func(2*proc() begin x:=2 end_proc()*y, x, y):
```

then when the function is called an error message appears. This is the result of x being substituted in the procedure so that the body of the procedure call can no longer be evaluated.

2.7.3   Operands of a Pure Function

A pure function contains five operands, some of which have a fixed value. Because pure functions are expressions of type *DOM_EXEC*, they also have the corresponding operands.

Example 135 *The input*

```
>> f := fun(2*args(1)):
   f(0) := 1:
   op(f);
```

yields the following expression sequence:

```
    112, NIL, "fun(...)", table(0=1), 2*args(1)
```

The first operand is the index of the C-function responsible for evaluating the function. For pure functions this C-function has the number 112.
The second operand always has the value **NIL**.
The third operand always has the type *DOM_STRING* and represents the name of the function.
The fourth operand has either the value **NIL** or is an expression of type *DOM_TABLE*. This is the remember table of the function. It determines the values of the function for the parameters in the table. In example 135 the table determines that the function returns the value 1 when called with the parameter 0.

The fifth operand is the function body. It is evaluated when the function is executed and the value has not already been obtained by a look up in the remember table.

2.7.4 Assignments to a Pure Function

As with all data of the type *DOM_EXEC* a pure function can be assigned a value by assignment.

Example 136 *The function defined in example 135 returns for the argument* 0

```
>> f(0);
```

the value 1.

2.8 Parallelism

One of the fundamental development aims of MuPAD is the efficient processing of very large amounts of data. This aim is realized by using, among other things, parallel computers. Processing large amounts of data in a sequential system is not recommended due to the low processing speed and the lack of memory space. The processing speed can be easily increased by using a shared memory machine. The memory space can be increased almost infinitely by using a network of computers.

MuPAD offers two different forms of parallelism, *micro-* and *macro-parallelism*. Micro-parallelism is easy to use because MuPAD solves most of the problems that occur during parallel programming itself. With macro-parallelism, a parallel program can be optimally adapted to the underlying hardware so that even in networks with high communication times an acceptable speed can be attained.

While using micro-parallelism the user needs neither to deal with the distribution of tasks to the available processes nor with inter-process communication. The user needs only to mark certain areas of a program as simultaneously executable and MuPAD then takes over the concrete parallel execution of these sections.

However, MuPAD cannot satisfactorily solve this distribution of tasks when inter-process communication is too expensive. To use MuPAD efficiently on such a system the use of macro-parallelism is recommended. Macro-parallelism offers a network of independently working clusters. Each cluster can consist of many processes again and can be programmed using micro-parallelism. The inter-process communication within a cluster should be fast enough for running micro-parallelism efficiently.

Comfortable message passing constructs and global variables are provided for communications between the clusters.

Using micro-parallelism the user can develop parallel programs very easily. Using macro-parallelism the user can optimize his/her program for all hardware-architectures, even for those with slow communication.

2.8.1   Micro-Parallelism

The micro-parallelism offered by MuPAD is remarkable for its simplicity of use. There are parallel statements with which program sections can be simultaneously executed. The user's only task is to mark these statements. Load balancing and process communication are carried out by the system itself without the user having to deal with it. This form of parallelism makes it possible for the user to implement parallel algorithms in MuPAD without extensive knowledge of parallel processing.

As micro-parallel statements MuPAD offers two parallel **for** loops as well as one statement each for the creation of parallel and sequential blocks.

The management of the program sections to be simultaneously executed is always carried out with the help of the so-called *problem heaps*. The *tasks* created during a run of a parallel construct are entered into the problem heap. A task is a request to evaluate data and contains the data to be evaluated and additional information about the context for evaluation. This includes information about the parent – the task that created this task – as well as about any other child tasks – tasks which have been created by the parent simultaneously with this one. The context is necessary for access to the global variables of the task. The return of the result of a parallel sequence is also realized with the help of the context information. Further informations stored in a task describe its private variables — a kind of local variables. No history-information need to be stored in the task because each task has its own history context (see section 2.11).

The evaluation of a parallel statement takes place as follows: at first, the tasks corresponding to the statement are created with the information contained and are written on the problem heap. As soon as the tasks are on the heap the idle processes of the relevant cluster begin to remove the tasks from the heap and process them. For this purpose they read the task information, establish the described context and evaluate the given datum. Even the process that writes the tasks in the heap takes a part in this processing of the heap. Only when all the created tasks have been processed this process does continue in its original context. Due to this waiting for the end of the processing of the tasks a synchronization point is created in the execution.

The individual constructs for controlling micro-parallelism shall be described in more detail in the following. The possibility of prematurely terminating tasks shall also be mentioned. The use of micro-parallelism on a sequential machine is described in section 2.8.1.5

2.8.1.1   The Parallel for Loop

MuPAD makes two versions of the parallel **for** loop available. As in the sequential case (see section 2.5.5) these versions differ in the range of their loop indices. There is a parallel form of the **for to** loop as well as one of the **for in** loop. Naturally, a differentiation of the running direction of the loop variable makes no sense in parallel cases. In both cases, syntactical differentiation takes place by replacing the keyword **do** by **parallel**. These loops have the expression types **_for_par** and **_for_in_par** respectively.

Syntax:

```
<stmt>        ::=   for <ident> from <expr> to <expr>
                      <steppart> parallel <prvt>
                      <stmtseq> end_for .
                |  for <ident> in <expr> parallel <prvt>
                      <stmtseq> end_for .

<steppart>    ::=   step <expr>
                |  .
```

In the first form, during the evaluation of a parallel **for** loop firstly the start and end values of the run variable and their step width are determined. In the second form of the loop the expression to be processed is evaluated and its operands are determined. Subsequently an individual task is created for each resulting value of the loop variable. This consists of, among other things, the body of the loop and the corresponding value of the loop variable. These tasks are then written in the problem heap. When all the tasks have been processed the statements following the loop are evaluated.

In contrast to the sequential loop, where the result of the last execution of the body is the result of the loop, the result of the parallel loop consists of an expression sequence, which contains the results of the individual tasks. The order of the values in this sequence corresponds to the order of the operands, or respectively, the natural order as given in the value range of the loop variable.

The declaration of private variables *<prvt>* (which is also possible in **for** loops) is described in section 2.8.1.3.

Example 137 *The elements of a list are to be squared and finally returned in the form of a set. This can be achieved by the following sequential loop:*

```
>> L := [1,2,3,4]: M := {}:
   for i in L do M := M union {i^2} end_for;
```

whereby at the start **M** *is initialized as an empty set and* **L** *is the list to be processed.*

Obviously all iterations are independent so that instead of the sequential form above the parallel form of the **for** *loop can be used. However a simple replacement of the sequential form by the parallel form of the* **for** *in loop is not possible here because the assignment*

```
>> M := M union {i^2};
```

could lead to an error if two tasks try to access **M** *simultaneously with different values of* i. *This is due to the fact that firstly both read the contents of* **M** *and then execute the union. The tasks would then calculate different sets which they would then assign as values to the variable* **M**. *The result of the task assigning its value first to* **M** *would be lost because the other task would delete this when assigning its own result to* **M**.

Thus the result of the **for** *loop must be used. The following statement leads to the desired result:*

```
>> M := {(for i in L parallel i^2 end_for)};
```

Here the **for** *loop returns the expression sequence* 1, 4, 9, 16 *for the list* L := [1, 2, 3, 4]. *With the curly brackets this expression sequence is then converted into a set.*

2.8.1.2 Parallel and Sequential Blocks

To be able to execute any statements or statement sequences simultaneously MuPAD offers the **parbegin** as well as the **seqbegin** statement.

Syntax:

```
parbegin <prvt> <stmt_seq> end_par .

seqbegin <stmt_seq> end_seq .
```

If a statement sequence is enclosed by the keywords **parbegin** and **end_par** it means that the individual statements of the sequence may be processed simultaneously. Here, the corresponding tasks are created during the evaluation of the blocks and written in the problem heap, too. The declaration of private variables is also possible.

The statements enclosed by **seqbegin** and **end_seq** are always executed sequentially. With this construct it is possible to define sequential blocks in a **parbegin** statement. The statements have the expression types **_parbegin** and **_seqbegin**.

Example 138 *The execution of the statement*

```
>> parbegin
      a := 1;
      b := 2;
      seqbegin
        c := 3;
        d := 2*c;
      end_seq;
      e := 4
   end_par;
   a + b + c + d + e;
```

has the consequence that four tasks may be carried out simultaneously. Each of the assignments a := 1, b := 2 *and* e := 4 *is carried out by a task and the fourth task executes sequentially the statements* c := 3 *and* d := 2*c. *The expression* a + b + c + d + e *is only evaluated after these four tasks are finished.*

2.8.1.3 Private Variables

If parallel tasks access common global variables for the purpose of writing, then their values are dependent on the random sequence of the processing of tasks. Even simple reading of variables can sequentialize tasks because for internal reasons only one task can read from a variable at a time. Because reading from the stack is very fast this sequentialization is only noticeable when the number of processes in a cluster is very large or when these global variables are accessed frequently. If variables are only necessary locally in a task it is useful to define them as local. The **private** declaration of variables enables the user to give local variables the same name for each task of a parallel statement. These cannot influence or block one another.

Syntax:

$$\langle prvt \rangle \quad ::= \quad \text{private } \langle identseq \rangle;$$

If a variable is declared as **private**, then it is considered as local the task corresponding to the parallel statement and is therefore multiply represented in memory. In this way a blocking access of the tasks is prevented. The index of a parallel loop is also a **private** variable which however is automatically declared as such.

Example 139 *The following procedure calculates the row sum norm of a matrix A. The procedure assumes that the matrix is separated into blocks which can be created by a call of the form*

```
>> A := array(1..n, 1..m, [1,m]);
```

This enables the user to work directly on the subarrays (in this case the rows). This prevents unnecessary access to the matrix A and consequently a possible sequentialization.

```
>> proc(A)
      local i;
   begin
      for i in A parallel
      #i runs through all rows of A#
         private sum, j;
         sum:=0;
         for j in i do
         #j runs through all elements of the row i#
            sum:=sum+abs(j);
         end_for;
```

```
        end_for;
        max(%);
      end_proc;
```

The calculation of each individual row sum is executed sequentially, but all the calculations of the row sums are carried out in parallel. If the variable sum *were not declared as a* private *variable then the calculations of the individual row sums would influence one another. In this example this behavior is not possible because the variable* sum *is local for each row. The same is valid for the variable* j.

The parallel for *loop returns an expression sequence with the sums of the absolute values of the elements of the individual rows. By calling* max *with* % *as its argument the row sum norm is returned.*

2.8.1.4   Termination of Parallel Statements

To terminate the statement, or the tasks created by it, the commands **break** and **next** are available. In parallel constructs these have an analogue meaning as in the sequential case:

If the **break** command is used in a parallel statement then the current task and all descendant tasks of this parent-task – called connected tasks – are terminated. As the result of the statement the last value calculated by the task interrupted with **break** is given. For all other terminated tasks a NIL is entered in the resulting expression sequence. Tasks which were finished before termination enter their usual return values in the expression sequence.

The **next** command, which terminates the current iteration and continues then with the next one in sequential loops, has an analogue effect on parallel constructs. The current task is terminated, but no other tasks are affected. As the result of the terminated task its last calculated value is entered in the resulting expression sequence.

A further possibility of terminating parallel statements is the function **return**. Its effect on parallel constructs does not differ from its behavior in sequential cases. The current procedure is exited and the argument is the result of the procedure. In addition all tasks created in this procedure and their successors are either terminated or deleted from the problem heap (as their results are not relevant).

While the **next** command is only there to make the termination of a task easier for the user, the **break** and **return** commands enable OR parallelism because all connected tasks can be terminated when the desired result is reached.

Example 140 *A function is given which checks for a prime number in a transferred list L. If this is the case it should return* TRUE, *otherwise* FALSE. *This function can be implemented as follows:*

```
>> proc(L)
   begin
     for i in L parallel
         if isprime(i) then
           return (TRUE);
         end_if;
       end_for;
       return (FALSE);
     end_proc;
```

This program tests simultaneously if the list elements are prime. As soon as a prime number is found all connected tasks are terminated and TRUE *is returned.*

2.8.1.5 Parallel Constructs in the Sequential MuPAD Version

It is also possible to use the parallel constructs of the MuPAD language on a sequential computer. Programs that have been written for a parallel version can also be checked on a sequential computer for correct syntax. Furthermore, some logical errors can be detected.

This is made possible by statement sequences being processed at random. In this manner data dependencies are revealed.

Example 141 *This can be seen when the following* for *loop is looked at:*

```
>> for i from 1 to 4 parallel print(i) end_for:
```

Multiple executions create ever changing output, such as

```
        1   2   3   4
        3   1   2   4
        3   4   1   2
```

etc., because the print *statements are executed in a random order.*

2.8.2 Macro-Parallelism

As the micro-parallelism in MuPAD can only efficiently work when the inter-process communication time is low, MuPAD additionally offers macro-parallelism. This form of parallelism has the advantage that it can be efficiently used in nets with slow communication and that all forms of parallelism can be emulated with it. Its disadvantage is that it is more complicated to use and therefore supposes that the user has knowledge of parallel processing. The user must, for instance, distribute the tasks himself in this form of parallelism. The user must also handle the load balancing and communication.

From the viewpoint of macro-parallelism MuPAD consists of a group of independent clusters which are numbered from 1 to n. This numbering is used to specify the individual clusters. Each cluster in its turn can be formed of many processes and micro-parallelism can be used to program a cluster. The processes of a cluster should be connected that tight that the automatic scheduling and the shared memory of micro-parallelism work efficiently. The tasks in the individual clusters are not visible to the other clusters. It is also not possible to address a certain task in a cluster from another cluster.

The individual clusters can communicate by sending messages. Furthermore there is a special sort of variable, *net variables*, which have the same value for all clusters and therefore can be used as shared memory.

Most of the following examples are only executable when more than one cluster is present. This is not the case in a sequential system. In a parallel system the user can determine during the start-up how many clusters MuPAD should consist of and how many processes each cluster should contain.

2.8.2.1   Net Variables

Apart from the usual variables of a cluster there are the *net variables* in Mu-PAD. These have the same value in all clusters so that they can be considered as shared memory.

The contents of a net variable `a` can be read-out with `global(a)`, whereby the read value is evaluated. `global` does not evaluate its argument but considers it as the name of the net variable.

`global(a, b)` assigns b to the net variable a. Again the first argument is not evaluated. In contrast the second argument is evaluated as usual. During the evaluation of the second operand, the net variable cannot be changed by other tasks. This means that with net variables atomic statements, and with these locks and semaphores, can be created too.

Net variables have nothing to do with the usual variables of a cluster despite the fact that the variables of a cluster can be used as the name of a net variable. The value of a local variable of a cluster is therefore independent of the value of the net variable of the same name.

Example 142 *In the net variable* **sum** *the numbers 1 to 10 are to be added. The following command sequence performs this:*

```
>> global(sum, 0);
   for i from 1 to 10 parallel
       global(sum, global(sum)+i);
   end_for;
```

If, in the example above, a usual variable had been used instead of a net variable then in a real parallel system the value of the variable at the end would not have been 55 but a much smaller value (see section 2.8.1.1).

2.8.2.2   Queues

Each cluster can have any number of *queues*. To distinguish them each queue has a name which can be any MuPAD expression. Each cluster has the possibility with the command **writequeue(a, i, value)** to write the expression **value** in the queue of the cluster **i** with the name **a**. This function returns **value** as result.

With the function call **readqueue(a)** a cluster can read the first element of its queue **a**. If this queue is empty then **readqueue** returns an object of the basic type *DOM_NULL*.

However, with the function call **readqueue(a, Block)**, the function waits until there is a value in the queue and then returns this.

Queues need not be declared before use and can be employed when a cluster collects information from multiple other clusters, if the order is irrelevant.

Example 143 *A cluster is to request values from the queue* "in" *continuously, square the read values and write the results into the queue* "out" *of the cluster 1. The following loop performs this:*

```
>> while TRUE do
       writequeue("out", 1, readqueue("in", Block)^2);
   end_while
```

2.8.2.3   The "work" Queues

No cluster has a task to perform when the system is initialized. The cluster with the number 1 communicates with the user and can get a task in this way.

Every other cluster constantly checks its queue with the name "work" when it has no work to do. If it reads an expression from this queue it evaluates it and starts checking the queue "work" again.

The cluster with the number 1 never checks if there is anything in its "work" queue. If this cluster has no more tasks then it waits for more work specified by the user.

Example 144 *The cluster 2 has to add the squares of the numbers 1 to 100 and write the sum in the net variable* **a**:

```
>> writequeue("work", 2, hold(global(a, (sum := 0;
                        for i from 1 to 100 do
                            sum := sum + i^2;
                        end_for
                        )))
            );
```

Because the function **writequeue** *evaluates its arguments before they are inserted, the statement to be sent must be protected by the function* **hold**. *The statement is sent to cluster 2 which reads it from the "work" queue and processes it as soon as its previous tasks have been processed.*

2.8.2.4 Pipes

With queues, messages can be sent to a certain cluster but undisturbed communication between clusters is not guaranteed because different clusters may write into the same queue. In this case the cluster that reads from this queue cannot be sure that the value is sent from the cluster it wants to receive a message from.

To solve this problem there is a further data structure organized as a queue; the so-called *pipe*. For two clusters **i** and **j** and a datum **a** there is exactly one pipe called **a** that runs from node **i** to node **j**. Pipes that have the same name but come from different clusters are really different so that data from different clusters are not mixed.

With **writepipe(a, j, value)** a cluster can write the expression **value** in the pipe of the cluster **j** called **a**.

With the function **readpipe(a, i)** a cluster can read an expression from the pipe of the cluster **i** called **a**. If this pipe is empty then the function returns an object of the basic type *DOM_NULL*.

However with the function call `readpipe(a, i, Block)` if the queue is empty the function waits until a value has been written in and then returns this as the function result.

Like queues, pipes can be used without having been declared previously.

Example 145 *A cluster is to continuously request data from the pipe called "in" coming from cluster 2. The data are to be squared and written in the pipe "out" of cluster 1:*

```
>> while TRUE do
       writepipe("out", 1, readpipe("in", 2, Block)^2);
   end_while;
```

The difference between this example and example 143 is that this program only reads data coming from cluster 2, i.e. data that other clusters send are not taken into account.

When choosing names for pipes and queues the user should note that the functions that access this data evaluate their arguments first. Unexpected errors can occur when names are chosen that do not evaluate to themselves.

2.8.2.5   The Function topology

The system function `topology` is used to give the user information about the organization of the actual MuPAD processes.

If this function is called without parameters then it returns the number of clusters in the system.

The call `topology(0)` returns the number of processes of MuPAD, i.e. the sum of the processes of the individual clusters.

If the function is called with an integer between 1 and `topology()` it returns the number of processes in the cluster with this number.

When called with the argument `Cluster` the function returns the index of the current cluster.

Example 146 *When used sequentially the calls*

```
>> topology();
   topology(0);
   topology(1);
```

```
topology(Cluster);
```

all return the value 1.

Example 147 *A function is to be written that adds-up the values f(1) to f(n) and returns the sum as result when f is a function dependent only on the variables defined during the booting of the system and their parameters. Hereby the function should use all available processes. This function can be defined as follows:*

```
>> proc(n, f)
   local l, i, sum;
   begin
      l := n div topology();
      global(l, l);
      global(f, f);
      #The values from l and f are written in the net variable
       of the same name so that the other clusters can
       read their values#

      for i from 2 to topology() do
         #Cluster i reads the value from net variable f#
         writequeue("work", i, hold((f:=global(f)))) ;

         #Cluster i reads the value from net variable l#
         writequeue("work", i, hold((l:=global(l))));

         #Cluster i calculates f((i-2)*l+1)+...+f((i-1)*l) and
          writes the sum in the queue "value" of cluster 1#

         writequeue("work", i,
            hold(writequeue("value", 1,
               (num := topology(Cluster) ;
                _plus(
                   (for i from (num-2)*l+1 to (num-1)*l parallel
                                f(i);
                   end_for))))))) ;
      end_for;
      #Cluster 1 adds-up the summands, which are not added-up by
       any other cluster#
      sum := _plus((for i from (topology()-1)*l+1 to n parallel
```

```
            f(i)
          end_for));

   #Cluster 1 adds-up the sums of the other clusters#
   for i from 2 to topology() do
      sum := sum + readqueue("value", Block);
   end_for;

   sum;
 end_proc;
```

The arguments of the function writequeue *containing statements are to be enclosed by* hold *in order to prevent their evaluation before sending (see example example 143).*

This procedure can only be called from cluster 1.

Programming with the tools provided by macro-parallelism is relatively time-consuming and complicated. However the parallel version of MuPAD will contain a library which makes working with macro-parallelism more comfortable.

2.9 Direct Use of Internal System Functions

One special characteristic of the MuPAD system is that for each operator and statement there is a functional equivalent. The identifiers of these so-called *underline functions* are made up of the name of the operator or statement with an underline (_) in front. Apart from this an expression type is associated with each system function. This describes the type of the unevaluated function call.

Example 148 *The system operator + used for the addition of numerical values can also be used as the function* _plus. *The entries*

```
>> a+b+1+2;
   _plus(a, b, 1, 2);
```

create identical expressions. The result a+b+3 *is a sum that can not be evaluated further and has the expression type* "_plus".

The arguments of a underline function corresponding to one of the systems operators represent the operands of this operator. In the case of the functional equivalent of a statement the arguments are created from the user specified parts of the statement. Generally these are the parts of the statement that contain no keyword. The arguments of the statements are to be given in the order they appear. An exact description of the operands of the individual system functions can be found in table A.2 in the appendix. In this table the expression type of each underline function is described.

In contrast to the usual treatment of operators and statements a comprehensive check of the operands is not carried out when using the functional equivalents. This is a consequence of the syntax for expressions and statements already implicitly restricting the number of operands and of types. Therefore, conditions concerning the operands of the underline functions need not be checked again. It follows that, for instance, according to the syntax an assignment must contain two operands, of which the first must have a certain form. In the corresponding function **_assign** these conditions are assumed to have already been checked and therefore are not checked again. Usage not conforming to the usual notation can lead to a program crash. Therefore, the user himself is responsible for the correct use of underline functions. 

Apart from this the functional notation is only another form of input. The output is independent of the input form by using the operator notation as well as the language constructs of the statements.

Through the assignment of a procedure definition to one of the described underline identifiers the user can change the functionality of the function or the operator or the statement, respectively. This should also be used with great care to avoid any unwanted side-effects.

The system kernel contains the following underline functions:

_and	_assign	_break	_case	_concat
_div	_equal	_exprseq	_fconcat	_for
_for_down	_for_in	_for_in_par	_for_par	_if
_index	_intersect	_leequal	_less	_minus
_mod	_mult	_next	_not	_or
_parbegin	_plus	_power	_procdef	_quit
_range	_repeat	_seqbegin	_seqgen	_stmtseq
_unequal	_union	_while		

Example 149 *The product* a*b*c *can be equivalently represented by*

```
>> _mult(a, b, c);
```

The generation of large sums with the aid of expression sequences can be effectively carried out by using the underline function _plus. Let

```
>> S := i $ i = 1..100;
```

be the expression sequence of the first one hundred natural numbers. The addition of the numbers is done by

```
>> _plus(S);
```

The underline function **_assign** *is the equivalent of a usual assignment. The function is called with two parameters. The first specifies the expression to which the value is to be assigned. The second parameter gives the value to be assigned. Thus, the assignment*

```
>> f(a) := b;
```

can also be written as

```
>> _assign(f(a), b);
```

More complex statements can also be entered in functional notation. Thus, the **for**-*loop*

```
>> for i from 1 to 5 step 2 do s := s+i end_for;
```

corresponds to the function call

```
>> _for(i, 1, 5, 2, (s:=s+i));
```

A complete usage of the functional notation results in

```
>> _for(i, 1, 5, 2, _assign(s, _plus(s, i)));
```

Even when a parameter in a statement is optional it cannot simply be left out when using the functional equivalent but must be replaced by **NIL**. Optional parameters of this type are, for instance, the step width in the sequential **for**-loop as well as the declaration of private variables in the parallel constructs. The user should also note that, when using the **_procdef** function instead of procedure definitions, **NIL** must be used to replace formal parameters, local variables or options when these are missing. The parameter of a procedure definition, which represents the corresponding remember table, must also be replaced by **NIL** if missing.

Example 150 *The procedure definition*

```
>> proc(a,b) local c; begin c := a*b; end_proc;
```

can be equivalently written as

```
>> _procdef((a, b), c, NIL, _assign(c, _mult(a, b)), NIL, NIL);
```

With the aid of the function **op** the user has the possibility of analyzing the exact structure of the operands of a statement. The parameters of the underline functions must exactly correspond to these operands in their order and type.

Example 151 *With*

```
>> EVAL_STMT := FALSE:
   op((if a=b then s:=1 else s:=2 end_if));
```

the user obtains the sequence a=b, (s:=1), (s:=2) *as the operands of the* if*-statement above. The variable* EVAL_STMT *is set to* FALSE *to prevent the premature evaluation of the* if*-statement. The operands reflect the structure of the* if*-statement. Hence, it can be also entered as*

```
>> _if( a=b, (s:=1), (s:=2));
```

2.10 Manipulation of Objects

One of the most important concepts of the MuPAD system is the complete equivalence of expressions and statements. This means that, apart from the difference in functionality, internally statements and expressions are treated identically. This is achieved by using identical data structures for statements and expressions.

Due to this fact expressions, statements and procedures (a procedure is nothing else but a MuPAD expression) can be created or changed interactively or in a program. To avoid ambiguous notation, we use the terms expression and datum synonymous with MuPAD expression and MuPAD statement in the following.

The functions **type**, **domtype** and **testtype** are available for determining the expression and basic types as well as the user-defined types. The functions

nops and **op** are used for analyzing any expression; the functions **subsop**, **subs** and **subsex** manipulate expressions and statements.

The first argument is not flattened during evaluation of the arguments of all the functions described in this section. This can be clearly seen in the following example.

Example 152 *The function* **nops** *may only be called with one argument. However, the call*

```
>> x := a, b, c;
   nops(x);
```

does not lead to an error, because the argument **x** *is considered as one parameter, and thus* **nops(x)** $\neq$ **nops(a,b,c)**.

2.10.1   Types and Expression Types

All MuPAD objects undergo a two-step typification. Firstly, each expression has a type which — depending on if it is an element of the system kernel or of a domain created by the user — is denoted as *basic type*, *user-defined type* or in short *type*.

For a finer structuring of the basic type *DOM_EXPR* the *expression typification* can be used. All expressions that can be formed by the system operators, all statements, procedure calls, calls of unknown functions as well as indexed expressions and domains are of the type **DOM_EXPR**. For all objects that do not belong to the basic types *DOM_EXPR* or *DOM_DOMAIN* the expression type and type are identical.

At this point we would like to remind that a declaration of identifiers is not necessary. It is allowed to assign the desired object to a variable without having previously declared the type of the variable.

2.10.2   The Functions type, domtype and testtype

The system functions **type** and **domtype** enable the user to determine the type or the expression type of any expression. Each function expects exactly one argument. This is evaluated and the type of the resulting expression is returned. The expression type is given in form of a string, the type in form of an identifier. No or more than one parameter results in an error.

Example 153 *The basic type of a numerical object is not distinguished by the functions* type *and* domtype.

```
>> domtype(1/2), type(1/2);

    DOM_RAT, DOM_RAT
```

In contrast to this the basic type and the type of the sum a+b *are different.*

```
>> domtype(a+b), type(a+b);

    DOM_EXPR, "_plus"
```

All the basic and expression types of the system are listed in the tables A.1, A.2 and A.3 in the appendix.

The function testtype checks if an expression belongs to a given type. For this purpose the function must be given two parameters expr1 and expr2 which are both evaluated before the actual comparison is executed. The comparison executed by testtype consists of several steps. Firstly, the basic type, to which expr1 belongs, is checked to see if it contains a method with the index "testtype". In this case this method is called with the arguments expr1 and expr2 and the result is that of this method. If the result is FAIL or no method is found then it is checked if the type specified by expr2 (it could be a name or any expression; in the latter case the type of expr2 is taken) contains a method "testtype". If yes, this one is called with the arguments expr1 and expr2 and the result is that of this method.

The corresponding methods are implemented for the basic types, so that by using testtype it can be checked whether an object belongs to one of the basic types.

```
>> testtype(a+b, DOM_EXPR);

    TRUE
```

```
>> testtype(1/3, DOM_RAT), testtype(4, DOM_STRING);

    TRUE, FALSE
```

In order to give an user-defined type as the second parameter of testtype a method testtype has to be implemented in the corresponding domain which executes the desired comparison. For instance, to check if an object belongs to a simple basic type i.e. one of the types *DOM_INT*, *DOM_RAT*, *DOM_FLOAT*, *DOM_COMPLEX*, *DOM_IDENT* or *DOM_STRING*, the user has to do the following. Firstly, a new domain SIMPLE must be defined:

```
>> SIMPLE := domain():
   SIMPLE::name := "SIMPLE":
```

To achieve the desired effect from **testtype** an appropriate method has to be implemented:

```
>> SIMPLE::testtype:= proc(x)
   begin
       case domtype(x)
           of DOM_INT do
           of DOM_RAT do
           of DOM_FLOAT do
           of DOM_COMPLEX do
           of DOM_IDENT do
           of DOM_STRING do
               TRUE;
               break;
           otherwise
               FALSE;
       end_case;
   end_proc:
```

With this definition the system function **testtype** is overloaded:

```
>> testtype("simple", SIMPLE), testtype(2.01, SIMPLE);

   TRUE, TRUE
```

```
>> testtype({a,b},SIMPLE), testtype((a,b),SIMPLE);

   FALSE, FALSE
```

Using this mechanism it is easy to extend the type testing mechanism.

In addition to the basic types there are several other type expressions which can be given as second parameter to **testtype**. These are implemented in the library domain **Type**, whose description can be found as a separate document in the MuPAD installation. This domain offers a comfortable type checking mechanism for expressions and compounded expressions, e.g. general expressions, lists or sets.

Example 154 *Checking whether a list only contains numerical values can be done in MuPAD with one call of* `testtype`:

```
>> list := [2, 2.4, 1/2, -3]:
   testtype(list, Type::ListOf(NUMERIC));

   TRUE
```

The result is TRUE *because every element in the list belongs to a numerical type (DOM_INT, DOM_RAT, DOM_FLOAT and DOM_COMPLEX).*

2.10.3   The Functions nops and op

The system functions **nops** and **op** are used to analyze an expression. With the help of **nops** the number of operands of an expression can be determined. Not only the operands of an expression formed by an operator are meant, e.g. a sum a+b+c with the three operands a, b, c, but also operands of compounded types, e.g. items in a table, elements of a set etc.. A complete list of these types can be found in the tables A.1, A.2 and A.3 in the appendix. These tables include the result when **nops** is applied to an object of the corresponding type.

The function **nops** should be called with one argument. This argument is evaluated and finally the number of operands is determined. The result is an integer.

Example 155 *With* nops *the number of operands in an expression can be determined. The function calls*

```
>> nops(a+b);
   nops({a, b});
   nops(a..b);
```

all return the value 2.

With the function **op** operands can be extracted from an expression. Like for the function **nops** operands are not only understood as the operands formed by an operator. What is exactly to be understood as the operands of an expression can be found in the tables A.1, A.2 and A.3 in the appendix. Further information can be found in the descriptions of the individual basic and expression types. If only one operand of an expression is requested this operand is returned. Otherwise the result is an expression sequence consisting of the specified operands.

Example 156 *The operands of a set or a list are the elements of the corresponding structure. Thus a call of* op *with a set or a list gives the elements of the set or the list in form of an expression sequence.*

```
>> op({a, b, c, d});
   op([a, b, c, d]);

   a, b, c, d
   a, b, c, d
```

An empty set or list has no operands. If op *is applied to such an expression the result is the element of the basic type DOM_NULL.*

Objects of the basic type *DOM_EXPR* are a special case. The function name or the operator name is considered to be the operator. It is not returned during a simple call of op, i.e. during a call of op with just one argument. nops is implemented in consistence with this.

Example 157 *The sum* a+b+c *consists of an operator and three operands, the identifier* _plus *and the arguments* a, b *and* c. *However, the function call*

```
>> op(a+b+c);
```

only returns the expression sequence a, b, c, *because it is an expression of the basic type DOM_EXPR and* _plus *is understood as the operator. Like for the function* nops *applied to this expression returns the integer* 3.

For objects of the basic type *DOM_ARRAY* the operator consists of an expression sequence which contains the dimensions, the ranges of the individual dimensions and if applicable the partitioning in subarrays (see section 2.3.12.2). The operator also differs from the other operands concerning evaluation. In contrast to the other operands it is returned not completely substituted (see example 162). In all other basic types numbering of the operands begins with one and by a call of op all operands are returned. The connection between the numbering of the operands and the manipulation functions shall be described in detail in the following.

Moreover, it is to be noticed that an expression of the form `-<expr>` is represented internally as `<expr>*(-1)` and that the operands are determined correspondingly. Similarly the expression `a/b` is treated like `a*(1/b)`. Of course, the result of `type` is consistent.

Example 158 `op` *and* `type` *work consistently when applied to signed expressions.*

```
>> op(-a); op(a/b);

   a, -1
   a, 1/b

>> type(-a), type(a/b);

   "_mult", "_mult"
```

For the basic types *DOM_RAT* and *DOM_COMPLEX* `op` works as follows. Although data of these types are treated internally as atomic objects, `op` applied to a rational or complex number still returns an expression sequence consisting of the numerator and denominator or respectively the real and imaginary part of the argument. In this way the handling of symbolic quotients and rational numbers as well as symbolic sums and complex numbers by `op` is different. 

Example 159 *Using the function* `op` *the numerator and denominator of a rational number can be determined. Thus the call*

```
>> op(3/4);
```

returns the result `3, 4`.

In order to extract specific operands from an expression, the user can specify the desired operand by giving a second argument to `op`. If the second parameter is a positive integer `i` then the call `op(expr, i)` returns the i-th operand of `expr`. The operator is, as already mentioned, only defined for objects of the basic types *DOM_EXPR* and *DOM_ARRAY*. It can be extracted by using 0 as the second argument in the call of `op`.

If an operand specified by an `op` call does not exist then the function returns `FAIL`.

Example 160 *As the product* a*b*c *is an expression of type DOM_EXPR its operator is defined.*

```
>> op(a*b*c, 0), op(a*b*c, 1);

    _mult, a
```

Similarly, a function call of an undefined function f *is an expression of the basic type DOM_EXPR and therefore the operator can be accessed.*

```
>> op(f(a, b, c), 0), op(f(a, b, c), 1);

    f, a
```

To extract more than one operand from an expression, op can be called with a range as its second argument. The range determines the positions of the desired operands. These are returned in form of an expression sequence.

Example 161 *To obtain the second to fourth operands of the sum* a+b+c+d+e *the following input is necessary:*

```
>> op(a+b+c+d+e, 2..4);

    b, c, d
```

Example 162 *The call of*

```
>> op(a+b, 0..2);
```

returns the result _plus, a, b. *The identifier* _plus *is an object of the basic type DOM_FUNC_ENV and thus, it can be decomposed further, as*

```
>> op(_plus);

    _plus, _plus, table("operator" = op_plusmult,
    "float"=_plus, "type" = "_plus")
```

shows. However, the output above looks a little bit strange to the inexperienced user, so that a _plus *is usually printed as* _plus. *The third operand of this output is the attribute table, in which, under the index* "operator" *the function to realize the operator functionality and under the index* "type" *the expression type are stored.*

Sets and tables form a special case in the numbering of operands. Due to the internal representation of sets and tables the numbering of their elements is random so that the i-th operand of two externally identical sets or tables can be different. The following invariant is valid: As long as a set or a table is not altered the order of its elements remains unchanged. Therefore, e.g. the i-th operand of an expression of type *DOM_SET* remains the same for two consecutive calls of op.

By entering a list as second argument the user can avoid nested op calls. This is necessary if the operands of an expression are to be decomposed further. The list may contain non-negative integers as well as ranges whose bounds also have to be non-negative integers. For the processing of such a call firstly the first element of the list is read and a call of op with this value as second argument is executed. Then all elements of the resulting expression sequence are processed individually. With each operand of the result a further call of op is executed with the second list element used as second argument. This strategy is carried on until the entire list has been processed. Thus, a call of the form op(op(op(expr, s1), s2), s3) is equivalent to op(expr, [s1, s2, s3]) whereby the si can be any non-negative integer. This equivalence does not apply to ranges because with the input of a list the operation is used recursively on each object created in each step.

Example 163 *To obtain the first factors of the product terms in* f+c*d+a*b*c *the call*

```
>> op(f+c*d+a*b*c, [2..3, 1]);
```

has to be entered. It returns the expression sequence

```
    c, a
```

as its result.

Example 164 *The input*

```
>> op(a*b*c+g*f(d,e), [2, 2, 1]);
```

is equivalent to the nested call

```
>> op(op(op(a*b*c+g*f(d, e), 2), 2), 1);
```

Both return d *as the result.*

2.10.4 The Function subsop

Using the **subsop** function the user has the possibility of substituting specific operands of an expression by other values. Before the function is evaluated all arguments are normalized. After the substitution has been executed the normalized form is no longer valid so that the final step in the execution of **subsop** is a simplification step. This last step can be prevented by using the option **Unsimplified** as the last argument of a **subsop** call.

The function **subsop** can be called with any number of arguments. The first argument is the expression in which the substitution is to take place. The other arguments are equations which specify the substitutions to be executed. The equations are processed in sequence from left to right, whereby the results obtained in the previous step are used. Between two such substitution steps, no simplification of the intermediate result is executed. The left hand side of an equation specifies the operand to be replaced, the right hand side is the expression that is to be inserted. Optionally, the last argument can be the identifier **Unsimplified**. The result of **subsop** is a new expression, i.e. the specified substitution is not executed in place but on a copy of the expression given as first argument.

It is to be noticed that due to the evaluation of the first argument the order of the operands can change (see example 167).

Like for **op** there are two ways to specify an object to be substituted: with a non-negative integer or a list of such numbers.

Example 165 *In the expression* a+b *the second operand is to be replaced by the expression* f(u).

```
>> subsop(a+b, 2=f(u));

    a+f(u)
```

Example 166 *Using a list the path of an operand to be substituted can be specified. The following call*

```
>> subsop(f(u)*2+g(u,v)*3, [2, 1, 1] = z);
```

replaces the first operand of the first factor in the second term of the sum, i.e. u *is replaced by* z. *Thus, the result is* f(u)*2+g(z,v)*3.

Example 167 *The first step in the execution of* subsop *is an evaluation of the arguments, therefore the call of* subsop *can lead to unexpected results. This can be seen by calling*

```
>> subsop(h(u,v)*g(u,v)^2+2*f(u,v), [1, 1, 1] = z);
```

which gives f(z,v)*2+g(u,v)^2*h(u,v). *The evaluation of the arguments transforms to a normalized form which includes sorting of the operands. For the expression above this means*

```
>> h(u,v)*g(u,v)^2+2*f(u,v);

    f(u, v)*2 + g(u, v)^2*h(u, v)
```

Thus u *is not substituted in* h(u,v) *but in* f(u,v).

Example 168 *The final step in the execution of a* subsop *call is a simplification step. The call*

```
>> subsop(a+b, 2 = a);
```

gives a*2. *By giving the option* Unsimplified *this step is prevented.*

```
>> subsop(a+b, 2=a, Unsimplified);

    a + a
```

By using subsop it is possible to generate a not fully evaluated expression. By using the function eval the user can force full evaluation.

Example 169 *After the assignment*

```
>> d := 2:
```

the call of

```
>> subsop(b*d + b*c, [2,2]=hold(d));
```

gives the result b*2 + b*d. *In the first step (evaluation)* d *is replaced by its value 2. The specified substitution replaces* c *by* d. *The final simplification step has no effect because the internal simplifier cannot apply any of its rules. Applying the* eval *function to the call above a complete evaluation can be achieved.*

```
>> eval(subsop(b*d + b*c, [2,2]=hold(d)));

   b * 4
```

Of course, it is possible to do substitution in more complex objects, e.g. tables, arrays, statements and procedures.

Example 170 *With*

```
>> Tab := table(a=x+y, b=y+z, c=x+z):
```

a table with three entries is created. The substitution of the first entry can be achieved with

```
>> subsop(Tab, 1=(a=z));
```

which gives `table(a=z, b=y+z, c=x+z)`. *Now let* A *be defined as a partitioned array (see section 2.3.12.1).*

```
>> A := array(1..4, 1..4, [2,2]):
   for i from 1 to 4 do
     for j from 1 to 4 do
       A[i,j] := a.(i.j)
     end_for;
   end_for:
```

The result is a 4×4 matrix with the identifier `aij` *at position* `A[i,j]`. *Using* **subsop** *it is possible to substitute single elements of the matrix or even complete subarrays*

```
>> subsop(A, 1=[1, 2, 3, 4]);

    +-                     -+
    |   1 , 2   | a13 , a14 |
    |           |           |
    |   3 , 4   | a23 , a24 |
    | ----------+---------- |
    | a31 , a32 | a33 , a34 |
    |           |           |
    | a41 , a42 | a43 , a44 |
    +-                     -+
```

If the left hand side of a substitution specification is a list of non-negative integers then it describes the operand to be replaced according to the functionality of **op**.

Example 171 *The first substitution of the example 170 could also be achieved by calling*

```
>> subsop(Tab, [1, 2]=z);
```

Example 172 *In order to replace both* dd *and* bb *in* ff+dd*ee+aa*bb*cc *by a numerical value the following call can be used:*

```
>> subsop(ff+dd*ee+aa*bb*cc, [2, 1]=11, [3, 2]=23);

   ee*11+ff+aa*cc*23
```

For objects of the basic type *DOM_EXPR* the manipulation of the operator is also permitted.

Example 173 *Consider the expression sequence* a, b, c, d, e. *One way for adding its elements is to substitute the expression sequence operator by the addition operator. The call*

```
>> Te := a, b, c, d, e: subsop(Te, 0=_plus);
```

gives a+b+c+d+e. *The same result can be achieved by using* _plus *directly, i.e. to call*

```
>> _plus(Te);
```

Manipulation of the operator of expressions of the basic type *DOM_ARRAY* is not permitted. In this case the function returns **FAIL**. 

A further usage of **subsop** is the deletion of operands. This can be achieved by substituting the operand by the element of type *DOM_NULL*.

Example 174 *Deleting the first element in the set* $\{a, b, c, d\}$ *by* subsop *can be done in the following way:*

```
>> M := {a, b, c, d}:
```

```
>> subsop(M, 1=null());

   {b, c, d}
```

With the function null *an object of the basic type DOM_NULL is created.*

The direct access to subexpressions allows to use also the assignment of NIL for deletion.

```
>> L := [a,b,c,d]: L[2] := NIL: L;

   [a,c,d]
```

2.10.5   The Functions subs and subsex

Like the function **subsop** the functions **subs** and **subsex** can also be used to manipulate expressions and statements. While the user has to know the exact position of the operand to be substituted in an object when using **subsop**, i.e. the user needs some knowledge of the internal data structure, with the functions **subs** and **subsex** the user can substitute objects without knowing their exact position.

The handling of the parameters by **subs** und **subsex** is similar to the handling of arguments by **subsop**. Before the actual substitution all arguments are fully evaluated. After the substitution the internal simplifier is called. Like in **subsop** this final simplification can be prevented by using the option **Unsimplified**.

The order of the arguments in **subs** and **subsex** is consistent with that of the other manipulation functions. The first argument is the expression to be modified. The other arguments specify the substitutions to be carried out. These are to be specified in form of equations whereby the left hand side is the expression to be substituted and the right hand side gives the expression to be inserted. Any number of equations can be given in one call. Optionally the last argument can be the identifier **Unsimplified**.

Example 175 *In polynomial expressions the variables can be substituted by* subs. *Firstly, define an univariate polynomial expression* P *by*

```
>> P := T^5+2*T^4-3*T+1:
```

and then by calling subs *the indeterminate* T *can be replaced*

```
>> subs(P, T=1);

    1


>> subs(P, T=z);

    z*(-3)+z^4*2+z^5+1
```

Example 176 *The option* Unsimplified *should be given to prevent the final simplification step, e.g. the call*

```
>> subs(a + b + c, b=a, c=a, Unsimplified);
```

gives a + a + a *instead of the simplified form* 3*a.

Example 177 *With* subs *not completely evaluated expressions can be created. A full evaluation can be achieved by calling* eval.

```
>> di := 3:
   subs(a + b*c, c=hold(di));
```

gives a + b*di. *The specified substitution replaces* c *by* di. *The final simplification step has no effect because the internal simplifier cannot apply any of its rules. Applying the* eval *function to the call above a complete evaluation can be achieved.*

```
>> eval(subs(a + b*c, c=hold(di)));

    a + b*3
```

As can be seen in the name of the function, **subsex** (which stands for *subs extended*), is an extended version of **subs**. In **subs** only those objects are found and replaced that represent operands in the sense of the function **op**. Every object that can be specified by a call of **op** without the use of a range, can be replaced by using **subs**.

Example 178 *The call of*

```
>> subs(a*b+d*e, a=x, d*e=y);
```

returns y+b*x. *The expressions to be substituted,* a *and* d*e, *are complete operands in the sense of* op: d*e *is the second operand of* a*b+d*e *(*d*e = op(a*b+d*e, 2)*);* a *is the first operand of* a*b, *and this in its turn is the first operand of* a*b+d*e *(*a = op(a*b+d*e, [1,1])*).*

```
>> subs(a*b*c+d*e, a*c=x);

   d*e+a*b*c;
```

a*c *is not an expression that can be directly extracted from* a*b*c+d*e *using* op. *Therefore the call of* subs *above does not lead to the desired substitution.*

A more complete, and as a consequence more complex substitution mechanism is realized with the function **subsex**. Here even those subexpressions that do not conform to the conditions above are found and replaced.

Example 179 *The substitution requested in the last example can be achieved by calling*

```
>> subsex(a*b*c+d*e, a*c=x);
```

which gives d*e+b*x.

However, **subsex** is not a fully fledged pattern matching mechanism. Only those expressions can be replaced, which are combinations of operands which belong to the same level with respect to the internal data structure. This can be seen in the following example.

Example 180 *The call of*

```
>> subsex(a*b^2, a*b=1);
```

returns the argument a*b^2 *unevaluated, because the internal representation is* a*(b^2), *and therefore* a *and* b *are not on the same level.*

This is the only difference between the two functions. With respect to the different possibilities for calling, both functions offer the same functionality so that here only the function **subs** shall be described.

Both functions offer two ways of substitution:

Sequential Substitution

This is the standard form of substitution that has already been used in example 178. The substitution specifications given are processed one after the other in their order.

Parallel Substitution

If this is the form of term replacement required the user has to specify the substitution equations in lists. All the substitutions given in the lists are logically carried out simultaneously.

Of course, both forms can be combined.

Example 181 *The call*

```
>> subs(a+b+c+d+x, b=x+y, x=z);
```

is evaluated internally as follows: the first substitution is applied to a+b+c+d+x. *The intermediate result is* a+c+d+y+x*2. *The second substitution is then applied to this. The result of this substitution and consequently the result of the entire function call is* a+c+d+y+z*2.

Example 182 *The call*

```
>> subs(a+b+c+d+x, [b=x+y, x=z]);
```

is, in contrast to example 181 evaluated as follows: both substitutions are executed simultaneously. The second equation is not applied to the result of the first term replacement but a special mechanism assures that replacements are not overwritten by any following parallel substitutions. The result of the call above is therefore a+c+d+x+y+z.

The simple examples above are only meant for demonstration purposes. They were used to show the application of the functions **subs** and **subsex** in a simple manner. The following example demonstrates an effective application of the substitution function.

Example 183 *With the substitution function* **subs** *it is easy to write a general Newton procedure for the fixed point problem in MuPAD. For a given function* f, *a value for* x *is searched, for which* $f(x) = x$ *holds.*

```
>> newton := proc(f, start, n)
      local x, y, i, df, DIGITS;
      begin

      DIGITS := 15;
      df := diff(f(y), y);
      x[0] := float(start);
      print(x[0]);
      for i from 0 to (n-1) do
          x[i+1] := subs((y*df-f(y))/(df-1), y=x[i]);
          print(x[i+1]);
      end_for;
   end_proc:
```

Firstly, the derivative of the function f *is determined using* diff. *To determine the value of this derivative at a specific point this point is substituted for the differentiation variable by using* subs.

```
>> ff := proc(x) begin x^2 - 1 end_proc:
   newton(ff, 1.0, 10);

      1.00000000000000
      2.00000000000000
      1.66666666666666
      1.61904761904761
      1.61803444782168
      1.61803398874998
      1.61803398874989
      1.61803398874989
      1.61803398874989
      1.61803398874989
      1.61803398874989
```

2.10.6   Manipulation of Statements and Programs

As already mentioned MuPAD does not distinguish expressions and objects like statements and procedure definitions. Therefore all functions for manipulation of expressions can also be applied to procedures and statements. It is important to note that the environment variable EVAL_STMT has to be set to FALSE beforehand to prevent a premature evaluation of the statements in the context of expressions. The operands of a statement can be analyzed using the function op (see also table A.3 in the appendix).

In the case of procedure definitions and statements there are several elements which are optionally contained in such an expression (see section 2.9). Unless otherwise stated these operands have the value NIL.

Example 184 *The operands of a procedure definition can be determined as follows:*

```
>> f := proc(a,b) option remember; begin a+b end_proc:
   op(f);

   (a,b), NIL, remember, a+b, NIL, f
```

The op call returns the operands of the procedure definition. The first operand is the sequence of formal parameters. The following NIL shows that this procedure has no local variable. The other elements of the expression sequence describe the options as well as the procedure body. The two final values represent the remember table (missing in this procedure) and the procedure name.

Example 185 *Using subsop a procedure can be given a remember table so that some values are assumed to be known. In this way the procedure fib for calculating the Fibonacci numbers can be realized without using an if statement:*

```
>> fib := proc(n)
   option remember;
   begin
     fib(n-2)+fib(n-1)
   end_proc:

>> T := table(0=0, 1=1):
   fib := subsop(fib, 5=T):
   fib(5);

   5
```

Because the function values fib(0) and fib(1) are stored in the remember table, the procedure body is not executed for the parameters 0 and 1. Therefore the recursion of fib is stopped automatically at this point.

Instead of substituting the entire remember table it can be built up step by step with assignments to a procedure call. For this the assignments `fib(0) := 0` *and* `fib(1) := 1` *have to be executed (see section 2.6.8).*

Example 186 *To change the condition of an* `if` *statement it can be substituted using the* `subsop` *function. As already mentioned the variable* `EVAL_STMT` *has to be set to* `FALSE` *to prevent a premature execution of the statement.*

```
>> EVAL_STMT := FALSE:
   s := (if a<b then TRUE else FALSE end_if):
   s := subsop(s, 1=(b<a)):
   s;

    if b < a then TRUE else FALSE end_if


>> a := 1:  b := 2: eval(s);

    FALSE
```

2.11   The History Mechanism

The history mechanism performs two tasks. Using the system function `last` or the per cent symbol followed by a positive integer the user can access previously calculated expressions. Furthermore, this mechanism enables the user to avoid the assignment of a value to an identifier as well as its re-evaluation.

Often in complex calculations subresults are calculated and then assigned to auxiliary variables in order to access the expression later. Alternatively, access to the expression can be done later without using any assignment using the function `last`. This means that, in certain cases, the amount of auxiliary variables can be reduced. Access to previously calculated values may be slower than carrying out an assignment and later access via a variable.

Example 187 *If the input*

```
>> a := b^2;
   f(a);
```

is replaced by the expressions

```
>> b^2;
   f(last(1));
```

an assignment to the identifier a *as well as a complete re-evaluation of a is avoided. However, this method can lead to errors, especially when in the meantime the value of the identifier* b *is changed. These changes are not taken into account by the history mechanism because a re-evaluation is not carried out.*

To explicitly evaluate a `last` call the system function `eval` should be used. This ensures that the entries of the history table are evaluated taking the value of LEVEL into account.

The entries of the history table can be viewed with the help of the system function `history`. The number of entries can be controlled with the environment variable HISTORY. That is a list containing two non-negative integers. The first element specifies the number of entries on the interactive level. The second element defines the number of entries while executing procedures. The default value is [20, 3].

In procedures there is a local history table for each procedure call. This means that values from outside the procedure can not be accessed via the history mechanism during the execution of the procedure and vice versa.

Not every result is inserted in the history table. Only the results of expressions and assignments that have been entered on the interactive level or are in the body of a procedure or statement are taken into account. The result of an assignment is the evaluation of its right hand side.

Example 188 *The variables* a, b, c *and* d *have the values* 1, 2, 3 *and* 4. *The identifier* e *has no value. The sequence*

```
>> e:=a*b;
   c+d+e: 10;
```

then inserts the values 2, 9 *and* 10 *in the history table.*

All other statements insert the values of their expressions and assignments evaluated during their execution. Only those expressions and assignments contained in the body of a statement are taken into account, for instance, the check of the condition in a `while`-loop is not inserted.

Apart from the number of stored elements the history mechanism does not differ between interactive input and the execution of procedures.

Example 189 *The interactive input*

```
>> if TRUE then
      sum := 0;
      for i from 1 to 3 do
         x.i := i^2;
         sum := sum+eval(x.i);
      end_for;
   end_if;
```

enters 0, 1, 1, 4, 5, 9 *and* 14 *in the history table. The result of the* if *statement is the final entered value:* 14. *This is not entered again as the result of the* if *statement. In the same way the condition of the* if *statement, the values of the range and the index of the* for *loop as well as its step width are not entered after their evaluation.*

Statements which occur in expressions have a special status. They also insert their results in the history table.

Example 190 *In contrast to the expression*

```
>> b + d;
```

after the evaluation of the expression

```
>> (a:=b) + (c:=d);
```

the values d *and* b *can be accessed with the help of the calls* last(2) *and* last(3). *The same is true, for example, for statements in the condition of an* if *statement. Thus after the statement*

```
>> if (a:=b; c; d := TRUE) then
      s := 1;
   end_if;
```

the expression

```
>> %1, %2, %3, %4;
```

returns the sequence 1, TRUE, c, b.

A further peculiarity concerns statements with signs. If these are isolated i.e. as complete input on the highest level or as the operand of a procedure body or other statement body, then the result of the statement is given the appropriate sign and, in addition, entered in the history table.

Example 191 *The* `if` *statement*

```
>> - ( if TRUE then -(a:=b); 2-(c:=d); e:=f; end_if );
```

leads to the values b, -b, d, -d+2, f *and* -f *being entered in the history table. Because the equation* `a:=b` *is signed and is also an operand of a statement body both its result* b *and the result with a sign* -b *are entered. The next entry causes the execution of the statement* `c:=d`. *The value with the negative sign is not entered here because this forms the operand of another expression. However, this expression is entered as the operand of a statement body which leads to the entry* -d+2. *The assignment* `e:=f` *enters the value* f *in the history table. This is also the value of the* `if` *statement. Because this has been entered on the interactive level and has a negative sign the value* -f *is entered too.*

Like procedures, every task produced by a parallel statement sets up its own local history table. Therefore, the history table of a parallel task and the table valid in the context outside do not affect each other. The result of a parallel statement is an expression sequence formed by the results of the individual tasks. This is the only value which is inserted in the history table for parallel statements. In the same way, the results calculated on the level of parallel statements or in other tasks, cannot be accessed from a parallel task with the help of the history mechanism.

Example 192 *The parallel* `for` *loop*

```
>> for i from 1 to 3 parallel
      i*2
   end_for;
```

only inserts its result, that is the sequence 2, 4, 6, *in the history table. Each of the three tasks created by the parallel loop sets up its own local history table and in this it enters its result of the evaluation of* `i*2`. *Therefore they each contain only one entry.*

Chapter 3

Debugging

3.1 Introduction

MuPAD offers two possibilities to observe the execution of a program in detail. The *trace mode* only generates a protocol of the execution, whereby the user has no control over the execution of the program. By comparison the *interactive, line oriented source code debugger* offers the following possibilities:

- User controllable program execution.
  (step by step or by setting breakpoints)

- Identification of those source code lines in which a run time error has occurred.

- Displaying and changing the values of variables.

This debugger mechanism, which is integrated in the MuPAD system kernel, has a character/line oriented user interface and offers, even on a terminal with a character screen, operational convenience and gives detailed information in textual form. An interface based on OpenWindows which offers even more convenience, is the program `mdx`, which is separated from the MuPAD system kernel (see section 3.3).

The debugger is activated by specifying options when calling MuPAD . There are three options available which can be also used simultaneously (see below).

A program, that is either to be executed under the control of the user (debug mode) or to be protocolled (trace mode), has to be a text file which contains a syntactically correct MuPAD program. This text file is either specified as usual when calling MuPAD or read with the system function `read`.

Calling procedures, defined in this text file, gives detailed information about their execution.

Under UNIX the debugger can be activated with the following command.

Syntax:

```
mupad [-g] [-t] [-v] [further MuPAD options]
```

These options have the following meaning:

−t : Turn on trace mode.

During a program execution the number of the line in the source code where the executed instruction is to be found is shown. In addition the place in the source code where a (user-defined) function is called and the corresponding arguments are also shown. The place where the (user defined) function is left and the return value of this function are given too.

Initially all procedures are protocolled by the trace mode. One can use the system function **trace** to protocol only selected procedures.

The output of the trace mode has the following form:

- **Eval at line** <*line*> **in file** <*name*>.
- **Enter procedure** <*procname*> **from line** <*line*> **in file** <*name*>**, args =** <*exprseq*> **, proc depth =** <*depth*>.
- **Exit procedure** <*procname*> **at line** <*line*> **in file** <*name*>**, result =** <*exprseq*>.

−v : Turn on verbose mode.

During the execution of a **read** instruction, the line numbers of the procedure definitions are displayed. Information about possible positions in the source code for the setting of breakpoints are displayed in the form *# line #* during the output of user defined functions.

The output in this mode is:

- **Entry point of procedure** <*procname*> **at line** <*line*> **in file** <*name*>.

−g : Turn on debug mode.

In this mode the user can enter commands with the keyboard to interactively control the debugger. A survey of the control commands is given in the following table. A detailed description of the commands can be found in section 3.3.1; for the exact command syntax see section 3.3.2.

Functions	Commands
Listing active procedures	(d)own, (u)p, (w)here
Displaying and changing of variables	(D)isplay, (e)xecute, (p)rint
Setting/deleting of breakpoints	(g)oto proc, (S)top at, (C)lear, clear (a)ll, (l)ist
Controlling the program execution	(c)ont, (n)ext, (q)uit, (s)tep, <Ctrl-C>

In contrast to the graphical user interface `mdx`, MuPAD offers the user no windows for displaying debug informations. It is given in the following form:

- `Stop at line` <*line*> `in file` <*name*>.
- `Enter procedure` <*procname*> `from line` <*line*> `in file` <*name*>`, args = ` <*exprseq*> `, proc depth = ` <*depth*>.
- `Exit procedure` <*procname*> `at line` <*line*> `in file` <*name*>, `result = ` <*exprseq*>.
- `Procedure` <*procname*> `at line` <*line*> `in file` <*name*> (Output with Up and Down).
- <*var*> `= ` <*value*> (Output with Print and Display).

3.2 Interaction between the User and the Debugger

If MuPAD is started with the debugger turned on, e.g. with `mupad -g` or `mdx`, then additional information for the debugger is produced internally. This information is transparent to the user. From the user's point of view there is no difference to the standard mode of MuPAD. The interactive debug mode is only activated by the system function **debug**. The argument of the function **debug**

can be any instruction. If this instruction contains a procedure call the prompt
mdx> appears and the user is now in the input mode of the debugger, in which
commands for controlling the program execution or commands for information
about the program status can be entered (see sections 3.3.1 and 3.3.2). During
the execution of the instruction only commands for the debugger can be given.
After the instruction have been executed the debugger reports **Execution**
completed, and the user is back to the input mode of MuPAD.

3.3   mdx: The X-Frontend of the Debugger

The program **mdx** is one of the graphical user interfaces of MuPAD. This
program requires either the X window system or OpenWindows. It improves
operational convenience by:

- Simplifying the input of debug commands.

- Visualizing the information given by the debugger.

- Simultaneously displaying the source code and the output of a program
  in separate windows.

For this reason **mdx** is divided into five windows (see figure 3.1):

- Status

  The status window contains general information about the current status. This includes, among other things, the line number in which the
  program was stopped and the line number of the file displayed in the
  source window.

- Source

  The source window displays the source code of the working file. After
  calling the debugger the window shows the first line of the file if a file has
  been specified in the call, otherwise the window remains empty. During
  the program execution, the part of the code which includes the line at
  which the program stopped, is displayed. The user can determine which
  part of the program is displayed by using the Goto proc button.

- Buttons

  Frequently used debugger commands can be executed by clicking on the
  corresponding button with the mouse. Commands that are not included
  here can be directly typed into the terminal window with the keyboard.

- Terminal

  All output produced by the execution of a program, as well as some output evoked by the debugger commands, appears in this window. Input, to both MuPAD and the debugger, has to be entered in this window. Input to the debugger has to follow the commands specified in section 3.3.2.

- Display

  The values of variables, that are to be permanently shown, appear in this window.

The source, terminal and display windows offer a scrollbar on the right-hand side, to enable the user to influence the displayed text.

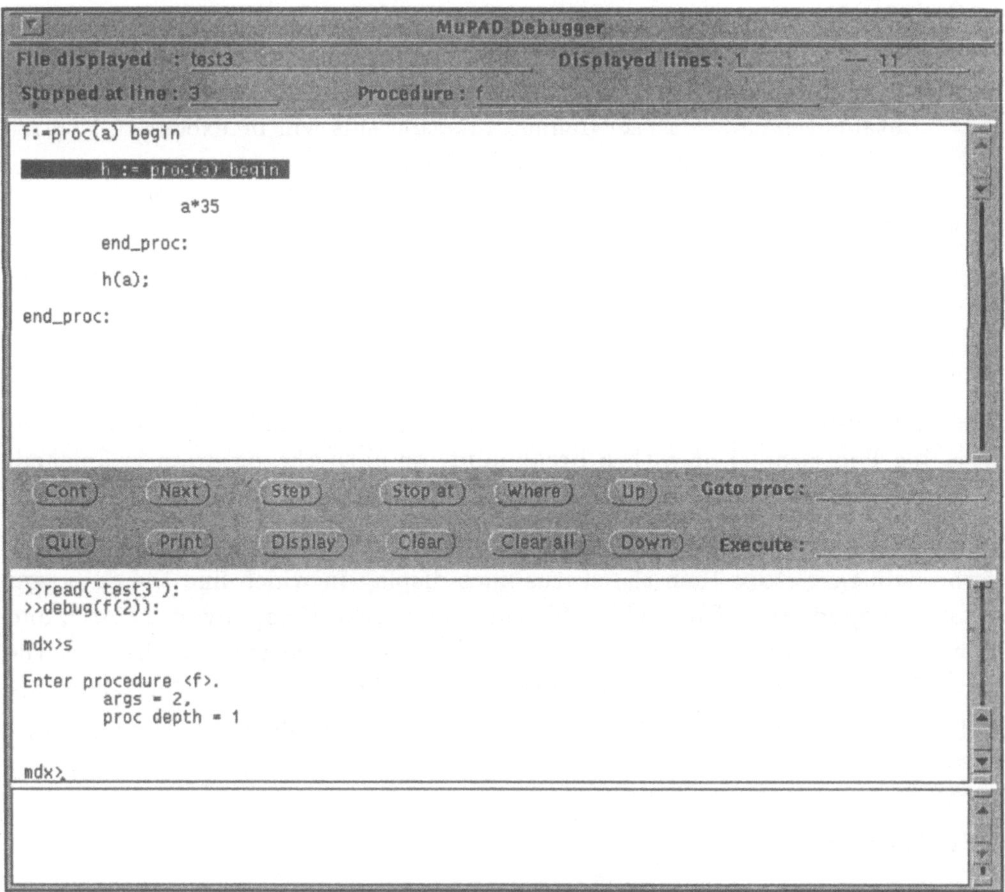

Figure 3.1: **mdx** Window

Syntax:

$$\boxed{\texttt{mdx [} -H\ h1,h2,h3\ \texttt{] [} \textit{MuPAD options} \texttt{]}}$$

With the option −H the height of the source, terminal and display windows can be specified. The unit is a pixel. The call `mdx` is equivalent to `mdx -H 250,150,50`.

MuPAD options contain all options that configurate the MuPAD kernel can be specified. However, the options −t and −g should not be used.

3.3.1 Description of the Commands

- Next

  With this command all instructions of the current line are executed. The debugger will stop at the next line. If the executed instruction contains the call of a user-defined function this will be executed without the debugger stopping inside the function.

- Step

  Analogue to Next, but the debugger stops inside the user-defined function.

- Cont (for Continue)

  The program is executed until an instruction is found which corresponding line is marked with a breakpoint, or until the program is finished. See also Stop at.

If the debugger stops then the status lines display in which line the debugger stops [`Stopped at:`], from which file this line is read [`Displayed file:`], and in which procedure it stops [`Procedure:`]. The source window displays the part of the program in which the corresponding line is found. Which lines are displayed is given by [`Lines:  ... -- ...`]. The line in which the debugger has stopped is shown inversely.

- Goto proc

  The specified procedure in the corresponding input field is displayed in the source window.

- Quit

  Terminates the execution of the MuPAD program if the debugger is in input mode. The debugger reports **Execution completed**, and the user is back to MuPAD input mode. A Quit in MuPAD input mode terminates the MuPAD session. The execution of a program can be stopped with <Ctrl-C>. After that the debugger is in input mode.

- Execute

  The MuPAD command specified in the corresponding input field is executed. The command needs not to be a single instructions but can be any sequence of instructions. One application is, for instance, the interactive influencing of a program execution by altering the values of variables (see also section 3.3.2).

- Where

  With this command all procedures that have been called from the beginning of the program execution until this point are shown in the terminal window. Procedure names and associated line numbers in the source code and the corresponding file names are displayed.

- Up

  This command serves for displaying the line in the source window from which the current procedure has been called. The status line of the debugger is also updated. If the procedure that the user has interactively entered, has been reached, then the message **Top level reached** appears in the terminal window.

- Down

  This command can only be used if at least one Up has been previously used. The procedure from which Up was called is shown in the source window. If the current procedure, i.e. the procedure in which the debugger stops, has already been reached, then the message **Bottom level reached** appears in the terminal window.

- Clear all

  With this command all breakpoints are deleted.

- Clear

  This command serves for deleting a specified breakpoint. For that the line in which a breakpoint was set is to be marked by a double mouse click. See also Stop at.

- Stop at

  The user can specify with a double mouse click in which line in the source window a breakpoint is to be set. In order to set a breakpoint inside a procedure the user must load this procedure into the source text window. (see also Goto proc). Breakpoints can also be combined with a condition. This causes execution to stop only if the condition is fullfilled. A condition can only be bound to a breakpoint when the whole command was typed with the keyboard.

- Print

  The user specifies the expression whose current value is to be displayed by highlighting the expression with a double mouse click in the source window. The value of the expression is then displayed in the terminal window (see also Execute and section 3.3.2).

- Display

  The user specifies the variable whose value is to be permanently displayed by highlighting the name of the variable with the mouse in the source window. Then the value of the variable is displayed in the display window. Afterwards the current value of the variable is permanently displayed. However, an update only takes place when the debugger is in input mode, i.e. when the program execution is halted.

Each of the previously described commands which can be executed by a mouse click on the appropriate button, can also be entered directly into the terminal window using the keyboard. The actual command syntax shall be explained in the following section.

3.3.2 Command Syntax

Button	Command
Next	n
Step	s
Cont	c
Quit	q
Where	w
Up	u
Down	d
Clear all	a
Clear	C $<$filename$>$ $<$line$>$
Goto proc	g $<$name$>$
Print	p $<$expr_1$>$ ... $<$expr_n$>$
Display	D $<$expr_1$>$ ... $<$expr_n$>$
Stop at	S $<$filename$>$ $<$line$>$ [$<$cond$>$]
Execute	e $<$expr$>$

$<$expr$>$ stands for a MuPAD expression, $<$name$>$ a name of a MuPAD procedure, $<$filename$>$ a file name and $<$line$>$ a positive integer. $<$cond$>$ stands for a MuPAD expression whose Boolean evaluation must return either TRUE or FALSE.

The following commands can only be entered using the keyboard.

Command	Description
l	List of all breakpoints
?	Brief description of all debugger commands

In contrast to the Print- and Display buttons, with a keyboard the user can specify several variables simultaneously.

The Execute, Display and Print commands all have an effect on the current program environment. If the user executes a procedure or requests that the value of a procedure is to be displayed with Print then this may lead to the further program execution being changed. The reason for that is that procedures can have side-effects, e.g. may change the values of global variables, that cannot be easily recognized by the user. The user should take note what effects the execution of statements or the displaying of expressions can have. 

Chapter 4

Graphics

4.1 Introduction

MuPAD offers the possibility to create, display, print and save two-dimensional (2D) and three-dimensional (3D) graphics. The calculation of graphical data can be done on any computer. A screen with graphics capabilities is only needed to depict the plot on the screen.

The creation of graphics in MuPAD is modelled by the process of photography. To gain a better understanding one should imagine that a photograph with a specific content is to be taken. For this *objects* with specific characteristics are grouped into a *scene*. A photograph is then taken of the scene. To take the photograph the user must decide which section is to be photographed (*Zooming*), which scale is to be preset (*Scaling*) and how this is to be marked (*Axes*, *Ticks*, *Labels*), where the camera is set up (*Perspective*) and with which lighting (*Lighting*) the picture is to be taken.

The characteristics of the objects are determined by the title, the parametrization, the parameter range, the plot style and the color.

This user's model reflects the internal basic structure of the MuPAD graphics. It is possible to integrate many objects with their individual characteristics into a scene with further characteristics which are in their turn valid for all the objects.

For the creation of new graphics the MuPAD commands `plot2d` and `plot3d` are available. An exact description of these can be found in the help pages.

However, it is also possible to create a new graphics completely mouse-driven. To do so in XMuPAD one has to enter `plot2d` or `plot3d` without arguments (or equivalently choose the item Graphics from the Tools menu in XMuPAD).

This opens the so-called base window with an empty canvas. The pull-down menu Scene offers the items create 2D-scene... and create 3D-scene... for creating new scenes and the item read scene... for reading files containing graphical data. MacMuPAD provides the same functionality for the interactive creation of graphics.

To calculate 2D or 3D graphics it is necessary to give the specifications for the co-ordinate functions of the graphical objects. Even explicitly defined curves or surfaces must be given in parameter form. We shall illustrate the use of a parametrization with the help of the following example.

Example 193 *The graph of the function $f(x) := \sin(x)$ shall be depicted. In this case we parametrize the x co-ordinate by $x(u) := u$ and the y co-ordinate $y(u) := \sin(u)$. Furthermore, the range within which the parameterization is to be calculated must be specified. Here the interval $[0, 2\pi]$ has been chosen. To create the graph the MuPAD command* `plot2d` *is needed:*

```
>> plot2d([Mode = Curve, [u, sin(u)], u = [0, 2*PI]]);
```

If this command is given in XMuPAD or MacMuPAD then the graphic shown in figure 4.1 is created.

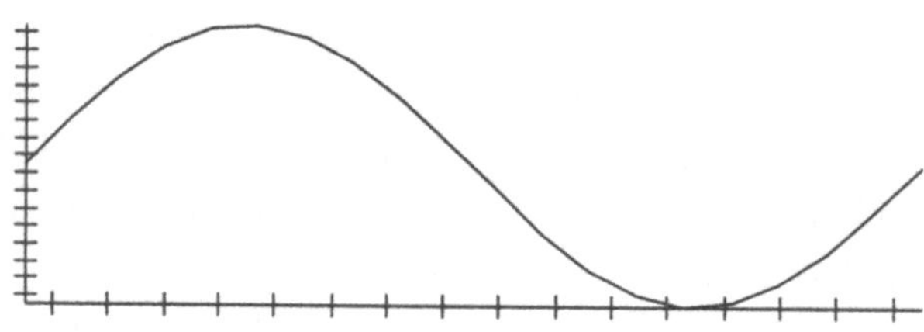

Figure 4.1: The Graph of the Function `sin`

In addition to function plots MuPAD 1.2.1 or later offers the possibility to plot lists of points or polygons. The following commands illustrate the use of the so-called graphical primitives `point` and `polygon`:

Example 194 *In a first step four points are defined which describe the corners of a quadrangle. Then this quadrangle is plotted as a list of two polygons.*

```
>> a := point( 0, 0, 0): b := point(-1, 1, 1):
```

```
c := point( 1, 1, 1): d := point( 0, 2, 2):

rect := [polygon(b, a, c), polygon(b, d, c)]:

plot3d(Axes = Box, Ticks = 0,
          [Mode = List, rect]);
```

*If this command is entered in XMuPAD or MacMuPAD the graphics shown
in figure 4.2 is created.*

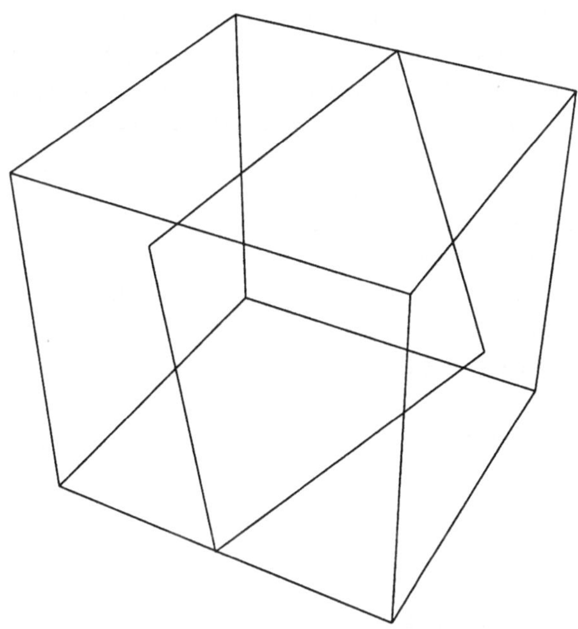

Figure 4.2: Use of Graphical Primitives

Beside the necessary declarations in a plot command, which are the mode of
an object, the parametrization, the parameter ranges or a list of graphical
primitives, a lot of further options exist which influence the appearance of
scenes and objects (see section 4.2). For any of these options a default value
is preset. If an option is not explicitly declared in a plot command then the
corresponding default value is taken. However, it is subsequently still possible
to change any option in VCam or MacMuPAD. Furthermore, the default values
can also be individually redefined (see section 4.4.6).

One specific option in a plot command is for instance the parameter *Grid*
which can be used to define the number of sample points in a plot. Apart
from that number MuPAD offers a further attribute, the so-called *Smooth-
ness-factor*. This factor determines the number of additional points to be
calculated between neighboring sample points. These are used to interpolate
a curve or surface more exactly. Hence, curves and surfaces can be depicted
more smoothly without too many parameter lines being drawn. The effect of
Smoothness is illustrated in the following example.

Example 195 *Three circles are to be drawn which were calculated with the
same number of sample points but different smoothness values. The circles are
parametrized with the help of polar co-ordinates, i.e. the x, y and z co-ordinates
are described by the functions $x(u,r) := r * \sin(u)$, $y(u,r) := r * \cos(u)$ and
$z(u,r) := 0$ The range of the radius r is the interval $[0,1]$ and for u we take
the interval $[-\pi, \pi]$. The corresponding MuPAD command is:*

```
>> plot3d(Axes = None,
        CameraPoint = [0.0, 1.0, 10.0],
        [Mode = Surface,
            [2.5+v*sin(u), v*cos(u), 0.0],
            u = [-PI, PI], v = [0, 1],
            Grid = [10, 10]
        ],
        [Mode = Surface,
            [v*sin(u), v*cos(u), 0.0],
            u = [-PI, PI], v = [0, 1],
            Grid = [28, 10]
        ],
        [Mode = Surface,
            [-2.5+v*sin(u), v*cos(u), 0.0],
            u = [-PI, PI], v = [0, 1],
            Grid = [10, 10],
            Smoothness = [2, 0]
        ]);
```

*This command produces the graphics in figure 4.3. The effect of the smoothness
factor can be clearly seen in the circle on the right. It contains exactly the same
number of parameter lines as the circle on the left but the radial parameter
lines of the circle on the right are curved. The middle circle shows how to
create curved parameter lines without using the smoothness factor. One has to
choose* **Grid = [28, 10]** *to obtain a reasonable result but with the effect that
28 radial parameter lines are drawn.*

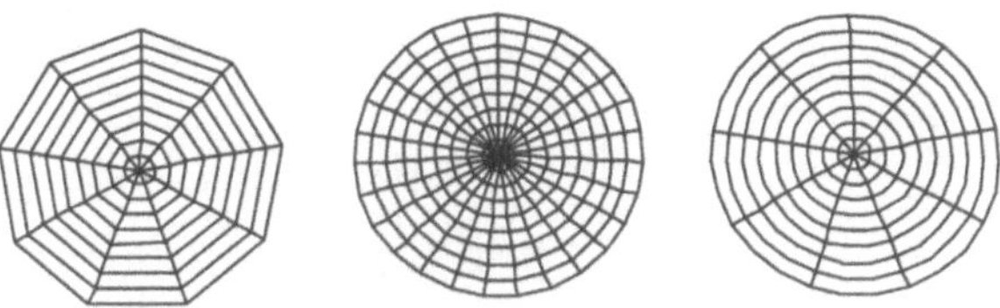

Figure 4.3: Demonstration of Smoothness

Apart from the options **Grid** and **Smoothness** MuPAD graphics offers a variety of different plot styles for various objects. The different plot styles available for curves are shown in the graphics in figure 4.4.

Example 196 *This graphics shows the different plot styles for a curve. They are produced by the following* plot3d *command:*

```
>> plot3d(Axes = Box, Ticks = 0,
        [Mode = Curve,
                [u, -PI, cos(u)], u = [-PI, PI],
                Grid = [40], Style = [Points]
        ],
        [Mode = Curve,
                [u, -1/3*PI, cos(u)], u = [-PI, PI],
                Grid = [40], Style = [Lines]
        ],
        [Mode = Curve,
                [u, 1/3*PI, cos(u)], u = [-PI, PI],
                Grid = [40], Style = [LinesPoints]
        ],
        [Mode = Curve,
                [u, PI, cos(u)], u = [-PI, PI],
                Grid = [40], Style = [Impulses]
        ]);
```

In addition to the plot style for surfaces it is possible to choose the parameter lines which are to be drawn. One can, for example, depict a sphere with only latitudinal degrees or only longitudinal degrees or both (see figure 4.5).

Example 197 *This example illustrates the different possibilities for drawing*

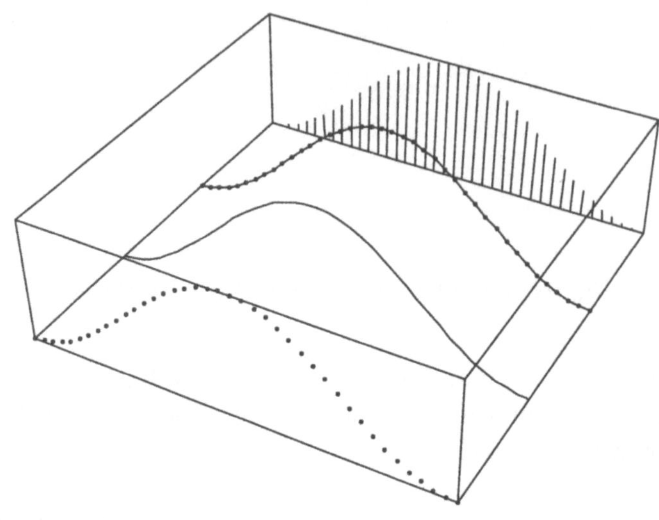

Figure 4.4: Plot Styles for Curves

parameter lines of surfaces. The graphics will be created by the following
`plot3d` *command:*

```
>> plot3d(Axes = Box, Ticks = 0, CameraPoint = [6.5, -21, 8],
        [Mode = Surface,
                [2.5+sin(u)*cos(v), sin(u)*sin(v), cos(u)],
                u = [0, PI], v = [0, 2*PI],
                Grid = [20, 20], Style = [HiddenLine, Mesh]
        ],
        [Mode = Surface,
                [sin(u)*cos(v), sin(u)*sin(v), cos(u)],
                u = [0, PI], v = [0, 2*PI],
                Grid = [20, 20], Style = [HiddenLine, ULine]
        ],
        [Mode = Surface,
                [-2.5+sin(u)*cos(v), sin(u)*sin(v), cos(u)],
                u = [0, PI], v = [0, 2*PI],
                Grid = [20, 20], Style = [HiddenLine, VLine]
        ]);
```

Further options (see also section 4.2) determine the appearance of the axes, the

Figure 4.5: Mesh Styles for Surfaces

perspective, the visible sections, the scaling or the title of the graphics. Furthermore, different possibilities of coloring graphical objects are implemented. Objects can be depicted in a freely chosen color or the colors of an object can be distributed according to the altitude. One can also define own MuPAD procedures for the coloring of objects. With this it is possible, for example, to project the Fatou set of a complex function on its absolute value or to simulate a temperature distribution on a surface. In future versions these functionalities are to be extended so that the colors are calculated according to the physical laws of a lighting model. There is also an automatic fine structure algorithm in the implementation phase with whose help the number of sample points used is adaptively calculated. A film tool integrated in the MuPAD graphics is also planned. With its help the user will be able to produce, store and re-play animation sequences.

In the following sections we shall describe the available options for the plot commands and the individual components of the graphical user's interface. The latter refers to the user's interface developed for the UNIX computer, the interface VCam — Virtual Camera. The Macintosh version will be described in section 5.3. This version has the same functionality as the graphics tool under UNIX but is designed as a typical Macintosh application with the "Look & Feel" of a Macintosh.

4.2 Plot Options

This section describes the options for a plot command with their admissible and their default values. The explicit syntax of a `plot2d` or a `plot3d` command can be found on the corresponding help pages.

Tables 4.1 and 4.2 show the options which can be used to specify a scene. The options of the first table are valid for 2D as well as for 3D scenes.

Option/Value	Meaning	Default Value
PlotDevice		Screen
Screen	The plot will be displayed on screen.	
"name"	The plot will be saved in binary format in the file named **name**.	
["name", format]	The plot will be saved in the format **format** in the file named **name**. **format** can have the following values: **Ascii**, **Binary**, **Raster**, **Gif** or **Postscript**.	
Title		""
"string"	**string** denotes the title of the entire plot.	
TitlePosition		Above
Above	The title will be placed above the plot.	
Below	The title will be placed below the plot.	
[f_1,f_2]	The real numbers **f_1** and **f_2** describe the position of the title. Admissible values lie between 0.0 and 10.0. [0.0,0.0] denotes the upper corner, whereas [0.0,10.0] stands for the lower corner both on the left-hand side.	
Axes		Box
None	No axes are drawn.	
Origin	Axes which cross in **AxesOrigin** are drawn.	
Corner	Axes are drawn beside the plot.	
Box	The plot will be framed in 2D or boxed in 3D, respectively.	

Table 4.1: Options for a Scene

Option/Value	Meaning	Default Value
Ticks		10
int	Every axis will be marked with the given number of ticks. Admissible values for `int` are integers and lie between 0 and 20.	
Arrows		FALSE
TRUE	Axes will be marked with an arrow.	
FALSE	Axes will not be marked with an arrow.	
Labeling		FALSE
TRUE	Axes will be displayed with a label and the tickmarks.	
FALSE	Axes will be displayed without a label and without tickmarks.	
Scaling		Constrained
Constrained	The plot will be scaled such that the ratio of the axes remains 1:1 in 2D or 1:1:1 in 3D, respectively.	
UnConstrained	The plot will be scaled such that the canvas can be optimally filled.	
ViewingBox		Automatic
Automatic	The measurement of the Viewing-Box (see section 4.4.9) is automatically determined such that the entire plot is visible.	
PointWidth		3
int	The integer `int` determines the width of a point. Admissible values lie between 1 and `max_long`.	
PointStyle		FilledSquares
Squares	Points will be displayed as squares.	

Table 4.1: Options for a Scene

Option/Value	Meaning	Default Value
FilledSquares	Points will be displayed as filled squares.	
Circles	Points will be displayed as circles.	
FilledCircles	Points will be displayed as filled circles.	
LineWidth int	The integer int determines the width of a line. Admissible values lie between 0 and max_long.	0
LineStyle SolidLines DashedLines	 Continuous lines are drawn. Dashed lines are drawn.	SolidLines
BackGround [r,g,b]	The real numbers r,g,b describe the color of the canvas. As above admissible values for r,g,b lie between 0.0 and 1.0.	[0.0,0.0,0.0]
ForeGround [r,g,b]	The real numbers r,g,b describe the color of the foreground in which the titles, the axes and the labels are drawn. Points and parameter lines of an object with plot style Transparent or ColorPatches and borderlines of filled polygons are also displayed in this color. Admissible values for r,g,b lie between 0.0 and 1.0 and determine the amount of red, green and blue in the RGB color model.	[1.0,1.0,1.0]
FontFamily		"lucida"

Table 4.1: Options for a Scene

Option/Value	Meaning	Default Value
`"string"`	The string `"string"` gives the font in which the titles and axes labels will be displayed.	
`FontStyle`		`"bold"`
`"string"`	The string `"string"` determines the desired font style.	
`FontSize`		10
`int`	The integer `int` determines the size of the font. Admissible values lie between 7 and 36.	

Table 4.1: Options for a Scene

In addition to the options above one can use the following attributes for a 3D scene:

Option/Values	Meaning	Default Value
`Labels`		`["x-axis",` `"y-axis",` `"z-axis"]`
`["x", "y", "z"]`	The strings `"x"`, `"y"` and `"z"` are used as labels for the x-, y- and z-axis.	
`AxesOrigin`		`Automatic`
`Automatic`	Axes cross in the middle of the plot.	
`[e_x,e_y,e_z]`	Axes cross in the point with co-ordinates `[e_x,e_y,e_z]`. Admissible values are real evaluable expressions and the identifiers `XMin`, `XMax` for `e_x`, `YMin`, `YMax` for `e_y`, and `ZMin`, `ZMax` for `e_z`. Here `XMin`, ..., `ZMax` denote the extremal co-ordinate values of the plot.	

Table 4.2: Additional Options of a 3D Scene

Option/Values	Meaning	Default Value
CameraPoint		Automatic
Automatic	The viewpoint of the camera is automatically determined depending on the size of the ViewingBox (see also section 4.4.9).	
[e_x,e_y,e_z]	The expressions e_x, e_y and e_z determine the co-ordinates of the viewpoint.	
FocalPoint		Automatic
Automatic	The camera is focused in the center of the viewing box.	
[e_x,e_y,e_z]	The expressions e_x, e_y and e_z determine the co-ordinates of the point in which the camera is focused.	

Table 4.2: Additional Options of a 3D Scene

For 2D scenes further options are the following:

Option/Value	Meaning	Default Value
Labels		["x-axis", "y-axis"]
["x", "y"]	The strings "x" and "y" are used as labels for the x- and y-axis.	
AxesOrigin		Automatic
Automatic	Axes cross in the middle of the plot.	
[e_x,e_y]	Axes cross in the point with coordinates [e_x,e_y]. Admissible values are real evaluable expressions and the identifiers XMin, XMax for e_x and YMin, YMax for e_y. Here XMin, ..., YMax denote the extremal co-ordinate values of the plot.	

Table 4.3: Additional Options for a 2D Scene

Like attributes for a scene there also exist different options for the objects.
Some options are admissible for all types of objects, some are only allowed for
special objects.

Options, which can be used for all objects are listed in table 4.4, whereas
table 4.5 and table 4.6 contain the additional attributes admissible for curves
and surfaces, respectively.

Option/Value	Meaning	Default Value
Title		""
"string"	string denotes the title of the object.	
TitlePosition		
[f_1,f_2]	The real numbers f_1 and f_2 describe the position of the title. Admissible values lie between 0.0 and 10.0. [0.0,0.0] denotes the upper corner, whereas [0.0,10.0] stands for the lower corner both on the left-hand side.	Depends on the number of the object.
Color		[Flat]
[Flat]	The object is colored with one color. This color is determined by the number of the object and the default color values.	
[Flat, [r,g,b]]	As above but the color in the RBG model is given by the real numbers r,g,b. Admissible values for them lie between 0.0 and 1.0.	
[Height]	The coloring is determined by the y co-ordinate in 2D and the z coordinate in 3D. The actual colors are calculated with the help of the number of the object and the default color values.	
[Height, [r,g,b], [r,g,b]]	As above but the starting color is given by the first RGB triple and the ending color by the second RGB triple.	

Table 4.4: Options for Objects

Option/Value	*Meaning*	*Default Value*
`[Function, function_name]`	The user-defined color function named `function_name` is used for the calculation of the coloring (see also section 4.5).	

Table 4.4: Options for Objects

In addition to the options in the table above one can choose for an object with `Mode=Curve` attributes from the following table:

Option/Values	*Meaning*	*Default Value*
`Grid`		`[20]`
`[i]`	The integer i defines the number of sample points of the curve. It must be equal or greater than 2.	
`Smoothness`		`[0]`
`[i]`	The integer i determines the number of additional interpolation points between two sample points. Admissible values lie between 0 and 20.	
`Style`		`[Lines]`
`[Points]`	Only the sample points of the curve are drawn.	
`[Lines]`	Only lines between sample points are drawn.	
`[LinesPoints]`	Sample points as well as the lines between them are drawn.	
`[Impulses]`	Sample points are drawn together with the corresponding section of the axis in y-direction in 2D or in z-direction in 3D.	

Table 4.5: Additional Options for a Curve

For an object with `Mode=Surface` one can use the options of table 4.4 and the following:

Option/Values	Meaning	Default Value
Grid		[20,20]
[i_1,i_2]	The integers i_1 and i_2 define the number of sample points in direction of the parameters **var_u** and **var_v**. They must be equal or greater than 2.	
Smoothness		[0,0]
[i_1,i_2]	The integers i_1 and i_2 determine the number of additional interpolation points between two sample points in direction of **var_u** and **var_v**. Admissible values lie between 0 and 20.	
Style		[WireFrame, Mesh]
[Points]	Only the sample points of the surface are drawn.	
[WireFrame, Mesh]	The surface is displayed as a wireframe model and both parameter lines in direction of **var_u** and **var_v** are drawn.	
[WireFrame, ULine]	As before but only parameter lines in direction of **var_u** are drawn.	
[WireFrame, VLine]	As before but only parameter lines in direction of **var_v** are drawn.	
[HiddenLine, Mesh]	The surface is displayed as a solid object with parameter lines in both directions being drawn.	
[HiddenLine, ULine]	As before but only parameter lines in direction of **var_u** are drawn.	
[HiddenLine, VLine]	As before but only parameter lines in direction of **var_v** are drawn.	

Table 4.6: Additional Options for Surfaces

Option/Values	Meaning	Default Value
`[ColorPatches, Only]`	The surface is displayed as a solid object, all surface patches are colored. No parameter lines are drawn.	
`[ColorPatches, AndMesh]`	As before but parameter lines in both directions are drawn.	
`[ColorPatches, AndULine]`	As before but only parameter lines in direction of **var_u** are drawn.	
`[ColorPatches, AndVLine]`	As before but only parameter lines in direction of **var_v** are drawn.	
`[Transparent, Only]`	As before but the patches are filled with patterns to produce a feeling of transparency. No parameter lines are drawn.	
`[Transparent, AndMesh]`	As before but parameter lines in both directions are drawn.	
`[Transparent, AndULine]`	As before but only parameter lines in direction of **var_u** are drawn.	
`[Transparent, AndVLine]`	As before but only parameter lines in direction of **var_v** are drawn.	

Table 4.6: Additional Options for Surfaces

4.3   Graphical Primitives

In MuPAD it is possible to plot lists consisting of graphical primitives. This is done in 2D as well as in 3D by use of the object mode **Mode = List**.

The simplest user-defined list contains only points which are generated by use of the function **point**. A detailed description of this function can be found on the corresponding help page. Depending on the dimension the list consists only of 2D or 3D points and an additional option to specify the color of the points.

Example 198 *The following commands serve for the definition of four points, which then are displayed by use of a* plot2d *command*

```
>> a := point(0, 0): b := point(0, 1):
   c := point(1, 1): d := point(1, 0):
```

```
plot2d(Axes = None,
          [Mode = List, [a, b, c, d]]);
```

Primitives of higher order can be created by use of the function `polygon`.
Polygons themselves are composed out of an arbitrary number of points. Furthermore there exist additional options. By use of these options the user is
able to influence the appearance of the polygon. These options are used to
define the color of a polygon and to determine whether the polygon is to be
closed and/or filled. The last two options are only available for triangles in
MuPAD 1.2.2, because polygons of higher order need to be triangulated before
drawing and the necessary algorithms are not yet implemented.

Example 199 *The following example shows how to draw a filled three-dimensional rectangle:*

```
>> a := point(  0,   0,   0):
   b := point(-0.5, 0.5, 0.5):
   c := point( 0.5, 0.5, 0.5):
   d := point(  0,   1,   1):

   plot3d([Mode = List,
             [polygon(a, b, c, Filled = TRUE),
              polygon(b, c, d, Filled = TRUE)]
         ]);
```

4.4   VCam — The Graphics' User's Interface

This section explains the components and the use of the graphics' interface
VCam. As an introduction some general remarks about the design of VCam
should be made:

- VCam was developed and implemented as an independent module of
  MuPAD. VCam can also be used without MuPAD as an external viewer
  to present graphical data (see section 4.6.2 and section 4.6.3).

- To enable easy learning and usage an equivalent of the 2D and 3D interface has been realized as far as possible.

- For each option that can be set mouse-driven in VCam there is an al-
  gorithmic attribute, which can be used in a plot command. Hence, the
  same plot can be produced by a `plot2d` or `plot3d` command.

- In contrast to other computer algebra systems it is possible to create new plots completely menu-driven.

- Some of the buttons and menus described in the following are inactive at certain times, i.e. clicking these buttons or menus has no effect. A deactivated interface object can be easily recognized; its contour changes from black to grey.

- In the actual version some options are not yet implemented. If one activates such an option a message will inform the user. After that the previous value of the option will be restored.

The graphics module VCam basically consists of a *base window* for displaying the plot and a corresponding *manipulation window* in which the menu-controlled options can be set. In addition, further components, like the *default window* in which the default values are shown and can be altered, are implemented. These ensure a user-friendly handling of the entire module.

4.4.1   The Base Window

The base window (see figure 4.6) is used to display plots. It contains a drawing area, in which the graphics is depicted, and various buttons and menus for controlling the creation and storage of plots. The base window can be activated by choosing the item Graphics in the Tools menu of XMuPAD or by entering `plot2d();` or `plot3d();` in XMuPAD. If one of these commands is given without arguments then the base window is opened and a plot can be created interactively. If, however, the command is given with a valid description of a scene (see the help pages for `plot2d` and `plot3d`) then the base window is opened and the corresponding plot is drawn. At the same time the manipulation window with the specified options is opened. In this window the plot options can be manipulated.

The base window contains the following buttons and menus:

- Scene
  This menu is used to create a new scene. It contains the following items:

  - create 2D-scene...
    With the help of this menu item new two-dimensional scenes can be created. Please note that this item can only be activated when no other scene is present. When this menu item is activated the

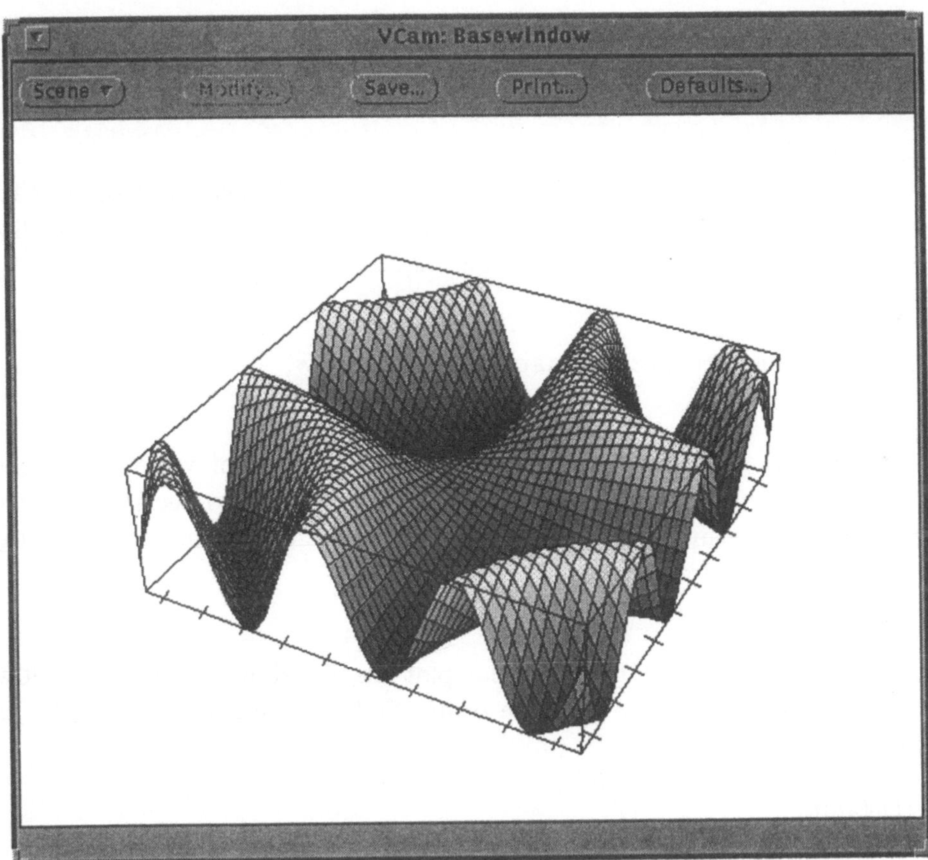

Figure 4.6: Base Window

manipulation window for a 2D scene is opened. In this the necessary arrangements for the new scene can be made.

- create 3D-scene...
  This is used to create a three-dimensional scene. Again this item can only be chosen when no other scene exists. After activating this item the manipulation window for a 3D scene in which the scene options can be chosen is opened.

- read scene...
  This menu item can be used to read graphical data from a file. A new scene can only be read when no other scene exists. After clicking this item the *read window* (see also section 4.4.3) in which the path and file name of the required file can be entered is opened.

- delete scene

This item is used to delete an existing scene in order to create a new scene with the menu items create 2D-scene..., create 3D-scene... or read scene....

– animate scene...
This item opens a further window in which the necessary options for animation of the current scene can be declared. This functionality is not implemented in version 1.2.2.

- Modify...
This button activates the manipulation window.

- Save...
The Save... button is used to store a plot in one of various formats. After clicking this button the *save window* (see also section 4.4.4) is activated. In this window the user can choose the format in which the plot is to be stored.

- Print...
This button is used to print the plot displayed in the base window. After activating this button the *print window* (see also section 4.4.5) is opened.

- Defaults...
This button is used to display and change the default values of the options of a plot command. After clicking on it the default window is opened. In this window the values can be changed by menus. More detailed information about the default values and the default window can be found in section 4.4.6.

4.4.2   The Manipulation Window

The manipulation window contains information about the status of the current plot and is used for entering and altering the graphics options. The options for 2D and 3D differ and therefore there are separate windows for 2D scenes and 3D scenes. The principle structure and the visual presentation are the same in both windows. The manipulation window is divided into two areas, one for setting the scene characteristics and the other for choosing the current object and for setting its characteristics. The characteristics of scenes and objects are arranged under main headings such as Axes for instance. Beneath these main headings there are menus for choosing from the possible options. The current choice is displayed.

In the following we shall firstly introduce the manipulation window for a three-dimensional plot and finally we shall explain the differences to the manipulation window for a two-dimensional plot.

4.4.2.1 The Description of a 3D Scene

As already mentioned the manipulation window for a 3D scene (see figure 4.7) is divided into two areas. In the upper part of the window the user can choose the options for a scene and in the lower part the current object and its values can be defined. In the following we shall firstly explain the specific options for a scene, then we shall introduce the options for the objects.
Please note that any changes to options only become visible after the Plot button is pressed. This enables the user to make a number of changes before the plot is redrawn with the altered options.

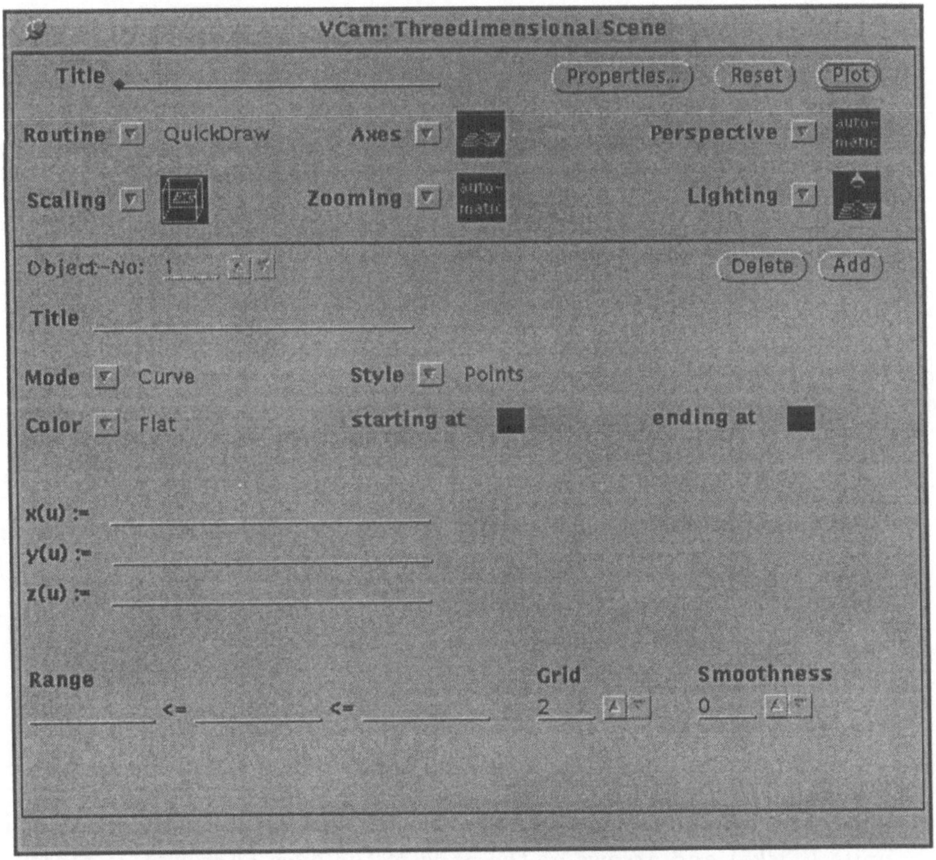

Figure 4.7: Manipulation Window for a 3D Scene

Attributes of Scenes

- Title
  Each scene can be given a title which is entered or altered here. If the
  title was declared in a plot command and is too long for the text field,
  then only a part of the title is displayed. In this case the title field
  is deactivated so that no interactive changes can be made. The title
  is placed in a predefined position in the base window and later can be
  moved by using the mouse. To do this the user must click on the title
  in the canvas and then move the mouse with the button pressed.

- Routine
  In this menu the user can decide which routine is to be used to calculate
  the graphical data. The following possibilities are available:

  - QuickDraw
    The parametrization is evaluated for the sample points defined by
    the parameters Grid and Smoothness.

  - QualityDraw
    In addition to the sample points defined by the parameters Grid and
    Smoothness further sample points, dependent on the curvature of
    the object, are evaluated using an adaptive procedure. This option
    is not yet implemented in version 1.2.2.

- Axes
  Four possibilities are available:

  - None
    No axes are drawn.

  - Origin
    Axes are drawn which cross in AxesOrigin. The default setting
    of AxesOrigin lies in the middle of the graphics, but this can be
    altered by using the *properties window,* see also section 4.4.7.

  - Corner
    The axes are drawn beneath the graphics.

  - Boxed
    The plot is framed by a rectangle in 2D or a box in 3D.

The presentation of the axes, like the use of axes labels (Labels), scale
units (Ticks) and arrows at the ends of the axes (Arrows), is preset by
the default values. It can be altered interactively by using the properties
window.

- Perspective
  Two possibilities are available.

  - Automatic
    The perspective is automatically chosen dependent on the size of the ViewingBox (see also section 4.4.9).

  - Manual
    If this option is chosen the *perspective window* is opened. In this window the user can change the perspective interactively (see also section 4.4.9) .

- Scaling
  With this menu the user can choose how the plot is scaled. The options are:

  - Constrained
    Scaling is carried out so that deformation only occurs as a result of perspective transformation. In this case spheres appear as spheres (and not as ellipsoids) in the base window. In drawing areas that are not square this may lead to the drawing area not being filled optimally.

  - UnConstrained
    In this case the plot is scaled so as to fill the drawing area optimally.

- Zooming
  With this menu the depicted section of the plot can be changed. The following options are available:

  - Automatic
    The section is chosen so that the entire plot is visible in the base window.

  - Manual
    If this item is chosen the *zoom window* (see section 4.4.8) is activated. In this window the depicted section can be interactively changed.

- Lighting
  It is planned to offer different lighting models for a scene in this menu item. This option is not implemented in version 1.2.2.

Further functions are given by the three buttons Properties..., Reset and Plot.

- Properties...
  The clicking of this button opens the properties window in which additional options, like the axes labels or the font for the title of the scene or object, can be changed. More detailed information can be found in section 4.4.7.

- Reset
  With this button the entries of the previous plot command are recalled, provided that a command has already been given. If this is not the case then the default values are used.

- Plot
  The current values of the attributes are entered in a plot command. This command is evaluated and the resulting plot is drawn in the already existing base window. At this point it should be mentioned again that changes in attributes only take effect after the Plot button has been pressed.

Attributes of Objects

This part of the manipulation window contains an area for choosing the current object and an area for entering the attributes of this object. The current object is determined by changing the number of the text field Object-No:. With the two buttons Add and Delete new objects can be added to a scene or existing objects can be deleted from a scene.

For the current object the following attributes can be set.

- Title
  Each object can be given a title which must be entered here. This title is drawn in a predefined position in the base window and can be moved with the mouse exactly like the title of the whole scene.

- Mode
  With mode the user can define the type of an object. The following possibilities are available.

    - Curve
      The current object is a curve.

    - Contour
      The contour diagram of a three-dimensional object is drawn. This option has not been implemented in version 1.2.2.

- Surface
  The current object is a surface.

- List
  An already stored list is read and drawn. Although user-defined lists can be drawn with a corresponding plot command, the interactive input is not implemented in version 1.2.2.

• Style
Here the user can determine how the current object is to be graphically depicted. The options are dependent on, and change with, the chosen object mode. If the user has chosen **Mode = Curve** then the following possibilities are available:

- Points
  Only the sample points are drawn.

- Lines
  Only connecting lines between sample points are drawn.

- LinesPoints
  Both the connecting lines and points are drawn.

- Impulses
  The abscissa of the y co-ordinates in 2D or the z co-ordinates in 3D are drawn.

For **Mode = Contour** the following is possible:

- Lines
  The lines are drawn.

In addition the user can also choose:

- Bottom
  The contour lines are drawn on the bottom of the ViewingBox.

- Attached
  The contour lines are drawn with respect to the height.

If **Mode = Surface** is set then the following is possible:

- Points
  Only the sample points are drawn.

- WireFrame
  A wireframe model of the object is drawn.

  – HiddenLine
    The surface is considered opaque and therefore hidden lines are not
    drawn.

  – ColorPatches
    Similar to HiddenLine, except that the patches which form a surface
    are colored.

  – Transparent
    Similar to ColorPatches, except that the patches are filled with
    patterns. By using different patterns for different objects a feeling
    of transparency can be produced.

In case of `Style=WireFrame` or `HiddenLine` the user can decide which
parameter lines are to be drawn.

  – Mesh
    All parameter lines are drawn.

  – ULine
    Only the parameter lines in direction of the first variable are drawn.

  – VLine
    Only the parameter lines in direction of the second variable are
    drawn.

With `Style=ColorPatches` or `Transparent` the options are as follows.

  – Only
    Only the colored patches are drawn.

  – AndMesh
    The patches and all the parameter lines are drawn.

  – AndULine
    The patches and the parameter lines in direction of the first variable
    are drawn.

  – AndVLine
    The patches and the parameter lines in direction of the second
    variable are drawn

In the properties window (see also 4.4.7) the user can also choose the
width of points and lines and the point and line styles.

• Color, starting at, ending at
  With this menu the user can determine the color(s) of the current object.
  The user is offered a color chart containing a maximum of 36 colors. In

this chart the user can choose two colors using the starting at and ending at buttons. By interpolation of the start value with the end value further colors are then internally calculated.

With the menu Color the user can choose a color distribution for an object from the following list:

- Flat

  The object is one-colored, namely the color chosen for starting at.

- Height

  The colors are calculated using the height (z co-ordinate for 3D objects and y co-ordinate for 2D objects) and the colors given by starting at and ending at.

- Physical

  The colors are calculated using the physical laws according to the lighting model. (This option has not been implemented in version 1.2.2.)

- Functional

  The colors are distributed according to a user-defined MuPAD procedure. After choosing this item the text field Function will be activated in which the name of the user-defined color function has to be entered (see also section 4.5).

If the buttons starting at or ending at are activated, then a further window, the so-called *color window* (see also 4.4.11) is opened. In this window the colors can be changed.

- x(u,v), y(u,v), z(u,v)

  Here the user can enter the functions which give the x, y and z co-ordinates of the current object. With `Mode = Surface` these are in general three functions dependent on two variables and with `Mode = Curve` three functions dependent on one variable. In the second case the three attributes are re-named x(u), y(u) and z(u). As with the titles it is possible that the character strings given in a MuPAD plot command for the parametrization are too long for the text fields. In this case the text fields are deactivated so that no interactive changes can be made.

- Range

  These text fields are used to give the names of the independent parameters and the range in which they are to be evaluated. The text fields are ordered in the form `min <= name <= max`. The names given here must agree with the free variables of the parametrization. The number of ranges and names to be given are dependent on the mode of the object. With `Mode = Curve`, only one range and one name need be given. With `Mode = Surface` two variables and two ranges must be given.

- Grid
  Here the user can choose the number of sample points in each parameter direction. The number can be entered either directly in the appropriate text field or by changing the counter.

- Smoothness
  Here the user can determine additional interpolation points between two visible sample points. With `Mode = Contour` the user can enter the number of contour lines to be drawn here.

4.4.2.2 Description of a 2D Scene

In this section we shall deal with the differences between the 2D manipulation window (see figure 4.8) and the window of a 3D scene. In the area of the options for the scene all the icons are replaced by their two-dimensional analogues. The Perspective and Lighting buttons are not included. All other buttons and menus have the same functionality as in the 3D window. Concerning the options for the objects the following changes are visible:

- Mode
  The user now only has the choice between the options `Mode = Curve` and `Mode = List`, whereby the interactive input of lists is not implemented in version 1.2.2.

- Style
  This includes only those possibilities for depicting curves as already described. These are Points, Lines, LinesPoints and Impulses.

- $x(u)$, $y(u)$
  To define a parametrization of a 2D object two functions dependent on one variable are needed.

- Range, Grid, Smoothness
  As the parametrization is dependent on only one variable, only one range and two numbers are required. These numbers give the number of sample points and the number of points for interpolating adjacent sample points.

In the next sections further components of the user's interface shall be introduced and described. These components are, as already mentioned, opened

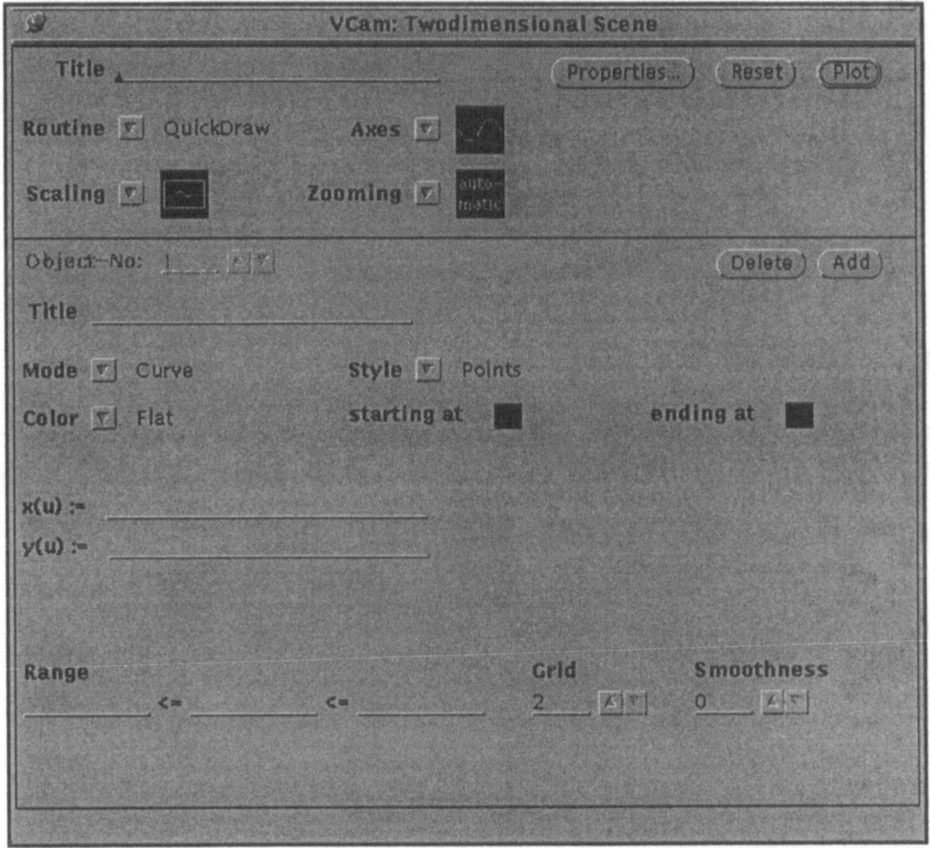

Figure 4.8: Manipulation Window of a 2D Scene

by using the appropriate buttons in the base window and the manipulation window.

4.4.3 The Read Window

The read window (see figure 4.9) is used to read a new scene. It is activated with the menu item read scene... in the Scene menu of the base window. This window contains two text fields for entering the path and file names of the file to be read and the button Read File to execute the reading. Files can be either in ASCII format or in MCode format. As usual the presetting of the path name is the path of the working directory. The user is given a message if the file cannot be found or cannot be read.

Figure 4.9: Read Window

4.4.4   The Save Window

The save window (see figure 4.10) is used for storing plots in various formats.
It is activated by the Save... button in the base window. It contains a pull-

Figure 4.10: Save Window

down menu for choosing the desired format and two text fields for entering
the path and file names.

- Save as
  This menu is used for choosing the format in which the graphics is to be
  stored. As yet the following possibilities are available:

  - Popup...
    With this the current plot in the base window is stored in a further
    window, the *saved plot window* (see also section 4.4.10). With the
    Replot button in that window the user can reproduce the stored
    plot in the base window.

  - Raster-File
    The plot depicted in the base window is stored as a raster file, where
    the path and file names of the file have to be entered.

- Gif-File
  The plot in the base window is stored as a GIF picture. Once again the path and file names have to be given.

- Postscript
  The plot in the base window is stored as a postscript file. Again the path and file names have to be entered. It has to be mentioned that the plot style **Transparent** is not supported in this format. Objects being plotted with this style are stored with plot style **ColorPatches**.

- ASCII-File
  The plot in the base window is stored as an ASCII file under the entered path and file names. Files in ASCII format can be re-read.

- MCode-File
  The plot in the base window is stored as a MCode file. Once again the path and file names must be given. MCode files can be re-read.

- Plot-Command (File)
  The plot command belonging to the plot depicted in the base window is calculated and stored in a file. Files containing this command can be re-read by MuPAD. Please note that this command only contains those options unequal to the related default value. If the default values are changed before such a command is re-read the resulting plot might differ from the earlier one.

- Plot-Command (XMuPAD):
  The plot command belonging to the plot depicted in the base window is copied into the XMuPAD window. There it can be changed and re-executed.

- Directory
  This text field is used to specify the path under which the plot is to be stored.

- File
  This text field is used to enter the name of the file.

- Save Scene
  With this button the chosen option is verified and the graphics is stored.

If options have been manipulated without pressing the Plot button afterwards, a warning will be displayed. In that case the plot is not stored.

For storing plots in the formats Raster and GIF MuPAD uses the package Netpbm based on Pbmplus (Copyright ©1989, 1991 by Jef Poskanzer), see also Appendix C.

4.4.5   The Print Window

The print window (see figure 4.11) is used to print a plot. It is activated by the Print... button in the base window. The print window contains the following

Figure 4.11: The Print Window

buttons and menus:

- Destination:
  With this the user can determine where the plot is to be printed. The following two possibilities are available:

  - Laserwriter
    The plot is printed on a laser printer.

  - Notebook
    The plot is inserted into a Notebook. This option has not been implemented in version 1.2.2.

- Printer:
  The name of the printer must be entered here. The printer name designated in the environment variable **LW_PRINTER** is preset here. However, this can easily be changed by typing any other printer name.

- Print Scene
  This button is used to verify the previously chosen options and to print the plot. If the user gives no printer name or options have been manipulated without pressing the Plot button afterwards, an error message is given.

4.4.6   The Default Window

With the default window (see figure 4.12) the user can change some of the preset default values. The default values are used to assign a given value to

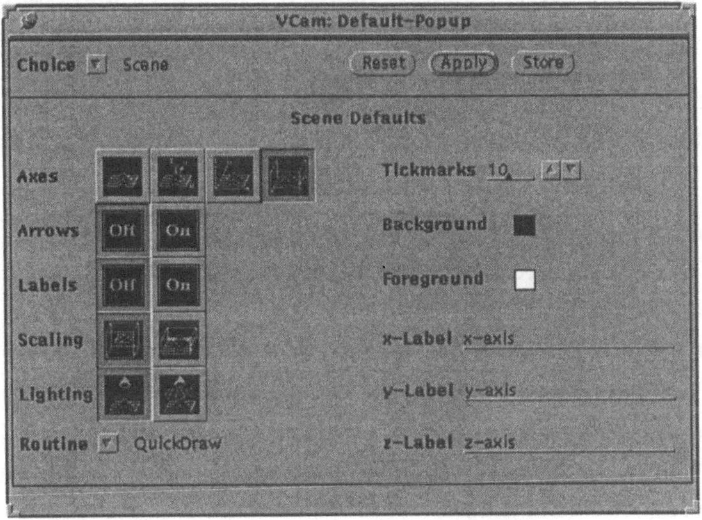

Figure 4.12: Default Window

the more important options of the plot commands. The graphics tool VCam proceeds as follows: The most important default values needed for a single plot are stored in a file named `.vcam_defaults`. This file is looked for in the user's home directory at the start of MuPAD. If it does not exist then the file is read from the path given by the environment variable `MuPAD_LIB`. These default values are then used for those options not explicitly specified in a `plot2d` or `plot3d` command.

In the default window the user can change the default values interactively. Furthermore, each user can define his own default values and store these in his home directory in the file `.vcam_defaults`. After changing the default values they are then available during the current session. On restarting VCam or XMuPAD the personal default values of the current user are used immediately.

The default window is used to show the chosen default values or to change them. It is activated by the Defaults... button in the base window.

The default values are divided into:

- default values for the scene,

- default values for the objects and

- default values that are used for drawing.

The default window contains two areas. One area contains a series of control functions in which the user can choose which default values are to be changed. In the other area, the current default values are displayed and can be changed. The control function area contains the following buttons and menus:

- Choice
  This list shows the headings under which the default values are categorized. One has the following choices:

  - Scene
    The default values for the attributes of a scene can be changed here (see figure 4.12).
  - Objects
    The default values for objects can be changed here (see figure 4.14).
  - Drawing
    The user can change the default values used by the drawing routines here (see figure 4.13).

- Reset
  With this button changes can be canceled. On activating this button all changes made since the last Apply command are reset.

- Apply
  With this button the changes are verified and are made available to the graphics internally as new default values for the current session.

- Store
  With this button the currently displayed default values are made available internally as new defaults for the current session and they are stored in the file .vcam_defaults in the user's HOME directory. With this we offer the user the chance to define his personal default values for attributes. At each new call of MuPAD these are then used.

The available options to change are dependent on the current value under Choice. In figure 4.12 the user can see the current default values for a scene. The following buttons, menus and text fields exist:

- Axes
  The values displayed here determine the default value for the Axes menu in the manipulation window.

- Arrows
  This value determines if arrows are drawn at the ends of the axes (On) or not (Off).

- Labels
  With the help of this menu the user can set if the axes labels and the axes markings are to be drawn (On) or not (Off).

- Scaling
  The values given here determine which scaling routine is to be used as a default for displaying the plot.

- Lighting
  This menu gives the default value of the lighting menu contained in the manipulation window. The user can choose between `Lighting = Off` and `Lighting = On` as a presetting. We should mention here that at the moment no light sources are supported in MuPAD. Due to this the choice `Lighting = On` will be ignored.

- Routine
  With this the user can change the default values of the Routine menu of the manipulation window. Please note that in the current version of MuPAD only `Routine = QuickDraw` for calculating the graphical data is implemented.

- Ticks
  The value shown here is the default value for the markings which are drawn on each axis.

- Background, Foreground
  The colors shown by these buttons are the default values of the fore- and background respectively. When one of these buttons is activated the so-called color window is opened (see also 4.4.11) .

- x-Label, y-Label, z-Label
  These text fields are used for entering the default character strings for the axes labels.

The default values which are used for the drawing routine (see figure 4.13), consist of the following entries:

- Font-Name, Font-Style
  These character strings give the name and the font style that are preset

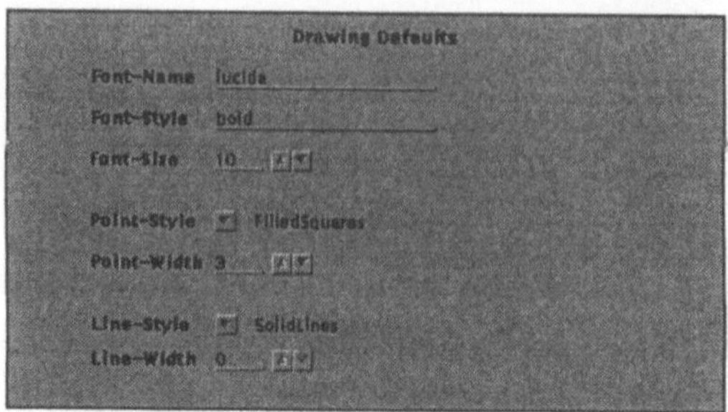

Figure 4.13: Drawing Default Values

for the title and the labeling of the axes. As the default value the font
lucida and the style bold are displayed. Other supported fonts are
charter, courier, helvetica and times for instance. As the font style
one can choose normal, medium, italic, bold for instance.

- Font-Size:
  This value (between 7 and 36) gives the size of the default font.
  If the user changes the values of the font options and verifies them with
  Apply or Store it is firstly checked if the newly specified font can be
  loaded. If not a message is given and the original font is reset.

- Point-Style
  With this attribute the user can decide how points are to be depicted as
  a standard. One has the choice of FilledSquares, Squares, FilledCircles
  and Circles.

- Point-Width
  This attribute gives the width of points in pixels. Only positive integers
  are admissible.

- Line-Style
  With this attribute the user can choose how lines are to be depicted as
  a standard. One can choose between SolidLines and DashedLines.

- Line-Width
  This attribute determines the width of lines in pixels. Lines can have a
  width of any positive integer or zero. The value 0 means that a line of
  width 1 is drawn but a faster algorithm is used.

With the aid of the window in figure 4.14 the user can change the default values
for two- and three-dimensional objects. For 2D objects the mode and the style
of an object can be changed. Additionally, for three-dimensional objects the

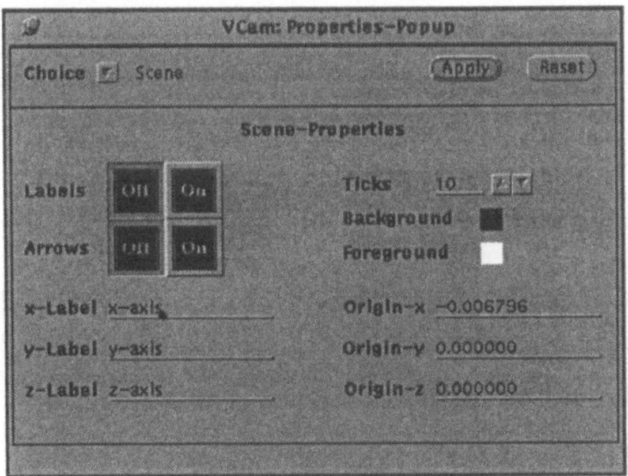

Figure 4.14: Object Default Values

user is able to determine a default value for the parameter lines to be drawn. Furthermore the user can determine the default values for the number of sample points (Grid) and the number of interpolating points (Smoothness). Additionally the user can set which color functionality (Color) and which names for the independent parameters of the parametrization (var_u and var_v) are to be used as defaults.

4.4.7 The Properties Window

The properties window (see figure 4.15) can be used to change further options. It is activated with the aid of the Properties... button in the manipulation

Figure 4.15: Properties Window

window. The user can change options belonging to the scene (e.g. axes labels) and options connected with the drawing routine (fonts, point and line styles). The properties window is similar to the default window in its construction:

There is an area in which the values to be changed can be chosen and verified and an area in which the changeable values are listed. The functions for choosing and verifying the changed values are:

- Choice
  With the aid of this menu the options to be changed can be chosen. The user has the choice between Scene (additional scene options can be changed) and Drawing (the drawing options can be changed).

- Apply
  This button is used to take over the changed values. They are then used when pressing the Plot button.

- Reset
  This button is used to reset all changes made since the last Apply command.

The options that can be changed in the properties window generally represent those in the default window (see also figures 4.12 and 4.13). The differences are concerned with the options of a scene. The properties window contains no possibility to change the axes style because this can be done in the manipulation window. The same is true for the scaling, routine and lighting options. In the properties window however the three text fields Origin-x, Origin-y and Origin-z exist which can be used for giving the co-ordinates of intersection of the axes.

Please note that these fields may only contain numerical values: MuPAD identifiers which represent real values are not admissible because they are not known by the graphics tool VCam.

4.4.8 The Zoom Window

The zoom window is used to change the visible section of a plot. This window (see figure 4.16) is activated by choosing the option Manual of the Zooming menu in the manipulation window. It contains the following buttons:

- Zoom In
  This button is used to enlarge the visible section of a plot. It is enlarged in all directions by a predefined factor.

- Zoom Out
  With the aid of this button the visible section is scaled down by a predefined factor.

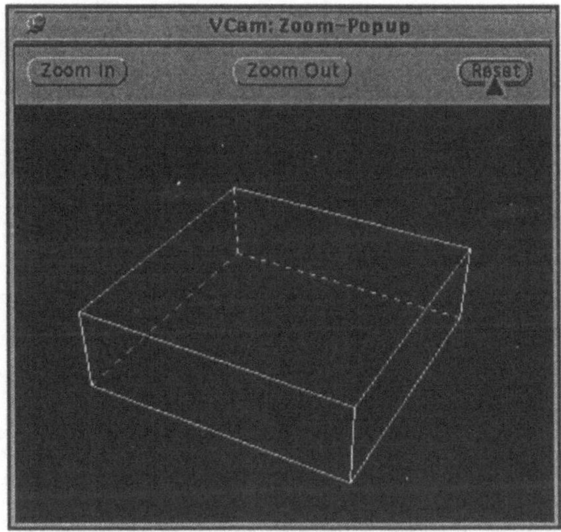

Figure 4.16: Zoom Window

- Reset
  This button is used to reset all the changes made since the last plot command.

By using the mouse the user also has the possibility to move the plot within the drawing area. To do this one must click the left mouse button in the drawing area and move the mouse with the button still pressed into the desired direction.

4.4.9 The Perspective Window

The perspective window (see figure 4.17) is used to change the perspective. This window is activated by choosing the option Manual of the Perspective menu in the manipulation window. Perspectives are calculated in MuPAD with two perspective parameters. These are the two three-dimensional points CameraPoint and FocalPoint. The CameraPoint gives the camera position and the FocalPoint gives the point at which the camera is pointed.

This window contains the usual Reset button for resetting the changed perspective parameters to the values that were valid at the last plot command. Furthermore, there is the possibility of explicitly giving values for the perspective parameters. For this purpose the text fields CameraPoint and FocalPoint and the sliders can be used. Furthermore, the perspective window contains a drawing area in which the *ViewingBox* is displayed in the current perspec-

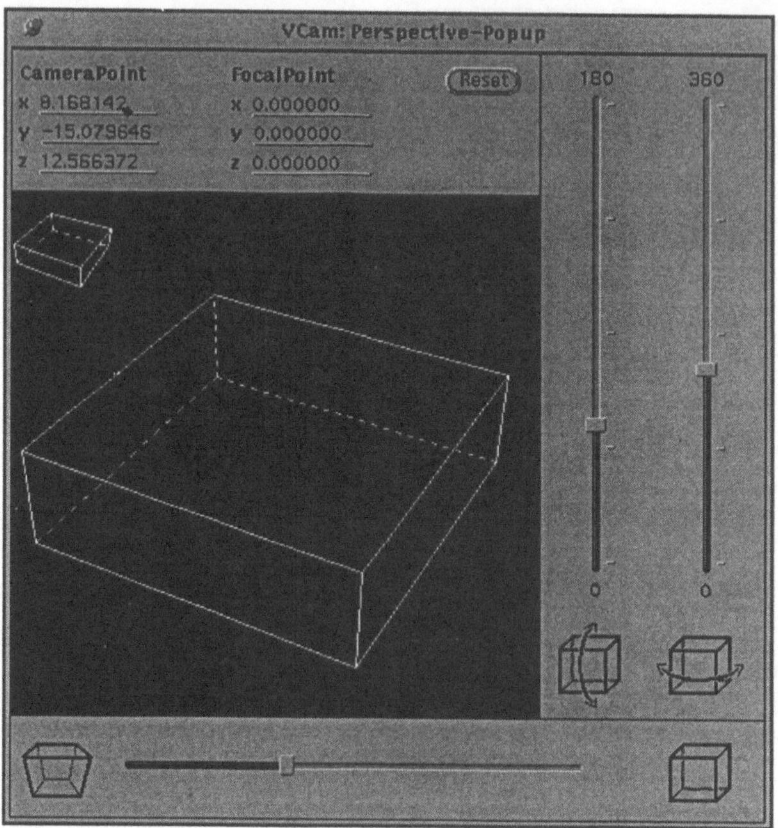

Figure 4.17: Perspective Window

tive. By ViewingBox we mean the minimal box which still contains all the objects of a scene. Changes in the perspective parameter are immediately displayed by the ViewingBox in the drawing area of the perspective window. The perspective parameters can be changed as follows:

- The user enters the explicit 3D co-ordinates in the text fields Camera-Point and FocalPoint and verifies the entry with <Return>.

  It is important that these text fields only contain numerical values; the user should not enter MuPAD identifiers which contain real values as these are not known by the graphics tool.

  The changes are displayed in the drawing area, i.e. the user can see the effects of the changes immediately. In addition to the ViewingBox in the new perspective a smaller ViewingBox is also shown in the original perspective to give the user a better idea of the effect his changes have. Furthermore when the explicit values for the CameraPoint and Focal-

Point are changed the corresponding values of the polar co-ordinates controlled by the sliders are adjusted.

- The second possibility open to the user is the changing of the polar co-ordinates with the mouse. For this purpose three sliders are provided. With these the user can rotate the depicted ViewingBox about its vertical and horizontal axis. With the lower slider the user can determine the distance from the CameraPoint to the ViewingBox. Once again, after each change the ViewingBox is displayed in the current perspective, and the 3D co-ordinates of the CameraPoint and FocalPoint are adjusted.

4.4.10 The Saved Plot Window

The saved plot window (see figure 4.18) is used to store a plot produced on the screen in a buffer. It is activated from the save window by choosing the item Popup... of the Save as menu. The plot displayed in figure 4.18 was produced at an earlier date with the aid of the graphics tool and then copied from the base window into this window. Internally this plot is stored temporarily as a MCode file. With the aid of the Replot button this plot can be re-read into the current session in order to change further options. It is then displayed in the drawing area of the base window. The corresponding options of the scene and the objects are displayed in the manipulation window.

4.4.11 The Color Window

The color window is used to assign color values to the scene and its objects. It is activated by clicking the Background or Foreground buttons in the properties window or default window, or when one of the buttons starting at or ending at in the manipulation window is pressed. The color window is depicted here (see figure 4.19), after the button starting at in the manipulation window has been pressed. This is indicated by the line `Choosing starting Color for Object 1`. The current color value has a white contour. This color can be changed by clicking another color with the mouse. The depicted color values are equivalent to the RGB values stored in the file `.vcam_defaults`.

4.5 User-defined Color Functions

In this section user-defined color functions are described. They allow to use MuPAD procedures for coloring graphical objects. Hence it is possible to

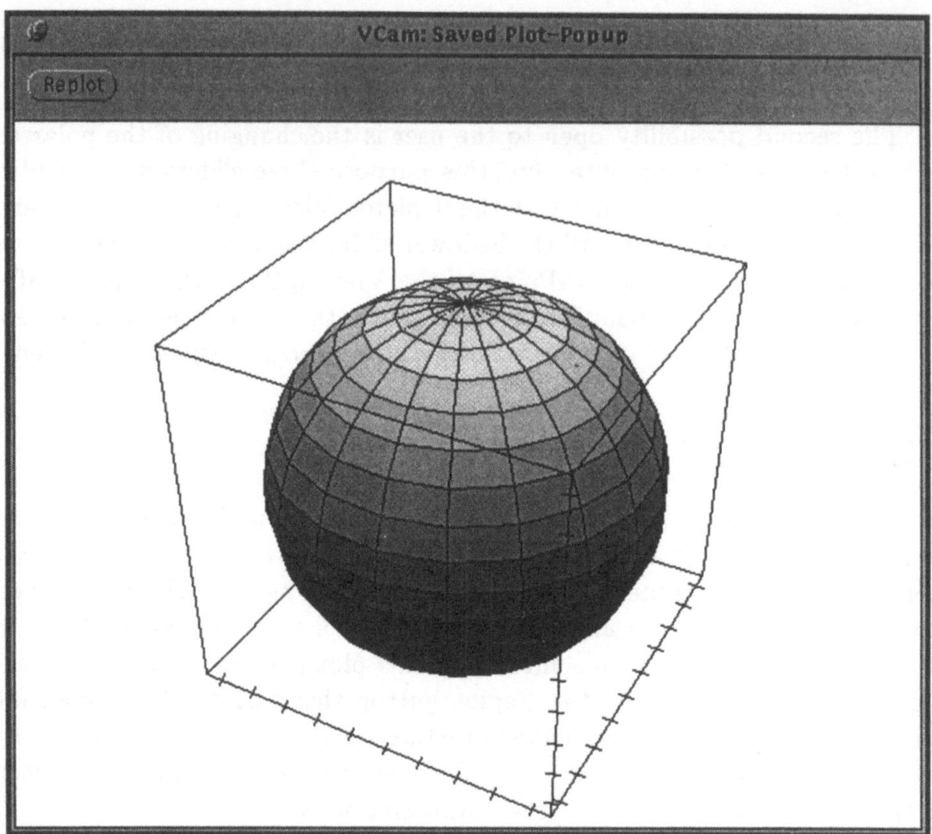

Figure 4.18: Saved Plot Window

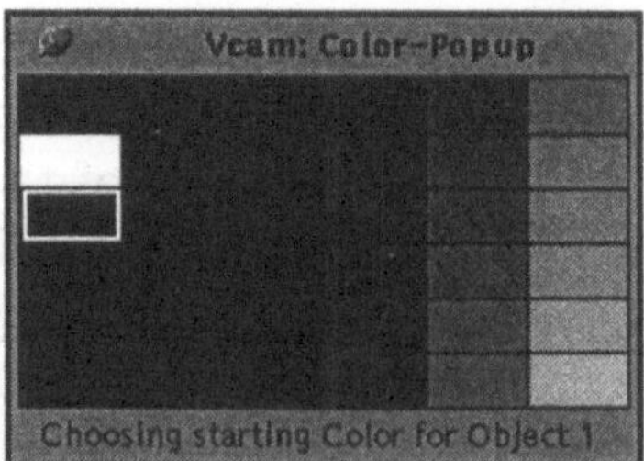

Figure 4.19: Color Window

calculate for instance the colors of an object with respect to the range of the variables or to project the basins of attraction of a complex function onto the surface described by the modulus of this function.

User-defined color functions are specified in a `plot2d` or `plot3d` command by use of the option `Color = [Function, function_name]`; `function_name` is the name of the user-defined function. The next sections explain how to work with user-defined color functions.

4.5.1 Arguments of Color Functions

In MuPAD graphical objects are represented by two- or three-dimensional co-ordinates, the so-called sample points. Therefore user-defined color functions are evaluated at these sample points. In MuPAD we distinguish between different types of sample points, which are dependent on the mode of the corresponding object.

- Sample points describing a two-dimensional curve, i.e. sample points of a two-dimensional object with `Mode = Curve`. In MuPAD such objects in common are described by two expressions depending on a single variable and a range for this variable. These two expressions, say `x(u)` and `y(u)`, are used to calculate the x and y co-ordinates of the object. Hence the range for the independent variable, say u, is equidistantly divided and the two expressions are evaluated at the equidistant points. So the sample points of a two-dimensional curve can be interpreted as a triple consisting of the current value `u_val` of the independent variable and the values of the above functions evaluated at the point `u_val`, i.e. `x_coord = x(u_val)` and `y_coord = y(u_val)`. For example, a circle is parametrized by the expressions `x(u) := sin(u)`, `y(u) := cos(u)` and the range `[-PI, PI]` for u. So the sample point at the left boundary of the range has the form: `(x_coord, y_coord, u_val) = (sin(-PI), cos(-PI), -PI)`.

- Sample points of two-dimensional lists:
  Sample points of two-dimensional lists consist of the two values describing the x and y co-ordinates of the corresponding points.

- Sample points of three-dimensional curves:
  In MuPAD a space curve is described by three expressions, say `x(u)`, `y(u)` and `z(u)`, depending on a single variable, say u, and a range for this independent parameter. Hence the sample points can be represented by the 4-tuple consisting of the values of the expressions evaluated at the different values for the independent variable, i.e. by `(x(u_val), y(u_val), z(u_val), u_val)`.

- Sample points of three-dimensional lists:
  As in the two-dimensional case the sample points of three-dimensional

lists consist of the different co-ordinates.

- Sample points of a surface:
  In MuPAD a surface is represented by three expressions depending on two independent variables, say u and v, and two ranges for these variables. The sample points of such an object are evaluated by defining a grid over the rectangular area defined by the two ranges and evaluating the three expressions at each grid point. So sample points of a surface can be represented by the 5-tuple (`x_coord`, `y_coord`, `z_coord`, `u_val`, `v_val`).

Because user-defined color functions have to be evaluated at the sample points of an object, the number of arguments of user-defined color functions depends on the mode of the corresponding object. In order to summarize the above explanations we give the explicit declarations of color functions for the different types of objects:

- two-dimensional curve:
  `cfun_curve2D := proc(x_coord, y_coord, u_val)`

- two-dimensional list:
  `cfun_list2D := proc(x_coord, y_coord)`

- three-dimensional curve:
  `cfun_curve3D := proc(x_coord, y_coord, z_coord, u_val)`

- three-dimensional list:
  `cfun_list3D := proc(x_coord, y_coord, z_coord)`

- surface:
  `cfun_surf := proc(x_coord, y_coord, z_coord, u_val, v_val)`

4.5.2 Results of Color Functions

The result of a user-defined color function has to be a list of the form [r, g, b], where every entry has to be a real value between 0.0 and 1.0. These values describe the amount of red, green and blue of the desired color.

4.5.3 Examples

In this section we want to give some examples for user-defined color functions.

Example 200 *The first example shows a color function for a two-dimensional curve. The object is colored with respect to the independent variable u, which is used to calculate the amount of red. At the lower boundary of the range this value will be 0 and at the upper boundary the amount of red will be 1.0:*

```
>> MIN_U          := -float(PI):
   MAX_U          :=  float(PI):

   cfun_curve2D := proc(x_coord, y_coord, u_val)
   local     value;
   begin
        value     := (u_val-MIN_U)/(MAX_U-MIN_U):
        [value, 0.5, 0.8]:
   end_proc:

   plot2d(Axes = Box, Ticks = 0,
          [Mode = Curve, [u, sin(u)], u = [MIN_U, MAX_U],
                Grid = [150], Color = [Function, cfun_curve2D],
                Style = [Impulses]
          ]);
```

Example 201 *The next example demonstrates a color function for a two-dimensional list. In a first step we define the procedure* sierpinski()*, which serves for generating the so-called Sierpinski-triangle of order n. The color function* cfun_list2D() *is used to color this triangle with respect to the x co-ordinates.*

```
>> sierpinski := proc(x1, y1, x2, y2, x3, y3, n)
   local nx1, ny1, nx2, ny2, nx3, ny3;
   begin
        if n = 0 then
             polygon(point(x1,y1), point(x2,y2), point(x3,y3),
                     Closed = TRUE):
        else
             nx1 := x1 + (x2 - x1)/2.0:
             ny1 := y1 + (y2 - y1)/2.0:
             nx2 := x2 + (x3 - x2)/2.0:
             ny2 := y2 + (y3 - y2)/2.0:
             nx3 := x3 + (x1 - x3)/2.0:
             ny3 := y3 + (y1 - y3)/2.0:
```

```
                 sierpinski(x1, y1, nx1, ny1, nx3, ny3, n-1),
                 sierpinski(nx1, ny1, x2, y2, nx2, ny2, n-1),
                 sierpinski(nx3, ny3, nx2, ny2, x3, y3, n-1):
        end_if:
end_proc:


XMIN := -1.0:
XMAX :=  1.0:


cfun_list2D  := proc(x_coord, y_coord)
local value;
begin
    value := (x_coord-XMIN)/(XMAX-XMIN);
    [value, 0.7, 1-value]:
end_proc:


plot2d(Axes = None, Scaling = UnConstrained,
       [Mode = List,
              [sierpinski(XMIN,0,XMAX,0,0,1,5)],
              Color = [Function, cfun_list2D]
        ]);
```

Example 202 *The third example shows a color function for a three-dimensional curve. The curve is colored with respect to the z co-ordinate of the sample points.*

```
>> MIN_Z         := -1.0:
   MAX_Z         :=  1.0:


   cfun_curve3D := proc(x_coord, y_coord, z_coord, u_val)
   local    value;
   begin
       value     := (z_coord-MIN_Z)/(MAX_Z-MIN_Z):
       [0, value, 1-value]:
   end_proc:


   plot3d(Axes = Box, Ticks = 0,
              [Mode = Curve,
                   [2*sin(u), 2*cos(u), cos(6*u)],
                   u = [-PI, PI], Grid = [125],
```

```
                        Color = [Function, cfun_curve3D]
            ]);
```

Example 203 *In this example the color function* cfun_list3D() *is used to color the points of a three-dimensional list. The procedure* tetra_rec() *creates a tetra-hedron, which is recursively refined. The colors of the single tetra-hedron are calculated with respect to the different co-ordinates, i.e. the x co-ordinate determines the amount of red, the y co-ordinate defines the amount of green and the z co-ordinate is used to determine the amount of blue of the current tetra-hedron.*

```
>> tetra_eder := proc(p1, p2, p3, p4)
   begin
        polygon(p1, p2, p4, Closed = TRUE, Filled = TRUE),
        polygon(p1, p3, p4, Closed = TRUE, Filled = TRUE),
        polygon(p2, p3, p4, Closed = TRUE, Filled = TRUE)
   end_proc:

   new_point := proc(p1, p2)
   local x, y, z;
   begin
        x := op(p1,1) + (op(p2,1)-op(p1,1))/2.0:
        y := op(p1,2) + (op(p2,2)-op(p1,2))/2.0:
        z := op(p1,3) + (op(p2,3)-op(p1,3))/2.0:
        point(x, y, z):
   end_proc:

   tetra_rec := proc(p1, p2, p3, p4, n)
   local np1, np2, np3, np4, np5, np6;
   begin
        if n = 0 then
            tetra_eder(p1, p2, p3, p4):
        else
            np1 := new_point(p1, p2):
            np2 := new_point(p2, p3):
            np3 := new_point(p3, p1):
            np4 := new_point(p1, p4):
            np5 := new_point(p2, p4):
            np6 := new_point(p3, p4):
            tetra_rec(p1, np1, np3, np4, n-1),
```

```
            tetra_rec(np1, p2, np2, np5, n-1),
            tetra_rec(np2, p3, np3, np6, n-1),
            tetra_rec(np4, np5, np6, p4, n-1):
      end_if:
end_proc:


XMIN  := -0.7:
XMAX  :=  1.0:
YMIN  := -0.5*sqrt(3.0):
YMAX  :=  0.5*sqrt(3.0):
ZMIN  :=  0.0:
ZMAX  :=  YMAX:


a := point(XMAX, 0, 0):
b := point(XMIN, YMAX, 0):
c := point(XMIN, YMIN, 0):
d := point(XMIN + (XMAX-XMIN)/2.0, 0, YMAX):


cfun_list3D := proc(x_coord, y_coord, z_coord)
local x_value, y_value, z_value;
begin
    x_value := (x_coord - XMIN)/(XMAX - XMIN):
    y_value := (y_coord - YMIN)/(YMAX - YMIN):
    z_value := (z_coord - ZMIN)/(ZMAX - ZMIN):
    [x_value, y_value, z_value]:
end_proc:


plot3d(Axes = None, CameraPoint = [0.15, -0.1, 7.0],
       ForeGround = [0,0,0],
       BackGround = [1,1,1],
       [Mode = List,
            [tetra_rec(a,b,c,d,3)],
            Color=[Function, cfun_list3D]]);
```

Example 204 *The last example is used to present a color function for a surface. In this example we plot the absolute value of a complex function. The color function then is used to mark the angle* phi *defined by*

$$z = abs(z)*exp(I*phi).$$

```
>> MIN_PHI     := -float(PI/2):
```

```
MAX_PHI      :=  float(PI/2):
EPS          :=  0.000000001:

complex_surf := proc(u, v)
begin
    1/((u+I*v)^3+1):
end_proc:

cfun_surf3D  := proc(x_coord, y_coord, z_coord, u_val, v_val)
local    erg, real, imag, phi;
begin
    erg       := complex_surf(u_val, v_val):
    real      := op(erg, 1):
    imag      := op(erg, 2):
    if abs(real) > EPS then
        phi   := atan(imag/abs(real)):
    else
        phi   := sign(imag)*PI/2:
    end_if:
    value     := (phi-MIN_PHI)/(MAX_PHI-MIN_PHI):
    [1-value, 0.5, value]:
end_proc:

plot3d(Axes = Box, Ticks = 0,
          [Mode = Surface,
              [u, v, min(abs(complex_surf(u, v)), 2)],
              u = [-2, 2], v = [-2, 2], Grid = [40, 40],
              Style = [ColorPatches, AndMesh],
              Color = [Function, cfun_surf3D]
          ]);
```

4.6 Graphical File Formats

As already mentioned, in the graphics tool VCam and in MuPAD itself several
file formats are used which we shall describe in this section. On the one hand
there is the file which contains the default values and on the other hand there
are the different formats in which a plot can be stored. These formats are:

- Raster format,

- GIF format,

- Postscript format,

- binary format and

- ASCII format.

From these formats only the last two are described here as, in contrast to the first three formats, they can be re-read from the graphics tool. But, firstly we shall describe the file which contains the default values.

4.6.1   The Format of the File `vcam_defaults`

The file **vcam_defaults** is used to store the default values used by the graphics tool and MuPAD. These values are read when VCam or MuPAD are called and are used for assigning predefined values to certain attributes. The file **vcam_defaults** is looked for as follows: Firstly a search is conducted in the user's **HOME** directory for the file **.vcam_defaults** — this file can be created with the help of the default window by using the Store button. If this file does not exist in the **HOME** directory it is read from the standard directory given by the environment variable **MuPAD_LIB**. With this users have the possibility of storing personal default values which are then available each time VCam or MuPAD are called. The two mentioned files **HOME/.vcam_defaults** and **MuPAD_LIB/vcam_defaults** are ASCII files. They are in form of tables, each consisting of three columns. The first entry gives the chosen value, the second column describes the attribute which is to be dealt with and the entries in the third column give the possible values which can be chosen as default values. This shall be shown in the following section of this file.

VCam-Default-Values		
Chosen Value	Meaning	Possible Values
0	Routine-Default	0 ==> QuickDraw (1) ==> QualityDraw
0	Scaling-Default	0 ==> Constrained 1 ==> Unconstrained

Please note that some of the values that can be chosen are not yet implemented. These values are indicated in the file above by being enclosed in brackets

(). If these values are chosen as default values the graphics tool displays a warning and the value is automatically reset to the first valid value.

At the end of the `HOME/.vcam_defaults` and `MuPAD_LIB/vcam_defaults` files 36 triples of real numbers between 0.0 and 1.0 are stored. These are the RGB values of which the basic color table of the graphics tool consists of . From these color values the user can choose the object colors and the fore- and background colors with the aid of the color window. These color values can be freely chosen and can be edited by the user. However the user should note that the first three values determine the background color and the next three values determine the foreground color.

4.6.2 Graphical Files in ASCII Format

In this section we shall describe a graphical file in ASCII format. These are files which contain data which can be read by the graphics tool and then displayed as a plot. However, these files do not contain all the possible attributes. They only contain the most important ones so that the file can be produced by hand or with the aid of an external program. The basic construction of such an ASCII file is organized as follows:

- It begins with the header `VCam-ASCII-Data`.

- Following this the most important scene options are to be specified. These consist of the scene dimension and the number of objects contained in a scene.

- For each object certain important object options must be defined. These options include:

  - The object mode; for a 2D scene this can only have the value 0 or 1(this is equivalent to `Mode = Curve` or `Mode = List`) but a 3D scene can have the value 0,1 or 2 (for `Mode = Curve`, `Mode = List` or `Mode = Surface`).

  - The parametrization, i.e. for a two-dimensional curve two character strings are expected whereas for three-dimensional objects three character strings are expected.

  - The number of sample points (`Grid`) and the extra points which are to be calculated in each parameter direction (`Smoothness`), i.e. the user needs two values for a curve and four values for a surface.

– Following this the free parameters with the ranges they traverse and in which the functions are evaluated must be entered. The user should note that for curves only one parameter and one range (i.e. one character string for the parameter and two character strings for the minimal and maximal value of the range) and for surfaces two parameters and two ranges must be given.

- After the object options have been entered in the file the actual graphical data can be entered in the file. These are the evaluated values of the parametrization which are arranged in lines i.e. each of the following lines contains two (for 3D scenes three) values, which represent the x and y co-ordinates (or the x, y and z co-ordinates). As an example we should like to make clear the order in which the graphical data for a surface must be given. They can be generated in the programming language C by using two nested loops:

```c
for ( i = 0; i < (vgrid-1)*(vsmooth+1); i++ )
{   /*
     * Calculate the value of the grid point v.
     */
    v     = vmin + i * delta_v;
    for ( j = 0; j < (ugrid-1)*(usmooth+1); j++ )
    {   /*
         * Calculate the value of the grid point u.
         */
        u     = umin + j * delta_u;
        /*
         * Calculate the values of the co-ordinate func-
         * tions.
         */
        x_coord = x(u,v);
        y_coord = y(u,v);
        z_coord = z(u,v);
        /*
         * Write the co-ordinates into the file.
         */
        fprintf(fp,"%f %f %f\n",x_coord,y_coord,z_coord);
    }
}
```

If one wants to store the following plane with the parametrization $x(u,v) = u$, $y(u,v) = v$ and $z(u,v) = u + v$ in the range $u = [-1, 1]$ and $v = [-1, 1]$ with the aid of an ASCII file, then one should produce the following file (here the chosen value for `Grid` is [2, 3] and for `Smoothness` is [0, 0]):

```
VCam-ASCII-Data
  3      1
  2
  u
  v
u+v
  2      0
  3      0
  u
 -1
  1
  v
 -1
  1
 -1     -1     -2
  1     -1      0
 -1      0     -1
  1      0      1
 -1      1      0
  1      1      2
```

Before creating ASCII files containing graphical data by external programs it is strongly recommended that one has a look at ASCII files which describe plots that were created in MuPAD by `plot2d` or `plot3d` and stored in ASCII format. 

4.6.3 Graphical Files in Binary Format

As already mentioned it is also possible in MuPAD to store plots in binary files. This can be done in the save window by choosing the option Save as MCode-File and saving the plot. These binary files can also be re-read by VCam and the scenes can then be changed. Contrary to the graphical files in ASCII format described above all the current options of the scene, objects and default values are stored, so that when re-read an identical plot is the result.

Please note that files containing binary graphical data must not be edited by hand, otherwise the re-reading of the file can lead to an error. 

Chapter 5

User Interfaces

In this chapter the front-end to the MuPAD-kernel under the X-Window system and on the Macintosh as well as the on-line hypertext help system shall be described. These interfaces are realized differently. As with the graphics tool the functionality of the tools has been kept as similar as possible: only the „Look & Feel" are dependent on the underlying system.

5.1 XMuPAD

In order to work with MuPAD no special input or output devices are necessary. A keyboard and a simple terminal are all that are needed. However, for a comfortable interactive use of the system XMuPAD was developed. It is a graphical user interface for MuPAD under the X-Window system. This interface is described in the following section.

XMuPAD is a completely independent program. It has no knowledge of computer algebra. It is implemented as an interface to the MuPAD kernel. XMuPAD is called with the command

```
xmupad [options]
```

which let appear a *basic window* (see figure 5.1) on the screen. This window consists of two components:

- a row of buttons and

- a text window for input and output.

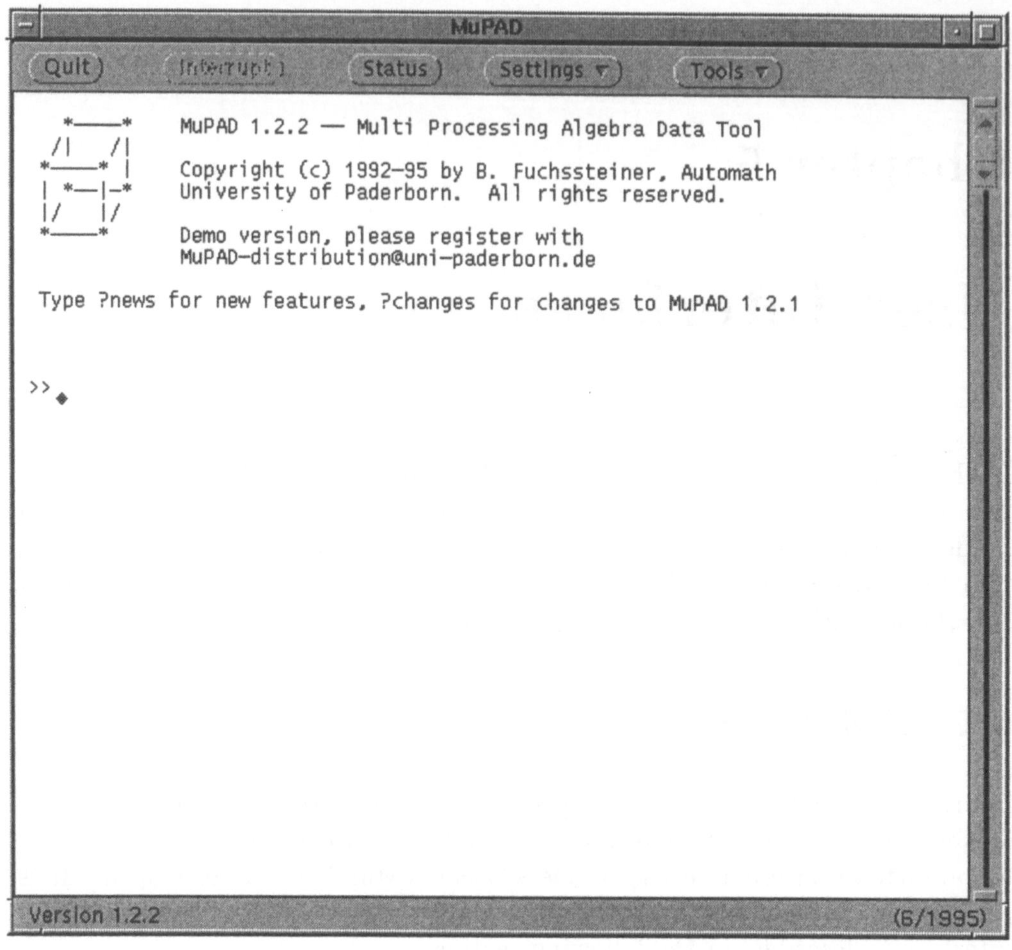

Figure 5.1: XMuPAD Basic Window

Internally, the kernel of MuPAD is started as a child process on the UNIX level.

Working with XMuPAD can be divided into three phases which are typical for interactive systems:

1. In the *input phase* the user enters commands in the text window. This stage is ended with the <Return> key.

2. In the *evaluation phase* the input is evaluated.

3. In the third and final stage the calculated result is displayed.

5.1.1 The Text Window

The text window is an usual window for the input of text. On the right hand side a scrollbar is located which allows to change the visible section of the text.

5.1.1.1 Input

Commands can be typed in with the keyboard at the current insert position, which is made visible by the cursor ▲. This insert position can be set anywhere in the text window by placing the mouse cursor at the desired position and press the left mouse button or by using the cursor keys ◄, ►, ▲ and ▼.

Any command has to be confirmed with the <Return> key. Then the current line is read and sent as a command to MuPAD for execution. Usually a command is not longer than one line. However, it is possible to enter longer commands which is described in section 5.1.1.2.

After the results being displayed the MuPAD prompt ("**>>**") will appear in a new line and will mask the current insert position.

After a command is sent to MuPAD, XMuPAD is in the evaluation phase. During this time the text window is inactive and no input can be given. This is indicated by a small clock being shown as the mouse cursor. To interrupt this stage, the buttons Quit (for termination of the complete session) and Interrupt can be used.

5.1.1.2 Special Features

Special features, which are explained in the following, make XMuPAD more comfortable to use than the terminal version of MuPAD.

- <Shift Return> or <Enter>
  To be able to enter commands spread over more than one line, a line can be ended with <Shift Return> or <Enter>. With this the current line is closed, a new line is inserted and the current insert position is set to the beginning of this line. However, the command just entered is not sent to MuPAD, i.e. it is not yet executed. In this way, a MuPAD command consisting of any number of lines, can be entered. The command is only executed when <Return> is entered. Hereby, it is unimportant where the current insert position is. All lines above and below the insert position are part of the input. In other applications the pressing of

<Return> usually enters a line break at the current insert position. In XMuPAD the line break is set at the end of the input.

Example 205 *In figure 5.2 the user has entered a command but not yet ended it with <Return>. The current insert position is in the second line after the x^i. If the user presses <Return> now, then the whole depicted*

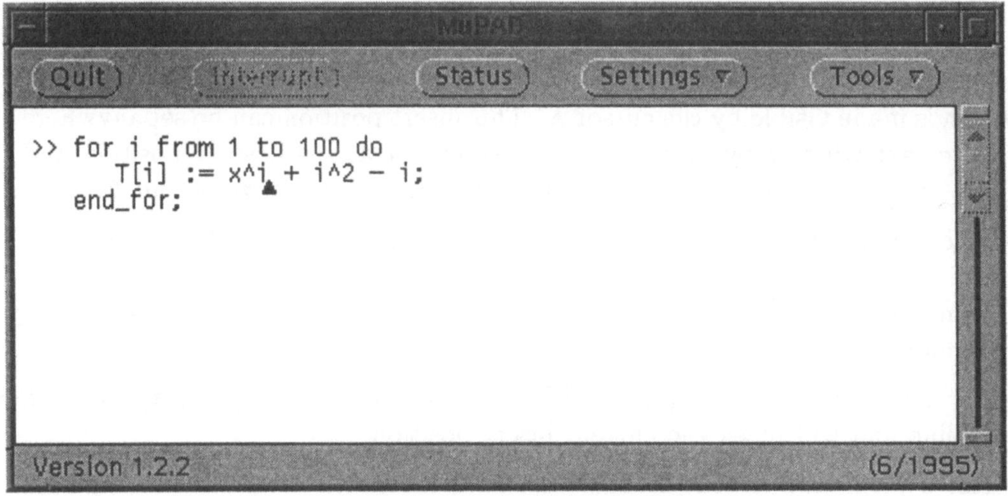

Figure 5.2: Text Window before <Return>

command, i.e. the entire **for** *loop, will be sent to MuPAD for execution. However the line break does not take place at the current insert position, but at the end of the command (see figure 5.3).*

- New Execution of Old Input
  By moving the current insert position into a previously given input and pressing <Return> once more, old input can be executed again. The insert position may be anywhere in the command even when this is longer than one line. It is also possible to modify previous input.

If Replace Mode is set to OFF in the pull-down menu Settings then the old output is retained and the new output is inserted in front of it. If Replace Mode is set to ON then the old output is overwritten. In this case, no new prompt is printed. Instead the current insert position moves to the end of the first line of the next command. If the Recalculate Mode is set to ALL, then not only the command at the current insert position is executed again, but also all commands until the end of the complete session. The order of the execution is determined by the displayed sequence of commands in the text window and not by the chronological

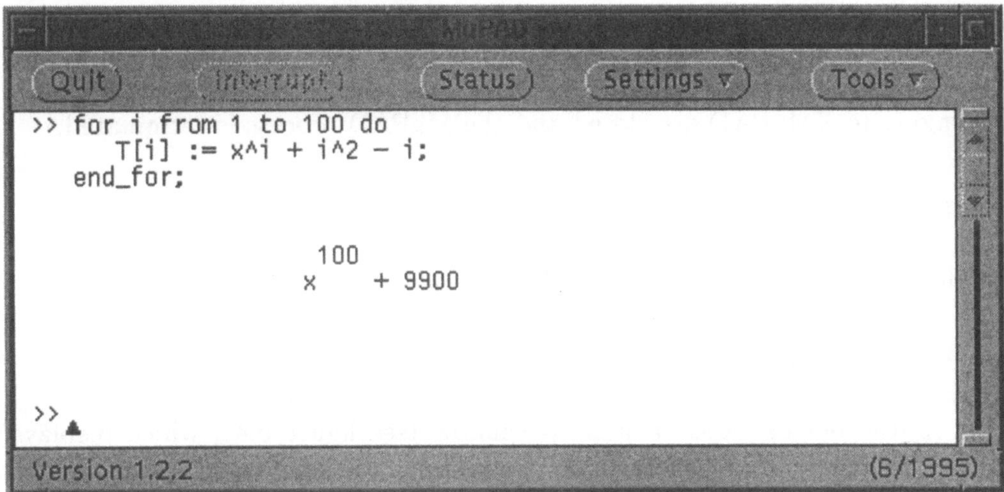

Figure 5.3: Text Window after <Return>

order of input. In this way it is easy to execute an interactive session again, if necessary with different start values.

- Insertion of new entries between old entries
  As described in section 5.1.1.1 the entering of a command can take place at any position in the text window. It is also possible to enter a new command between the output of an old command and the input of the next command. With this the order of all given commands is influenced which is of importance for the realization of the Recalculate Mode.

Example 206 *The user has entered the commands* cmd_1, cmd_2 *and* cmd_3 *one after the other, i.e. each in its own line. Now s/he inserts a further command* cmd_4 *between the output of* cmd_1 *and the input of* cmd_2. *If the Recalculate Mode is set to ALL and the user executes* cmd_1 *again then all four commands are evaluated again in the order* cmd_1, cmd_4, cmd_2, cmd_3.

5.1.2   The Buttons in the Basic Window

The buttons in the basic window are ordered horizontally (see figure 5.1). Sometimes certain buttons are deactivated, i.e. clicking these buttons has no effect. This is indicated by the usually black contour and label of the button changing to light grey. In the following the individual buttons are described in more detail.

5.1.2.1 The Quit Button

The Quit button terminates the XMuPAD session. With this all the windows belonging to XMuPAD are closed and the MuPAD process is terminated.

5.1.2.2 The Interrupt Button

With this button the current execution is interrupted.

5.1.2.3 The Status Button

The Status button opens a pop-up window (see figure 5.4), which displays status information about MuPAD. The information contained is the number of used and allocated bytes and the time in milliseconds passed since the start of the session (see help pages for the commands **bytes** and **time**).

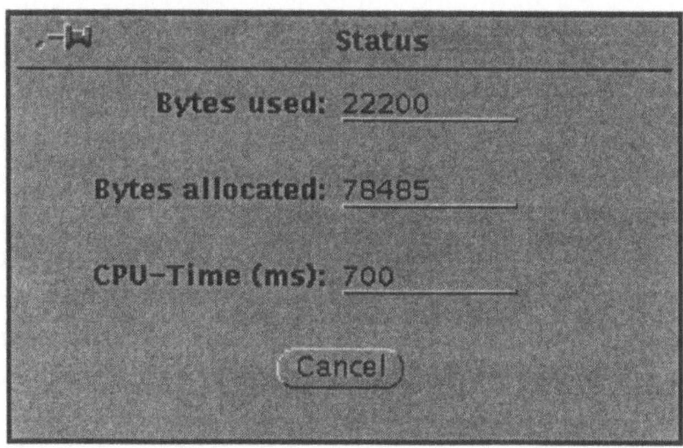

Figure 5.4: Status Popup-Window

5.1.2.4 The Settings Button

By clicking this button a pull-down menu is opened (see figure 5.5), whose menu items enable the user to execute the following commands:

- Delete Last Output
  This menu item deletes the output of the last command. As the first entry in the pull-down menu this is also defined as the default action. Thus, if the Settings button is pressed without the menu being pulled down this action is automatically executed.

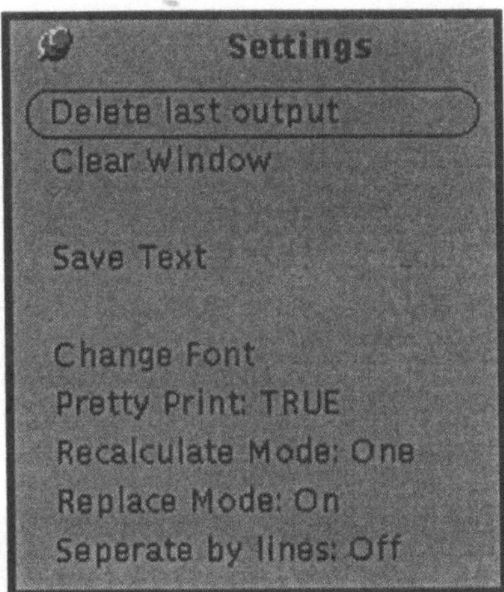

Figure 5.5: Settings Pull-down Menu

- Clear Window
  This command deletes the whole contents of the text window. After being activated the only text remaining in the text window is the MuPAD prompt ">>".

- Save Text
  The menu item Save Text opens the save pop-up window (see figure 5.6). The upper input field is used for entering the name of a directory. In the lower field the name of the actual file is entered. The name entered under "Directory" needs not to be ended with a /. It is automatically inserted between the two names. If <Return> is pressed in the "Directory" input field then the input cursor jumps to the "File" input field. With the OK button the contents of the text window are saved in the specified file. If a file with the same name already exists, a back-up of the old file is made before it is overwritten. For this back-up the ending % is added to the name of the old file. Between two calls of the save pop-up window the entered names are retained.

- Change Font
  Using the menu item Change Font a further pop-up window is opened (see figure 5.7). In this window the user can change the font of the text window. The user can choose the *font family*, the *font style* and the *font size*.

With the following menu items, the various modes for output in the text window can be set. These are binary switches whose status can be changed

Figure 5.6: Save Pop-upWindow

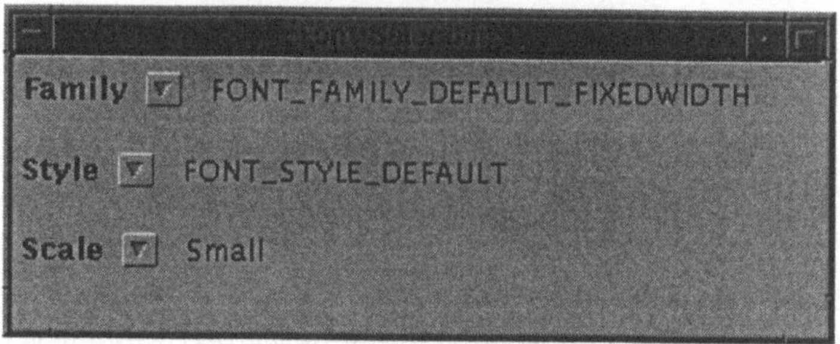

Figure 5.7: Font Selection Window

by clicking them. The current status is shown in the menu.

- Pretty Print
  With this command the two-dimensional output, the *Pretty-Printer*, can
  be turned on and off.

- Recalculate Mode
  In XMuPAD it is possible to execute already executed commands again.
  For this purpose the input cursor is moved into the old command and
  then executed by pressing <Return>. If the Recalculate Mode is set to
  ON then only the current command is executed again. If the Recalculate
  Mode is set to ALL then not only this command is executed, but also
  all the following commands in the text window.

- Replace Mode
  With this menu item the replace mode can be turned on and off. This
  mode is described in section 5.1.1.2.

- Separate by Lines
  With this menu item the user has the possibility to separate an output
  from the next input and vice versa with a line.

5.1.2.5 The Tools Button

The Tools button opens the tools pull-down menu (see figure 5.8). This menu
contains items for starting other programs. In MuPAD 1.2.2 items for calling
the debugger frontend mdx, the graphics tool VCam and the help system are
contained.

Figure 5.8: Tools Pull-down Menu

5.1.3 Special Cases – Differences to MuPAD

In a few points the functionality of XMuPAD differs to that of MuPAD.

5.1.3.1 UNIX Shell Command

With the command **system** (short form **!**) it is possible to enter commands
to a UNIX shell in MuPAD. Giving this command in XMuPAD for the first
time a terminal window is opened in which the user's default shell is started.
The parameter of the **system** command is then executed in this shell.

Using the Done button this window can be made invisible. However, the
contents of the terminal window is preserved. With a new call of **system** the
terminal window with its contents becomes visible again.

Due to implementation of XMuPAD with the XView library, the execution of
the **system** command in XMuPAD returns no result in the terminal window.
Hence, the user gets no information whether the command has been successfully executed or even terminated. Of course, in an interactive session the user
can see the result in the terminal window.

Furthermore the command **system** is has no blocking effect. If **system** is called
in a statement sequence then the execution of the following command does
not take place when the execution of the shell command has been terminated,
but both commands are processed "simultaneously" as two processes (one
command is carried out by the shell, the other by MuPAD).

5.1.3.2 User Interrupt

In MuPAD, the user has the possibility to interrupt the current evaluation with <Ctrl-C>. A menu appears in which

- abort – i.e. the termination of the current evaluation,

- continue – i.e. the continuation of the current evaluation, or

- quit – i.e the termination of the MuPAD session

can be selected.

This choice does not exist in XMuPAD since a <Ctrl-C> is interpreted as an usual text input in the text window. This is, however, not a great disadvantage as two of the three possible choices are covered by the buttons Quit and Interrupt.

5.1.3.3 Execution of `textinput`

With the function `textinput` the user is able to enter text interactively. In this case a further window is opened which offers the user a text editor. After the desired text has been entered and the Done button has been activated the text is executed by the MuPAD kernel.

5.2 HyTEX

HyTEX is a hypertext system, which has been developed by the group of N. KÖCKLER at Paderborn University. A modified version of this program is used in XMuPAD. As can be seen by the name, HyTEX is based on the TEX system of D.E. KNUTH. Apart from a special *Previewer* it makes some new TEX commands available. HyTEX is used for the on-line documentation. This includes the MuPAD user's manual, the MuPAD tutorial, an extended demo and the help pages for each MuPAD command.

After calling XMuPAD HyTEX is only visible as an icon on the screen. Only on an explicit request from the user the corresponding window is opened. This can be achieved in various ways. On the one hand, by selecting Help in the tools menu (see section 5.1.2.5) the MuPAD manual is opened and the table of contents is displayed. A second possibility is the command `help` (short form ?) in MuPAD. With this the help page for the command specified by the user is opened (see section 5.2.2). Of course, the help icon can also be directly clicked. In this case the cover sheet of the manual is opened.

5.2.1 The HyTeX Window

The HyTeX window (see figure 5.9) contains a LaTeX-Previewer, which is equiped with hypertext functions. It offers the user different possibilities of navigating through the document. These are described in the following.

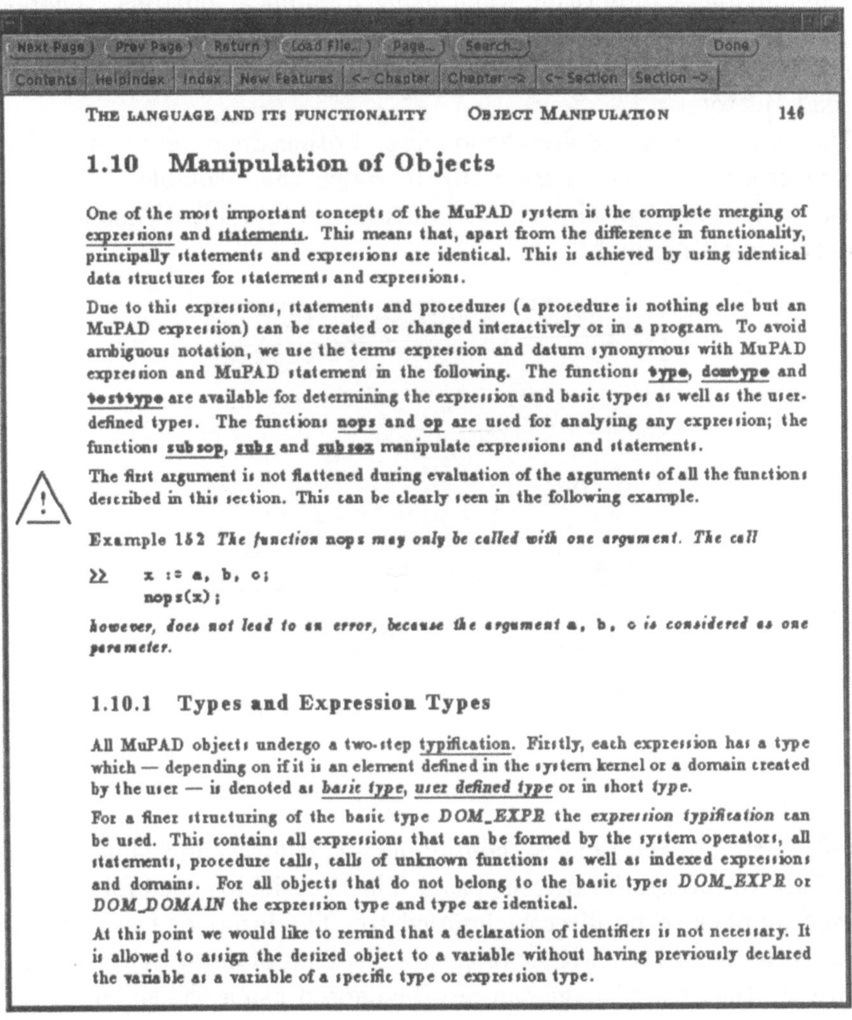

THE LANGUAGE AND ITS FUNCTIONALITY   OBJECT MANIPULATION   146

1.10 Manipulation of Objects

One of the most important concepts of the MuPAD system is the complete merging of expressions and statements. This means that, apart from the difference in functionality, principally statements and expressions are identical. This is achieved by using identical data structures for statements and expressions.

Due to this expressions, statements and procedures (a procedure is nothing else but an MuPAD expression) can be created or changed interactively or in a program. To avoid ambiguous notation, we use the terms expression and datum synonymous with MuPAD expression and MuPAD statement in the following. The functions **type**, **domtype** and **testtype** are available for determining the expression and basic types as well as the user-defined types. The functions **nops** and **op** are used for analysing any expression; the functions **subsop**, **subs** and **subsex** manipulate expressions and statements.

The first argument is not flattened during evaluation of the arguments of all the functions described in this section. This can be clearly seen in the following example.

Example 152 *The function nops may only be called with one argument. The call*

```
>>    x := a, b, c;
      nops(x);
```

however, does not lead to an error, because the argument a, b, c is considered as one parameter.

1.10.1 Types and Expression Types

All MuPAD objects undergo a two-step typification. Firstly, each expression has a type which — depending on if it is an element defined in the system kernel or a domain created by the user — is denoted as *basic type*, *user defined type* or in short *type*.

For a finer structuring of the basic type *DOM_EXPR* the *expression typification* can be used. This contains all expressions that can be formed by the system operators, all statements, procedure calls, calls of unknown functions as well as indexed expressions and domains. For all objects that do not belong to the basic types *DOM_EXPR* or *DOM_DOMAIN* the expression type and type are identical.

At this point we would like to remind that a declaration of identifiers is not necessary. It is allowed to assign the desired object to a variable without having previously declared the variable as a variable of a specific type or expression type.

Figure 5.9: HyTeX Window

5.2.1.1 Statical Buttons

Statical buttons are always visible in the HyTeX window irrespective of the displayed page. They are all to be found in the upper part of the window.

- Next and Prev Buttons
  With these two buttons the document can be leafed through page-by-page. Next moves one page forward, Prev one page backward.

- Return Button
  This button is important when using dynamical buttons. Therefore it shall be described later.

- Load Button
  The Load button enables the loading of other documents. A click opens a control panel (see figure 5.10), in which the available documents are listed. This list contains all DVI documents found in the same directory as the user's manual.

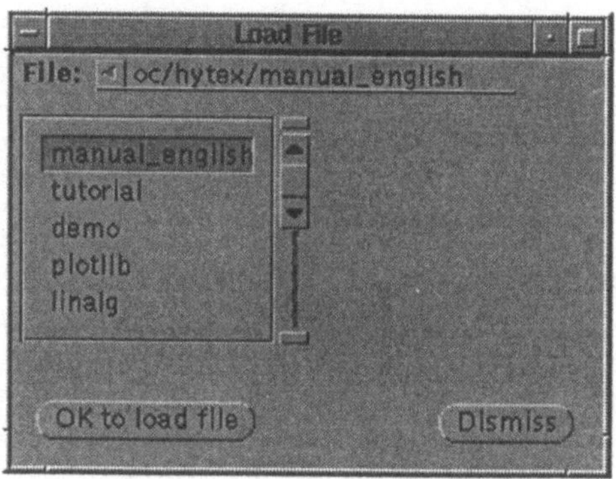

Figure 5.10: Load Control Panel

- Page Button
  This button opens a dialog, containing a text field and a slider, which can be used to choose the current page. The text field is used for entering the page number to be directly jumped to. The input of the page number has to be confirmed with <Return>. Instead of entering a page number directly the desired page can also be chosen using the slider.

- Search Button
  This button is used to search in the document. After it is activated the search window appears (see figure 5.11), in which the user can enter a text pattern. This is searched for in the entire document starting from the current page (wrapping around at the end of the text). Furthermore the user can stipulate if capital or small letters are to be taken into account.

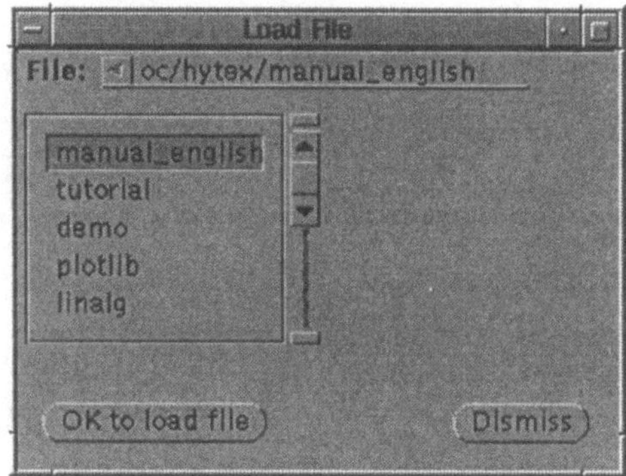

Figure 5.11: Search Window

- **Done Button**
  If the user clicks this button the HyTEX window disappears. With the **help** function or the Help button in XMuPAD it can be made visible again.

5.2.1.2 Dynamical Buttons

While writing HyTEX documents, one is able to mark sections with so-called nodes and references to these nodes. The nodes can be seen as internal marks in the text and are not visible. However, the references to the nodes (hyperlinks) are visible. In the HyTEX window they can be seen as underlined or boxed words (see figure 5.9). If such a word is clicked then the relevant node is automatically jumped to. In this way the user has the possibility to navigate through the document according to her/his personal interests.

In the MuPAD documentation there are two kinds of *dynamical buttons*. On the one hand there are those to be found in the control field in the upper part of the window. With these buttons it is, for instance, possible to jump to the next or previous chapter or section. These buttons can change from page to page. On the other hand, there are the dynamical buttons in the text. These buttons are references to linked sections or topics.

If the user has jumped to a page s/he will of course want to return to the start page as quickly as possible, especially when following hyperlinks. The Return button is used for this purpose. When using the dynamical buttons to jump to a page the previous page is internally stored. A click on Return has the effect of returning to the page from which the last jump was made.

Further to this there are also many examples in the on-line documentation marked with >> or **>>**. By clicking these symbols the example is passed to the XMuPAD window. The user can now study the example, change it if necessary and execute it with <Return>. Using this kind of "Learning by doing" it is easy to gain knowledge of the correct syntax to be used.

5.2.2   The Help Pages

As already mentioned above apart from the MuPAD user's manual, an on-line help system is also integrated in HyTEX. For example, the user can request information about the system function `igcd` with the command

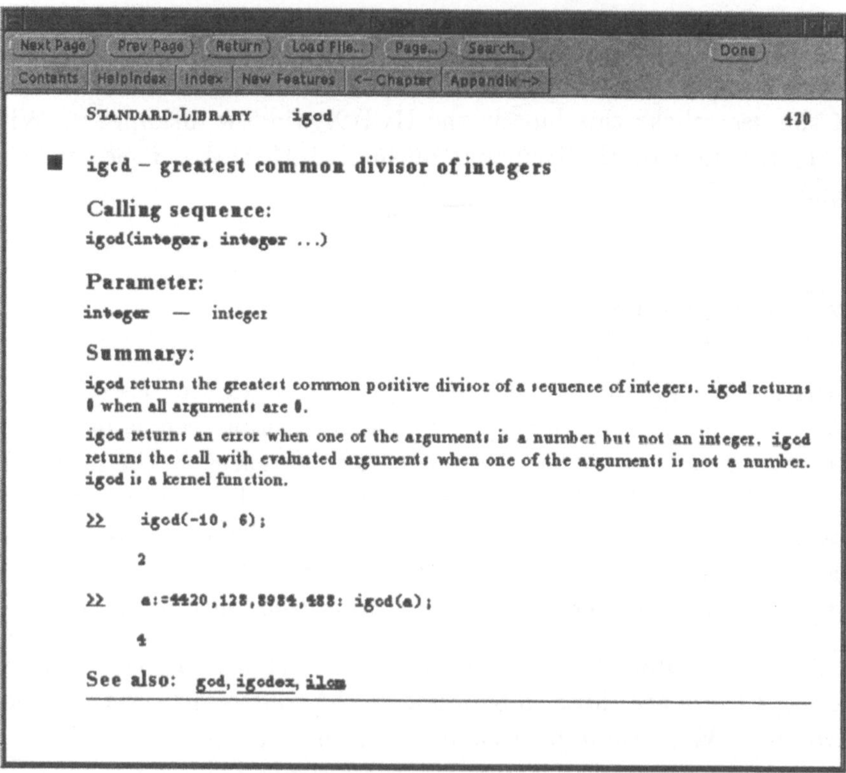

Figure 5.12: Help Page for `igcd`

```
>> help("igcd");
```

or

```
>> ?igcd
```

The corresponding help page is then opened in the HyTEX window (see figure 5.12). The exact syntax of the requested command is described on such a help page. Furthermore, it is explained how the command works. Also some examples are shown for its application, and hyperlinks to the help pages of related functions or commands are given. Each example is marked with a dynamical button. In contrast to the dynamical buttons in the text, which are used to provide cross references, by clicking these buttons the example is automatically entered as a command in XMuPAD.

5.3 MacMuPAD — MuPAD on the Macintosh

5.3.1 Introduction

This chapter describes the Macintosh version of MuPAD and explains its special features. Aspects such as hard- and software requirements, installation and configurability shall be dealt with.

Figure 5.13: MacMuPAD

5.3.2 Installation

MacMuPAD requires Macintosh system software 7.0 or higher, as well as at least 3 MB free memory. However, more memory is recommended for more complex tasks: as the saying goes; the more, the merrier.

To install MacMuPAD one should copy the program as well as the folder "lib" into the desired folder. MacMuPAD is immediately executable and can be started as usual with a double click.

Registration of MacMuPAD   Before starting MacMuPAD one should register the program. Unregistered versions of MacMuPAD are subject to limitations of the usable memory, so that the functionality of MuPAD can only partially be used. To register MacMuPAD select the menu item *Registration* in the *File* menu. A dialog is opened in which you should enter your user name and the registration code. These should be exactly the same as the name and code found in the MuPAD licence agreement.

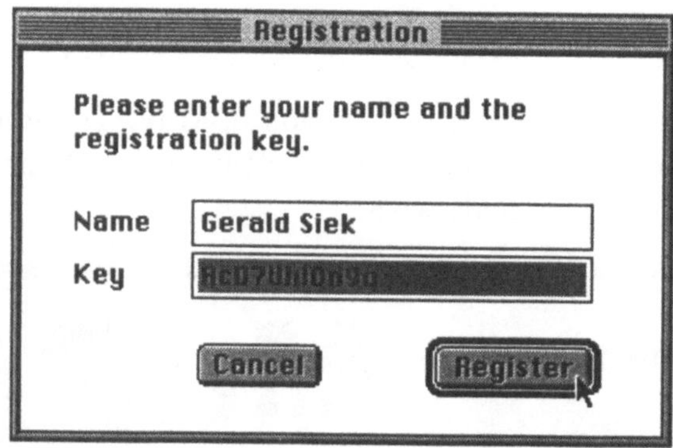

Figure 5.14: Registration of MacMuPAD

To become a registered user of MacMuPAD send a mail to

`mupad-distribution@uni-paderborn.de`.

After the licence agreement has been sent, one should receive the personal registration code.

5.3.3   A First Session

After starting MacMuPAD the *session window* is opened. Here, all MuPAD calculations and programs, are entered and displayed. MacMuPAD greets the user with a prompt and then waits for user input. One may now enter a simple expression, e.g.

`1+1;`

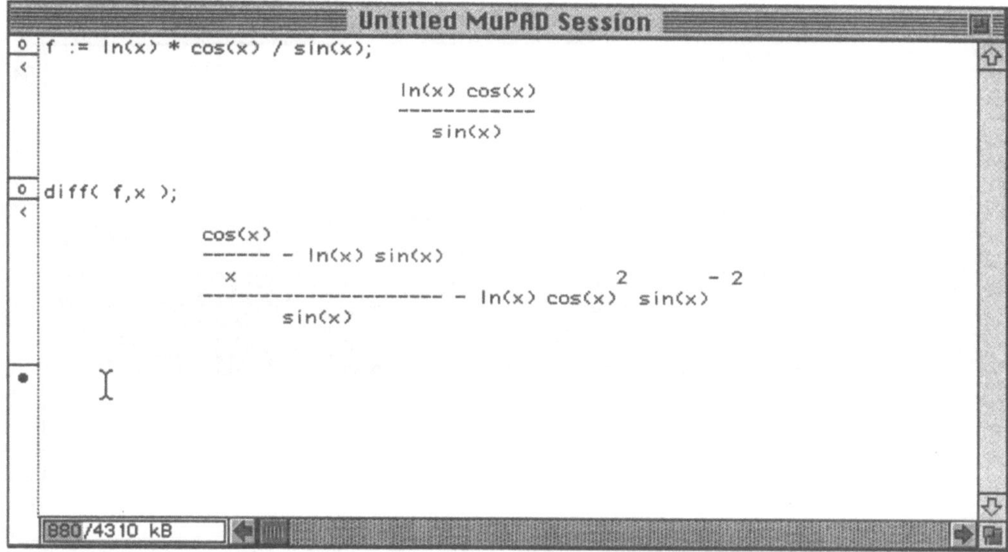

Figure 5.15: MacMuPAD Session Window

This should be closed with <Enter> on the keyboard which is found in the numeric keypad. For computers without numeric keypads, the input is closed with <Control>-<Return> or <Shift>-<Return>. Then MacMuPAD begins calculating the task and returns the result immediately.

The session window is optically divided into two parts. The larger part of the window is the actual text window, with the so-called *type bar* on the left. The type bar is used to separate the various input and output areas of the text window and to show their types. In the example created above , three different text types can be seen. The already evaluated input line "1+1;" has the type *Evaluated Input*. The result has the type *Output* and the following input line the type *New Input*. The different text colors and the symbols in the type bar are meant as orientation help in the document. Furthermore, you can see that MacMuPAD separates the various areas with horizontal bars in the type bar: this is a further orientation help. These related areas are called *Fields* in the following. If one clicks with the mouse in one of the areas of the type bar, then the corresponding section in the text window is highlighted.

Most of the time, MacMuPAD manages the fields automatically. In the example above one could see how the output of a result creates a new field of the type *Output* and how the input is converted into the type *Evaluated Input* after evaluation. It is possible to change old input and evaluate it again. Doing that, the input field is once more colored blue and the symbol in the bar changes to indicate that the changed input has not yet been evaluated. If

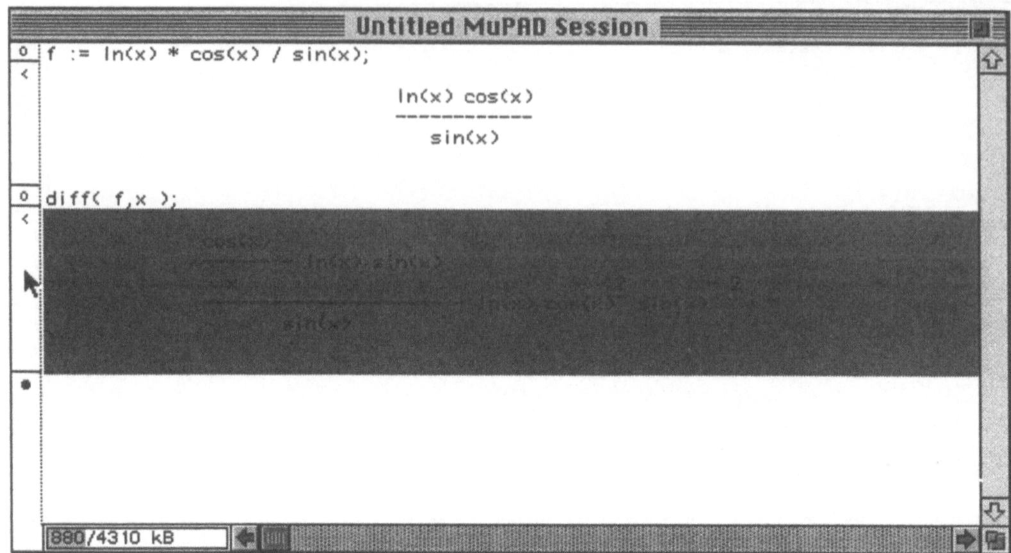

Figure 5.16: MacMuPAD

the input is once again closed with <Enter> then the input field is again converted into the type *Evaluated Input*. Furthermore, MacMuPAD overwrites the old output with the new result. This behavior can be influenced by the menu item *Replace old Results* in the *Session* menu (compare section 5.3.6.3).

Not every text in the text window can be changed. If one tries to change the output created by MacMuPAD then a signal tone is given as reply to each key depression to show that output may not be edited. The philosophy behind this is that only text created by MacMuPAD during calculation can be of the type *Output*. If the user wants to manipulate the output then the relevant field must be converted into the type *Text/Commentary* or *New Input* to indicate that the new text has been created by the user and not the system.

The use of the <Enter> key is subject to a similar limitation. If the cursor is placed in a text or output field and the <Enter> key is pressed then MacMuPAD once more replies with a signal tone to report a forbidden action. In a MuPAD session only input fields can be evaluated, i.e. evaluated by the MuPAD kernel. This limitation is appropriate because when the <Enter> key is pressed, MacMuPAD considers the whole field input and evaluates it. If the user presses the <Enter> key by mistake, for instance, in a large output field, then the evaluation of the whole field can under certain circumstances, have an unwanted effect on the internal state of MuPAD. Changes to the internal state are not reversible so that in the worst case results can be destroyed or falsified.

Further elementary operation of MacMuPAD is easy to grasp and is based on the usual Macintosh conventions. The cursor can be freely positioned, text can be cut, copied into the clipboard and pasted to the window. Scrolling of texts takes place in the usual way by using the scroll bar in the window.

MacMuPAD supports the on-line help-system of the Macintosh ("Balloon Help^{TM}"). If turned on, a balloon with a help text appears whenever the user moves the mouse cursor to a menu or a dialog item. Details of Macintosh Balloon Help can be found in the Macintosh's User Manual.

5.3.4   The Windows

5.3.4.1   The Session Window

The session window is used for input and output of data. The text is organized into *Fields* (sometimes also called *Blocks* or *Sections*). A field can contain any number of coherent lines. Fields of different types cannot be combined, but the type of a field can always be changed into *New Input* or *Text/Commentary*. Furthermore, fields cannot be split, i.e. coherent output always remains coherent. New input and text fields can be inserted at any position with the relevant menu items: *New Input* and *New Text* in the *Session* menu.

5.3.4.2   The Type Bar

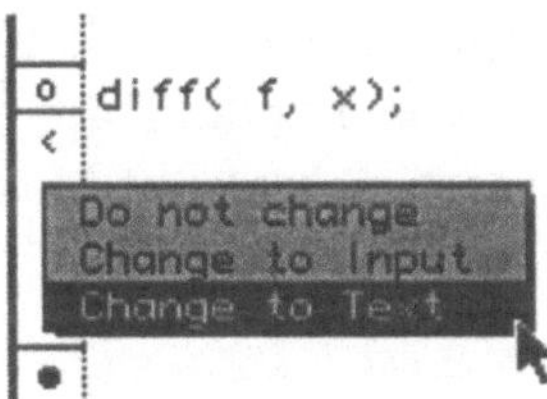

Figure 5.17: Pop-Up Menu for Type Changing

Primarily, the type bar is used for showing the various fields and their types. Furthermore by clicking in the individual fields of the type bar the user can select the corresponding text fields. These selected sections can be deleted with the <Delete> key. Selecting can be extended in the usual Macintosh way, i.e. by simultaneously pressing the shift key. Furthermore, the type bar can be used to change the type of a section. By clicking in a field of the type bar and simultaneously pressing the option key, a pop-up menu is activated. With this the selected field can be changed into the types *New Input* or *Text/Commentary*.

5.3.4.3 Field Types

At the moment MacMuPAD differentiates between the following five text types:

New Input: This type is used for text entered by the user for evaluation. Every evaluable text that has not yet been sent to the kernel is of this type. New input is symbolized in the type bar with a filled circle ($\bullet$); in the session window it is by default depicted in blue.

Evaluated Input: This type is used for text entered by the user and sent to the kernel for evaluation. Evaluated input is symbolized by a circle (o); in the session window it is by default depicted in green.

MuPAD Output: This type is used for text, results and error messages returned by the MuPAD kernel. MuPAD output is indicated by the smaller-symbol ($<$) in the type bar and is by default depicted in red in the session window. Output text cannot be changed by the user.

Text/Commentary: This type is used for text entered by the user that is not to be evaluated. This kind of text is primarily used for commenting on results in the compilation of spreadsheets. Commentaries are symbolized by a double tilde ($\approx$) in the type bar and are by default depicted in black in the session window. The start message of MacMuPAD is of the type Text/Commentary.

Debugger Input: This type is used for input to the debugger. There is no difference between evaluated and unevaluated debugger input. It is symbolized by a double arrow ($\gg$) and is normally depicted in violet in the session window.

5.3.4.4 The Memory Display

The memory status of MacMuPAD is displayed next to the horizontal scroll bar of the session window. The already occupied memory as well as the memory still available are given. This concerns only the memory which is used by the kernel in contrast to the memory used by the frontend. For further information see section 5.3.12.

5.3.4.5 The Edit Window

Beside the session window in which all input, calculations and output takes place, MacMuPAD offers a second type of document: the *edit window*. This can be created, manipulated and changed as usual for text documents. The edit window should be used as a "scratchpad" for the purpose of, for instance, storing results or entering programs which are to be copied later into the session window. A new edit window can be created with the menu item *New* in the *File* menu, existing documents can be opened by using the menu item *Open*. These documents can be processed as usual and saved for future use. With the menu item *Transfer to Session* in the *Edit* menu one can copy the current selection or the entire contents of the edit window into the current input field of the session window.

5.3.5 Documents

At present, MacMuPAD supports five different document types:

- session documents

- text documents

- help documents

- M-Code documents

- Graphical documents

Session Documents: If the contents of the session window is saved by using the *Save* or *Save as...* command then MacMuPAD saves the text including the type information, i.e. the text blocks and their types are saved. Documents of this type can be re-read into the MuPAD session. When saving a session, the user may choose to save Input and Text fields only by selecting the corresponding box in the save dialog.

To save a session document without any type information the menu item *Export...* in the *File* menu must be chosen.

MacMuPAD only stores the contents of the window, calculations and variables are not saved. Hence, if a session document is loaded after starting, all the calculations must be carried out again. An evaluation of all input fields in a document can be carried out with the menu item *Evaluate all* in the *Session* menu.

Text Documents: The MacMuPAD edit window contains usual text which is also saved as such. Documents of this can be read-in and processed with any standard editor, like, e.g. SimpleText or BBEdit. Session documents can be imported as text with the menu item *Open Session as Text*.

Help Documents: The MuPAD manual, tutorial and other documentation are available as help documents. These documents can not be edited but opened by the MacMuPAD help tool.

M-Code Documents: These documents are created by functions such as `fprint` or `fopen` and can contain any MuPAD data. These documents cannot be read by the frontend but have to be read by using suitable MuPAD commands, like `fread`.

Graphical Documents: These documents are used by plots which were created by the graphic tool. Plots stored in this way are saved in standard Macintosh Picture ('PICT') format and can be opened by almost every program for manipulating pictures on the Macintosh.

5.3.6 Working with the Session Window

5.3.6.1 Input of MuPAD Commands

New MuPAD commands are always entered in fields of the type *New Input*. If already evaluated input is changed then the type of the relevant field is converted to the type *New Input*. Input fields can be of any size. By pressing the <Return> key at the end of a line a further input line is inserted in the current field. Only when the <Enter> key is pressed then MacMuPAD sends the *entire* input field to the MuPAD kernel for evaluation. The position of the cursor in the input field is unimportant; the whole field is evaluated.

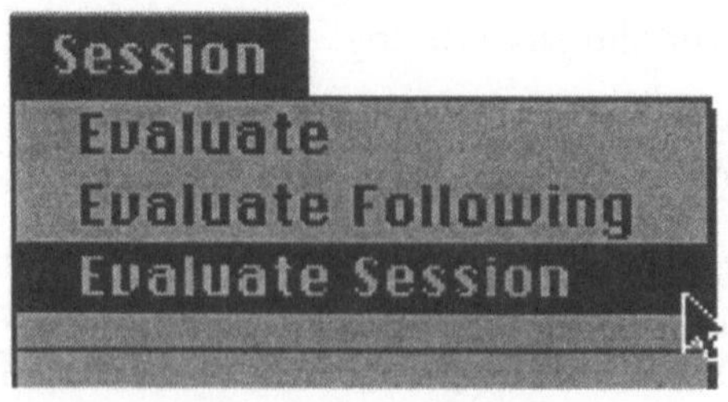

Figure 5.18: *Evaluate* Menu Items

Usually, one will only want to evaluate one input field at a time, however, sometimes it is desirable to also evaluate following input fields, for instance, to newly evaluate a series of calculations that are based on one another. For this purpose, MacMuPAD offers the items *Evaluate following* and *Evaluate all* in the *Session* menu.

Evaluate following evaluates the current input field and all following input fields. *Evaluate all* newly evaluates all input fields in the session window: that is from top to bottom.

5.3.6.2   Connecting Lines

Successive lines are interpreted by MacMuPAD as if they were separated by blanks. E.g. if one enters

```
a :=
1234;
```

followed by <Enter>, then this input is accepted by MacMuPAD because the separation between the assignment := and the number 1234 is syntactically correct. However, it is sometimes desirable that lines are not seen as separate, e.g. when entering very long numbers. For example

```
a := 1234
5678;
```

leads to an error. The separation of the lines is equivalent to:

```
a := 1234 5678;
```

which is syntactically incorrect. If this separation is to be suppressed then a backslash (\) must be entered at the end of the first line. E.g.

```
a := 1234\
5678;
```

5.3.6.3   Controlling Output

Replacing Old Output: Like input, MuPAD output is also organized as fields. Output fields cannot be changed or partially deleted. Only entire output fields can be deleted (see section 5.3.4.2). If already evaluated input is changed then with the subsequent evaluation the old output is replaced by the new. This behavior can be changed through the menu item *Replace old Results* in such a way that the new input is inserted in front of the old. This may be desirable, for instance for comparing results.

Separation Lines: If wanted, MacMuPAD can enter a separation line after each output to more clearly separate the results. This is activated or deactivated with the menu item *Separate by Lines*.

Pretty-Printer: MuPAD output can take place either formatted ("Pretty-Printed") or unformatted. The Pretty Printer is controlled by the environment variable `PRETTY_PRINT`. In MacMuPAD this variable can also be set to `TRUE` or `FALSE` with the menu item *Pretty Print* in the *Session* menu. Of course, the variable can also be set in the usual way through an assignment.

Text Width: A further possibility of controlling the MuPAD output is by regulating the text width with the environment variable `TEXTWIDTH`. The text width influences the appearance of the output in such a way that when the Pretty-Printer is active the output is centered according to this width and when the Pretty-Printer is deactivated the output is overrun at this width. In addition to the change of the variable `TEXTWIDTH` by assigning it a new value, MacMuPAD offers the possibility of carrying this out by using a menu. The item *Set Textwidth...* in the *Session* menu makes it possible to either directly give the desired width or to adapt it to the current window size.

Note that `TEXTWIDTH` does not change automatically as the user resizes the session window. You must use the appropriate menu-item or MuPAD command to change its value.

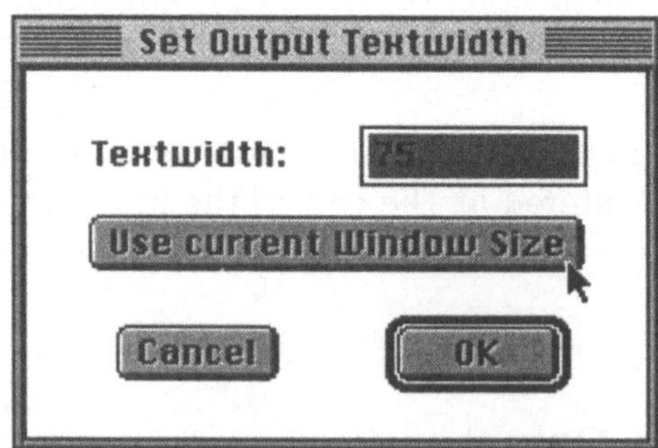

Figure 5.19: Dialog for Setting the Text Width

5.3.7 Text Editing

All document types in MacMuPAD can be edited in the usual Macintosh way with the functions *Cut, Copy, Paste, Clear* and *Select All*. The last input or change in the text can be taken back with the menu item *Undo* in the *Edit* menu as long as no evaluation has taken place. Also, *Replace All* cannot be undone (see below).

Furthermore, in every window, font type, size, style as well as the desired tabulator width can be chosen. This takes place with the menu items *Set Tabs & Font...* in the *Edit* menu.

5.3.7.1 Search and Replace

MacMuPAD offers the user the possibility of searching documents for character strings and of replacing them with other character strings. The item *Find...* in the *Search* menu calls the dialog for searching and replacing.

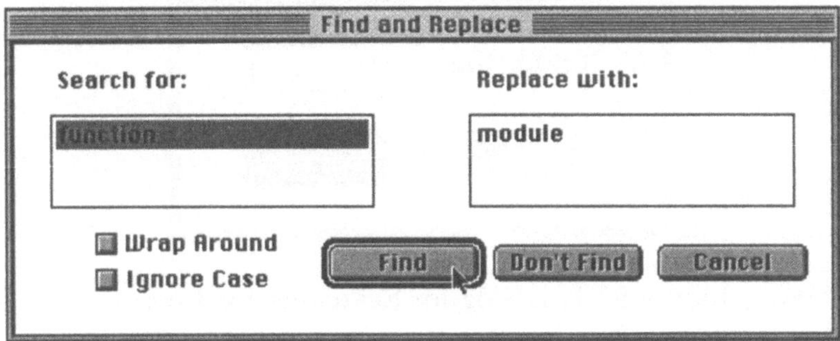

Figure 5.20: Find & Replace Dialog

Here, the user can specify a search and a replace pattern, whereby the replace pattern can also be empty. Furthermore, two parameters can be specified in the relevant check boxes. If *Wrap around* is *not* checked searching ends at the end of the document otherwise searching will continue at the start of the document and end at the current cursor position. If *Ignore Case* is checked small letters match capital letters and vice versa. If the search dialog is closed with *Find* then the search is started using the specified search- and replace patterns. *Don't Find* does not begin searching but remembers the search- and replace patterns. If the search pattern is not found then MacMuPAD replies with a signal tone. Otherwise, the first found search pattern is selected. The user now has the following possibilities:

- **Find again:** Searches for the next occurrence of the search pattern.

- **Replace:** Replaces the selection with the replace pattern.

- **Replace & Find again:** Replaces the selection with the replace pattern and resumes the search.

- **Replace all:** Replaces every occurrence of the search pattern with the replace pattern. A *Replace All* cannot be undone.

Generally, with a single *Replace* the current selection is substituted by the replace pattern. It is unimportant if the selection was made by a preceding search or by hand. Conversely, the user can declare the current selection as the search pattern with the menu item *Enter Selection*. In this manner it is easy to find any further occurrences of a given pattern.

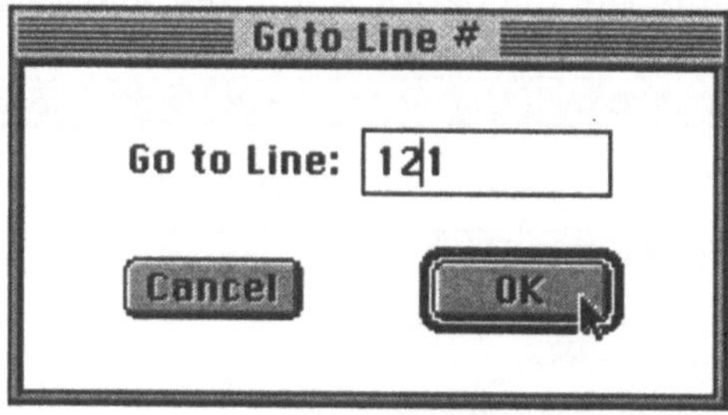

Figure 5.21: Dialog for Searching for Lines

Finally, the *Search* menu offers the possibility of jumping to specific lines by using the item *Goto Line*.

5.3.8 Graphics

The graphic tool integrated in MuPAD enables the user to comfortably plot mathematical functions, curves, surfaces and other objects and to manipulate them. This can be done either directly with the MuPAD commands `plot2d` and `plot3d` or interactively with the MuPAD graphic tool.

To use the graphic tool interactively one should choose one of the menu items *Create 3D Scene* or *Create 2D Scene* in the *Graphic* menu. The same can be done by entering `plot3d()` or `plot2d()` in the session window. In both cases a new scene is generated and a dialog for creating graphical objects is opened. With this dialog, the so called "object manager", all 2D and 3D objects are created and edited. Of course, in 2D scenes only 2D objects can be generated and in 3D scenes 3D objects.

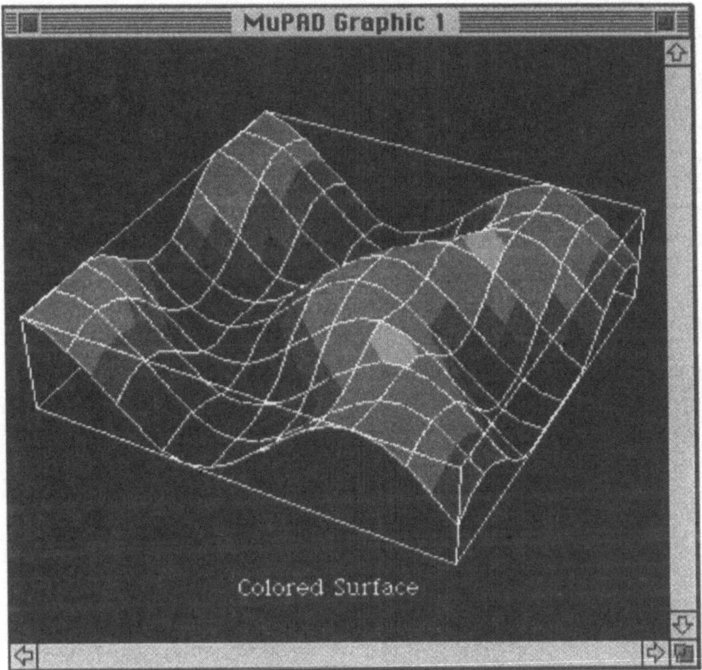

Figure 5.22: Plot of a Surface

5.3.8.1   The Object Manager

This dialog (see figure 5.23) is the main component of the MacMuPAD graphic
tool. It enables the user to manipulate all parameters of a graphical object in
a clear manner. A graphic object is specified by the following parameters:

- Object Type (Mode)
  2D objects can be curves or lists of so-called graphical primitives, i.e.
  points and polygons. 3D objects may be curves surfaces or lists. Specifying a list is not supported when creating objects interactively. See
  section 4. for details on graphical lists.

- Object Style (Style)
  With this, depending on the type of object, the plotting style is set.
  Thus, 3D surfaces can be depicted in the types points, wireframe, hid-
  denline, colorpatch and transparent.

- Mesh (Mesh)
  With this menu the appearance of the surface mesh for 3D objects can
  be set.

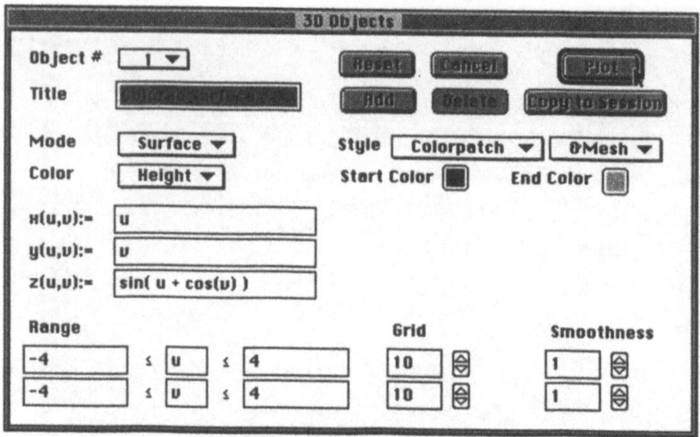

Figure 5.23: Parameter of a 3D Object

- Color (Color)
  This menu specifies how the current object is colored.

- Start and End Color (Startcolor, Endcolor)
  These items determine the color, or respectively the color range of the
  current object.

- Functional Specification
  Here the mathematical definition of the object has to be given in parameterized form.

- Value Range (Range)
  Here the value ranges of the parameter(s) u and v have to be given. If
  desired the parameter identifiers u and v can be changed according to
  the name of the identifiers which are used as parameters in the parameterization.

- Sample points (Grid)
  The number of sample points to be evaluated has to be entered here.

- Smoothness Factor (Smoothness)
  This number determines how many intermediate points should be calculated between two grid points to obtain a smoother representation.

- Object Titel (Title)
  Here a title for the object can be given.

All these parameters can be chosen in the usual manner, i.e. by choice in the pop-up menus or respectively by entry in the relevant fields of the dialog.

It may happen when changing the object parameters that some fields or menus of the object manager change. For instance, the *Style* pop-up menu has different entries for curves and surfaces. Also, the fields for the second parameter v are only available when a surface object is being edited. Furthermore, some fields will be deactivated when they are of no importance for the current object. Thus, the end color cannot be set for an object when it is drawn in "flat" color mode, because in this mode only one color, not a color range is needed.

We shall forgo a detailed explanation of the individual object parameters at this point, because the MuPAD graphics and their usage have already been introduced and explained in detail in chapter 4.

Setting Up and Deletion of Objects: In the object manager, with the buttons "Add" and "Delete" one can create new objects or respectively delete existing objects. The number of objects in a scene is only limited by the amount of memory available, whereby a scene must contain at least one object. If an existing object is deleted then a warning is given that enables the user to abort the deletion process. Once an object is deleted it cannot be retrieved again.

Restoration of the Previous Specification: After changing the definition of an object, it is still possible to restore the previous setting with the button "Revert". If an object is newly created then this action enters a default object. This restoration of an old definition is only possible as long as the new scene is not plotted.

Plotting an Object: After all objects have been created and their specifications set, the calculation of the corresponding plot is started by clicking the button "Plot". The object specifications are sent to the MuPAD kernel and the sample points are evaluated there. If one or more objects are incorrectly defined then a window is opened in which an error message is given. In this window there is the possibility of re-calling the object manager in order to correct the object definition or to abort the editing of the scene.

Creation of a MuPAD Plot Command: All graphical scenes can also be created using the commands plot2d() and plot3d(). MacMuPAD offers the user the possibility of converting a scene created using the object manager,

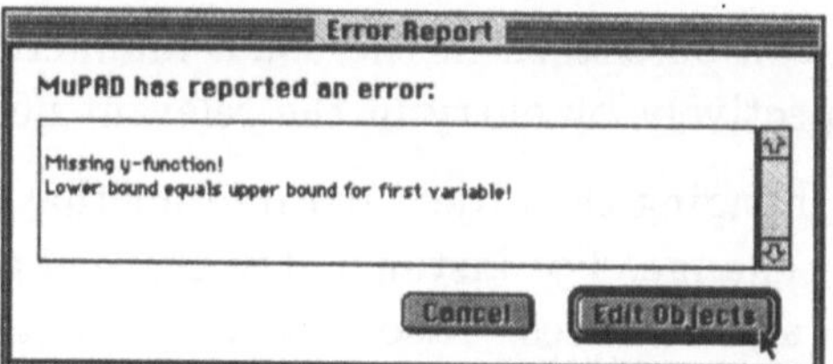

Figure 5.24: Error Message of the Graphic

into an equivalent plot command and then to output this. In this way one can, for instance, get to know the various applications of the command or one can save created plots in form of the corresponding plot commands. Clicking the button "Copy to Session" in the object manager creates a plot command that describes the current scene and which is subsequently copied into the session window. However, this is only possible when an input line is selected in the session window.

5.3.8.2 Manipulation of the Scene

The most important parameters of a graphical scene are:

- The "Constrained" or "Unconstrained" Mode.

- Drawing style (thickness of lines, point width, font, foreground and background color)

- Axes style (type of axes and labeling)

- Perspective (viewpoint of the user - only for 3D graphics)

- Zooming (visible section of the plot)

These parameters are modified through the corresponding dialogs. The dialogs are called through the four menu items *Drawing*, *Axes & Labels*, *Perspective* and *Zooming* in the *Scene* submenu of the *Graphic* menu. In the following the individual dialogs shall be explained in brief.

The *Drawing* Dialog: With this dialog all the basic drawing parameters are set:

- Font, Style and Size
  Here the font in which all labeling of the plot is made is set, This includes
  the labeling of the axes as well as the titles.

- Point Style and Point Width
  The appearance of individual points and their size can be given here.
  This is of importance for objects drawn in the style "Points". or "Lines-
  Points"

- Line Style and Line Width
  In this menu the drawing style of lines and their thickness can be defined.
  This is used everywhere a line is drawn, e.g. the axes or grid lines of an
  object.

- Foreground Color and Background Color
  The axes, their labels and the parameter mesh of objects are drawn in
  the foreground color. The background color is used as background color
  in the whole graphic window.

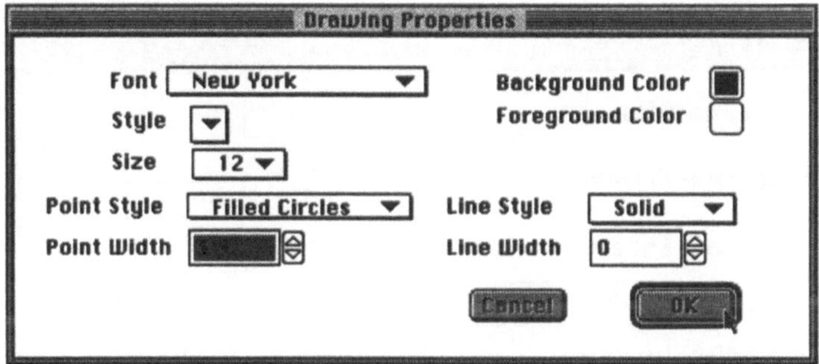

Figure 5.25: Drawing Dialog

All of these parameters are chosen in the corresponding fields in the dialog or
pop-up menu. The new values can then be either immediately accepted by
clicking the "Plot" button or discarded with "Cancel".

The *Axes & Labels* **Dialog:** This dialog sets the axes style and regulates
the labeling of the axes and the scene. The following options are present:

- Scene Title (Title)
  Here the title of the whole scene can be set. It is drawn below the scene
  in the chosen font.

- Axes Style (Axes)
  With this the user chooses if and how the axes are to be drawn. There is a choice between: no axes, axes cross at the origin, axes cross in the corner of the plot and box around the plot.

- Axes Marks (Ticks)
  This parameter sets if and how many tick marks should be drawn on the axes.

- Draw Arrows
  Here the user can specify if the axes are to be drawn with or without arrowheads.

- Axes Labeling (Labels)
  These text fields allow the user to label the axes as desired.

- Axes Origin (Origin)
  Here the co-ordinates at which the axes would cross can be specified.

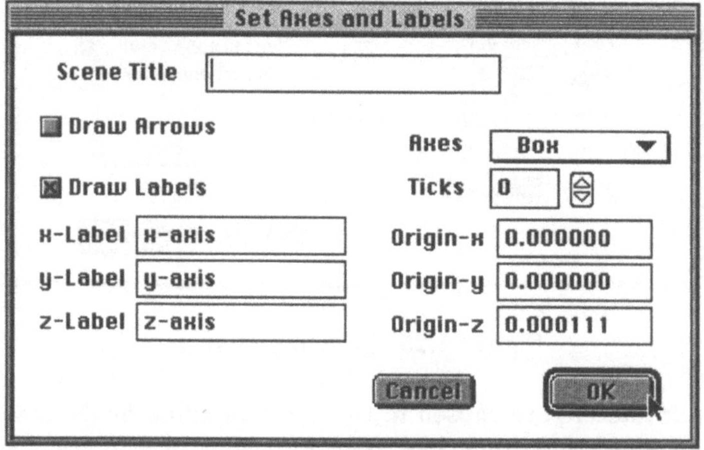

Figure 5.26: Axes and Labeling Dialog

The fields for entering the labels are only visible when the labeling of the axes is chosen, i.e. when the relevant box in the dialog is checked. In the same way the axes origin can only be entered when the axes style "Origin" has been selected.

The button "Plot" accepts the changes, the button "Cancel" discards all changes and the scene remains as it was.

The *Perspective* **Dialog:**   This dialog is used for setting the perspective of the scene, i.e. to freely rotate the plot and to change the distance between the viewer and the ViewingBox. This takes place through the relevant scroll bar in the *Perspective* dialog. With this the ViewingBox can be easily interactively rotated horizontally and vertically.

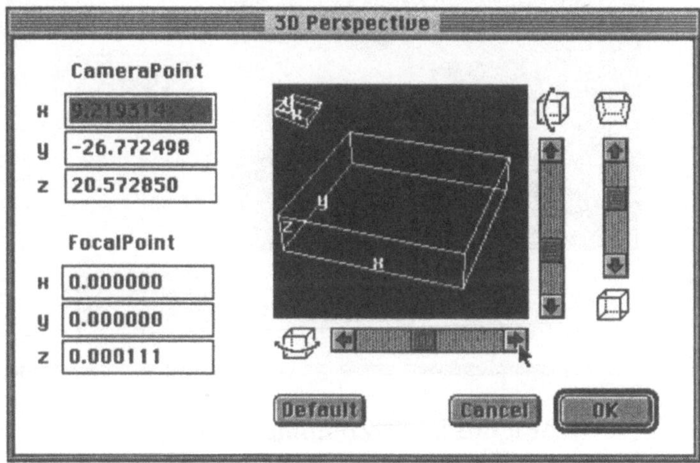

Figure 5.27: Perspective Dialog

With the perspective scroll bar on the right in the perspective dialog the distance to the ViewingBox is set. The nearer one moves toward the ViewingBox, the more the perspective is distorted. In contrast, the ViewingBox appears as an idealized parallelepiped at the maximum distance.

The most vital parameters when looking at the ViewingBox are the camera position ("Camera Point"), also called the viewer's position, and the position to which the camera points ("Focal Point"). As a default, the focal point is set to the centre of the ViewingBox and the user changes the camera point by moving around the scene with the scroll bars. Both points can also be changed by hand, by entering the desired values in the relevant fields. After the new values have been entered the ViewingBox is immediately re-drawn with the new perspective parameters.

The *Zooming* **Dialog:**   With this dialog the user can interactively change the visible section of a scene. Firstly, there is the possibility of enlarging the visible section ("Zoom in") or diminishing it ("Zoom out"). This takes place by simply clicking the relevant button in the *Zooming* dialog.

Furthermore, it is possible to move the plot horizontally and vertically. For this the mouse pointer is placed on the box which shows the visible section

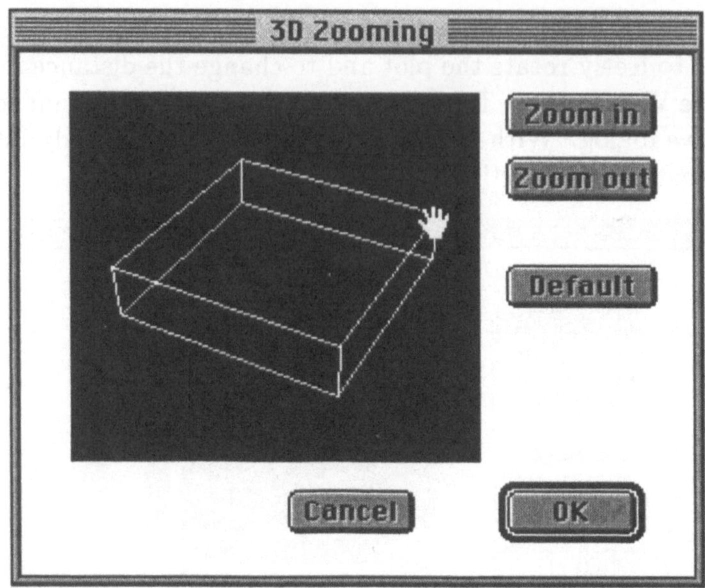

Figure 5.28: Zooming Dialog

and, while holding the mouse button pressed, this boxs is moved to the desired position.

Clicking the button "Plot" accepts the changes and draws the new scene. "Cancel" aborts the dialog without changing the visible section.

Constrained/Unconstrained Mode: Graphical scenes can be drawn in two different ways:

In the "Constrained" mode the scene is drawn in its real proportions. Circles appear as circles and squares as squares.

In the "Unconstrained" mode the scene is scaled so that the size of the graphic window is utilized to the full, i.e. the graphic is transformed to the window size. For instance, if the window is twice as wide as it is high then a circle appears as a flat ellipse that fills the whole window.

Saving Graphics: In MacMuPAD plots can be saved in two ways. A plot command that describes the plot can be created and is then saved as text in a text file. For this the button "Copy to Session" in the object manager has to be clicked. The command equivalent to the plot is then copied into the session window and can either be executed from there or copied to a text window with Cut&Paste and saved there.

Furthermore, a plot can be saved to a file in the Macintosh Picture ('PICT') format. For this, the menu item *Save* or *Save as* in the *File* menu has to be selected. So-created plots can be imported and processed further by all known programs which deal with graphics on the Macintosh.

5.3.9 The Help System

MacMuPAD offers the user an integrated help tool which gives extensive support for all features of MuPAD. The help tool is organized in the form of a hypertext system and includes the complete MuPAD documentation. At the moment this is the MuPAD user's manual, the tutorial, an exhaustive demo and a number of other documents. Further documents will be released during the next months. The help system can be opened in two ways in MacMuPAD.

The help system can be directly activated with the menu item *Open Manual* in the *Help* menu or by selecting the appropriate Manual name in the *Manuals* submenu. MacMuPAD then displays the contents of the manual. The user can browse through the manual page-by-page or with the use of cross references (hyperlinks).

Apart from this it is possible in MacMuPAD to request help to a specific keyword. This takes place by entering the command

```
help("<Keyword>"); or ?<Keyword>
```

in the session window. If a help page to the given keyword is available then this is shown.

Usually, the MuPAD manual is opened by the procedure above. This, and other documentation files, are stored in a folder "Manuals" which can be found in the same directory as MacMuPAD. On starting, MacMuPAD searches this folder and enters all found documentation files in the submenu of the menu item *Select Manual* in the *Help* menu. To open a different help document the desired manual has to be selected in this submenu.

5.3.9.1 Browsing through the Help System

If the help system is activated then one can leaf through the help document by using various commands of the *Help* menu. The menu items *Next Page* and *Previous Page* select the next, or respectively, the previous page. Instead of using these menu items the two arrows above the scroll bar in the help window can be used. With the menu items *First Page* and *Last Page* one can jump to

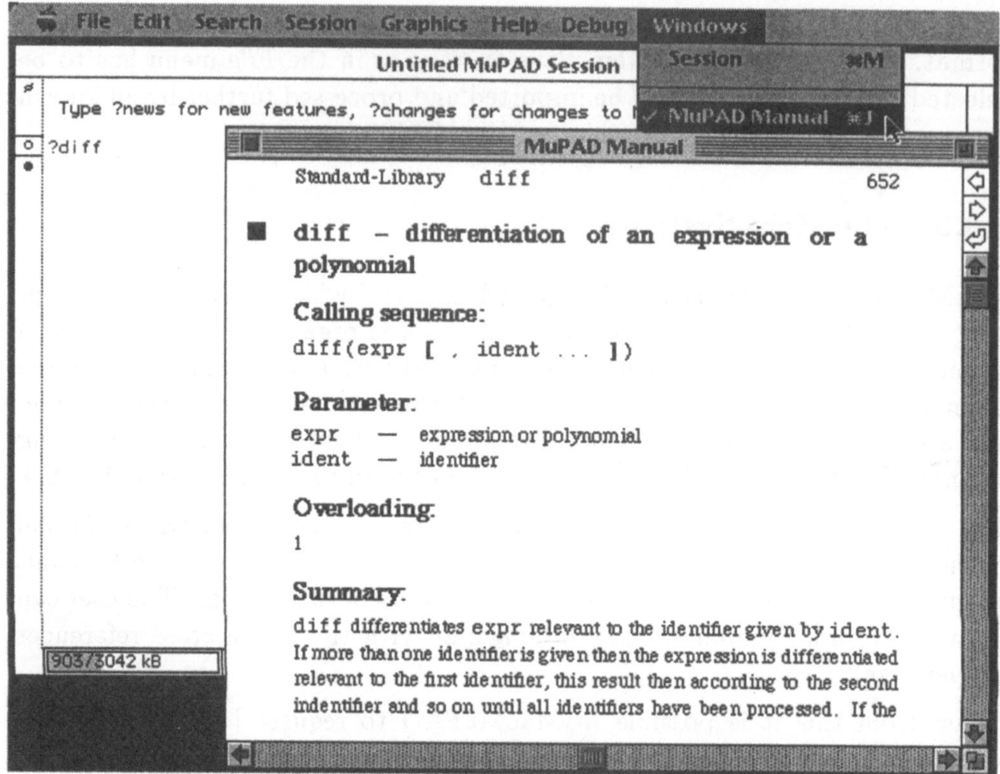

Figure 5.29: The MacMuPAD Help System

the first or last page of the help document. The command *Goto Page* opens a
dialog with which a specific page can be reached.

While leafing through the manual, MacMuPAD registers the pages already
read. With the menu item *Go Back* one can leaf backward to the previously
chosen page. This is especially useful in connection with cross references (see
below). The command *Go Back* can also be activated with the arrow pointing
backward above the scroll bar in the help window.

Cross References in the Text (Hyperlinks):   In the text important keywords are underlined. These denote cross references to the pages, on which
the relevant keyword is explained. These underlined words behave similarly
to buttons, i.e. one can click them with the mouse. Doing this MacMuPAD
automatically turns to the relevant page. Here MacMuPAD also registers the
previous page. Thus, it is easy to look up additional references or explanations
quickly and then to return to the actual starting page.

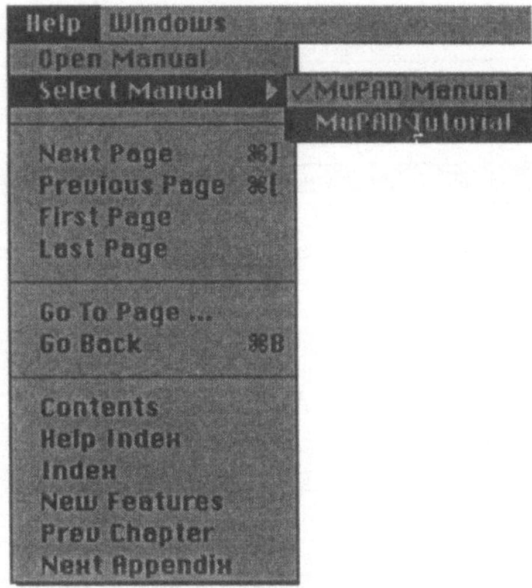

Figure 5.30: Selecting a new manual document

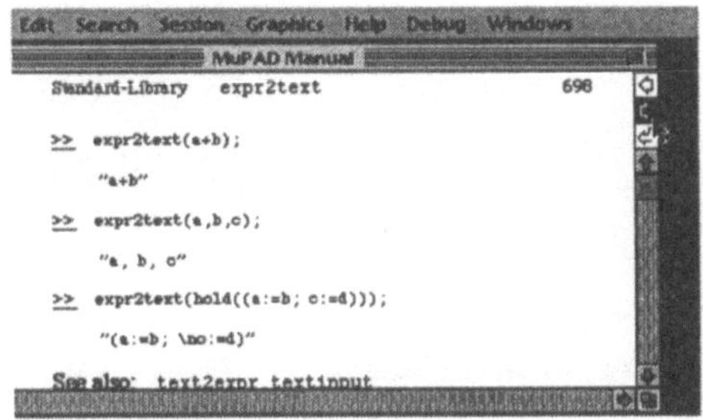

Figure 5.31: Browsing forward in the Help Tool

Cross References in the Menu: In addition to the hyperlinks in the text, additional cross references can be found in the *Help* menu. These cross references can be selected through the relevant menu items. While the hyperlinks in the text always refer to keywords in the text, the cross references in the menu refer to important passages in the text, e.g. the index register or the next chapter. Here, too, one can jump back to the previous page with the command *Go Back*.

Examples: All sections of the MuPAD documentation are illustrated by examples that can be directly copied to the session window and executed. The examples are underlined like the keywords in the text and can be clicked like buttons. When clicking on an example then the relevant example text is automatically copied into the session window and can be executed or changed

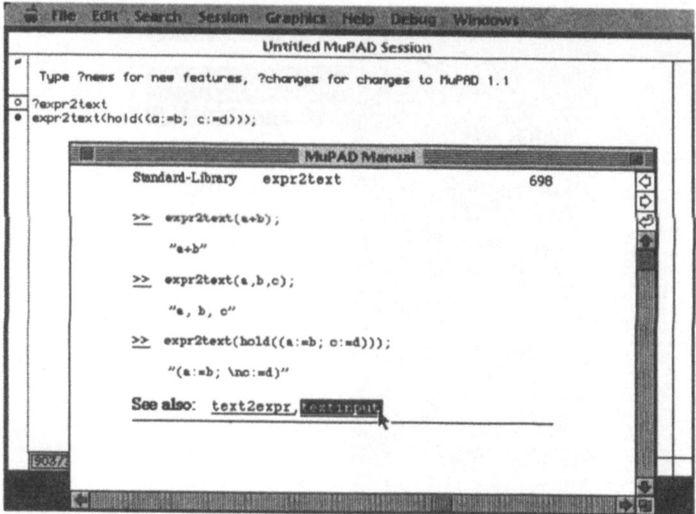

Figure 5.32: Clicking a Cross Reference

there.

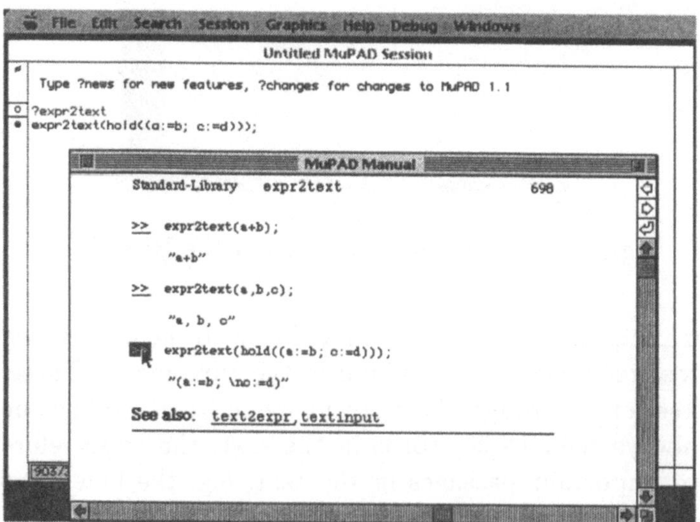

Figure 5.33: Clicking an Example

5.3.10 The Source Level Debugger

If the MuPAD debugger is turned on the user may choose to run his or her
MuPAD programs using the "debug()" command. Once the user starts a

function or program with this command the debugger windows are opened
and the MacMuPAD debugger frontend is ready.

The debugger displays two windows, the so called source window which contains the source text of the program and the data window which is used to
display the values of MuPAD variables and expressions.

The Source Window: The source window contains the source code of the
program which is currently executed, the statement markers, the name of the
current function and an arrow which marks the current statement.

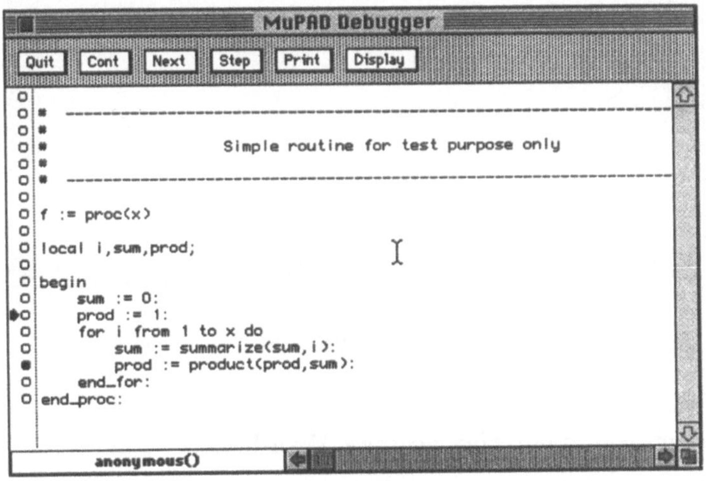

Figure 5.34: Debugger Source Window

The top of the source window has a six-button status panel. These buttons
control the execution of the current procedure. The names of these buttons
match the commands of the internal MuPAD debugger.

The column of circles along the left side of the source text are statement markers. The user can set and delete breakpoints by clicking on the corresponding
statement markers.

The arrow to the right of the statement markers is the current statement
arrow. It points to the statement that is to be executed next. When you start
a debugger session the arrow points to the first statement in the function to
be debugged.

The Data Window: With this window the user can examine MuPAD variables and expressions. All output done by the debugger is displayed in this

window. The top of the data window has a three-button panel. These buttons can be used to issue the debugger commands "Print", "Display" and "Undisplay"

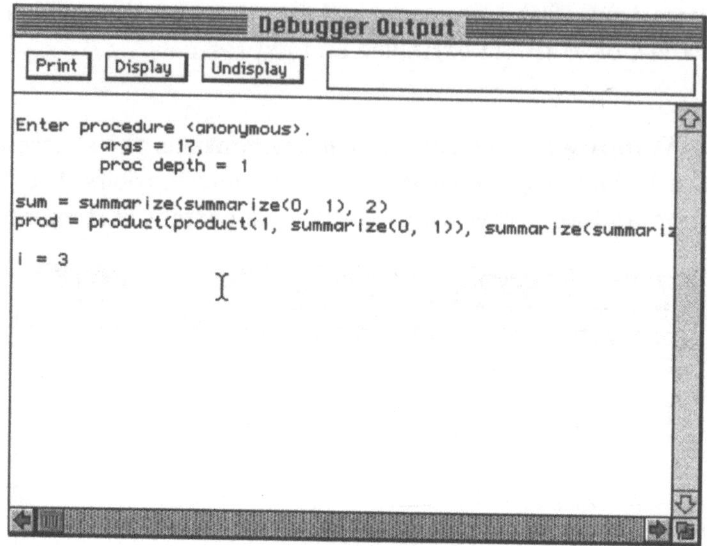

Figure 5.35: Debugger Data Window

5.3.10.1 Working with the Source Window

The source code of the procedure or function which is currently debugged is displayed in the source window. If other functions are called when stepping through the code the debugger loads the appropriate source file and displays it in the source window.

The debugger can only display code which was loaded from a file. Functions or procedures which were entered interactively cannot be debugged this way.

Setting Breakpoints: The MuPAD debugger lets the user set breakpoints by clicking on a statement marker circle. The circle then changes from hollow to filled, indicating that a breakpoint is set. To clear the breakpoint one has to click once again on the now filled circle. It will change to a hollow circle to show that the breakpoint is now cleared.

In the current version of MuPAD is it possible to set breakpoints in lines that are never executed (e.g. comment lines). Execution will not stop if a breakpoint is set in one of these lines so the user must make sure to set breakpoints only in lines that contain executable statements.

The Current Function Indicator: The current function indicator at the lower left of the source window displays the name of the function in which the current statement is found. When one clicks and holds the mouse button on the current function indicator, a pop-up menu will appear that shows the chain of calls. If then an item is chosen, the debugger opens the file that contains the selected function and selects the line from which the following function was called.

Controlling Execution: The source window shows six buttons which are used to control execution and to display the values of expressions and variables. All debugger commands are explained in the debugger section of this manual so we just give a brief overview.

- Quit: This aborts the current computation and ends the disables the debugger.

- Cont: This starts execution or continues after the program has stopped because of a breakpoint.

- Step: This executes the current statement and stops after that. If the function calls another function, the source code of the called function will be displayed.

- Next: This executes the current statement and stops after that. If the current statement is a function call, the function is executed and the current statement arrow advances to the next statement in the current function.

- Print: This prints (i.e. evaluates) arbitrary MuPAD variables or expressions. To print an expression, the user must select the relevant expression in the source window and click this button. The expression will be evaluated and the result will be displayed in the data window. To enter an expression by hand one must use the data input field in the data window.

- Display: This works like "Print" but displays the selected expression every time the debugger stops. This command is useful to monitor how the value of a variable changes during execution without clicking on "Print" after every step.

5.3.10.2   Working with the Data Window

In the data window all debugger output is displayed. In the entry field of the data window one can type any legal MuPAD expression which is evaluated after typing <RETURN> or clicking on one of the buttons "Print" or "Display". These buttons just work like in the source window except that the expressions to be printed (displayed) are not taken from the source text but from the data window's entry field. Furthermore the button "Undisplay" can be used to stop displaying a variable or expression every step.

5.3.10.3   The "Debug" menu

The "Debug" menu contains all those debugger commands that can be issued with one of the buttons in the debugger window. In addition to these the following menu items can be selected:

- Up: This goes up one level in the call-chain, i.e. displays the function that called the current function (if available).

- Down: This goes down one level in the call-chain, i.e. displays the function that was called by the function which is currently displayed.

- Go to Proc...: This allows the user to specify an arbitrary MuPAD function to be displayed. Of course the function can only be displayed if the source code is available, i.e. that the function was not typed interactively.

- Execute...: This allows the user to execute an arbitrary MuPAD command.

- Clear all breakpoints: This clears all breakpoints in all source files. This is useful when starting a new debugger session.

- Arrange windows: This arranges the debuggers source and data window on the screen.

5.3.11   Configurating MacMuPAD

MacMuPAD offers the user the possibility of definin own preferences. These settings are made with the various subitems of the *Preferences* menu under *File.*

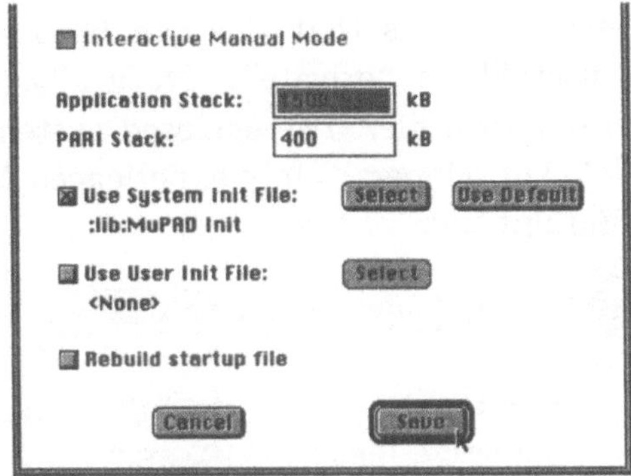

Figure 5.36: MuPAD Presettings

5.3.11.1   MuPAD Kernel

With this dialog the parameters that influence the behaviour of the MuPAD kernel are set. MacMuPAD must be re-started for these settings to come into effect. The following parameters can be changed:

Application Stack: With this the maximum size of the application stack can be set. The larger the application stack, the larger the recursion depth that MuPAD can attain. Should MuPAD crash during a very extensive calculation then this can be due to a stack overflow.

PARI Stack: With this the maximum size of the stack for PARI calculations is configurated. The PARI routines carry out all numerical operations. If problems arise during calculations with very long numbers then with this item the memory space of PARI can be enlarged.

The memory occupied by the application and the PARI stack will be subtracted from the memory available for frontend and kernel. 

System Init File, User Init File: MuPAD offers the user the possibility of reading data and programs or loading library functions directly after starting. This occurs with the help of two files, the *System Init File* and the *User Init File*. By using the relevant check boxes the user can set which files are to be read first. The system init file is a standard file which takes over the important initialization tasks. The user init file is intended for configurating personal settings.

The idea behind the system file is that this can be used by all users in a network, whereas the user file is "private". Naturally, the system file can also be replaced by a users own file. The enclosed system file represents the suggestions of the MuPAD developers. Only experienced MuPADusers should choose a different system init file.

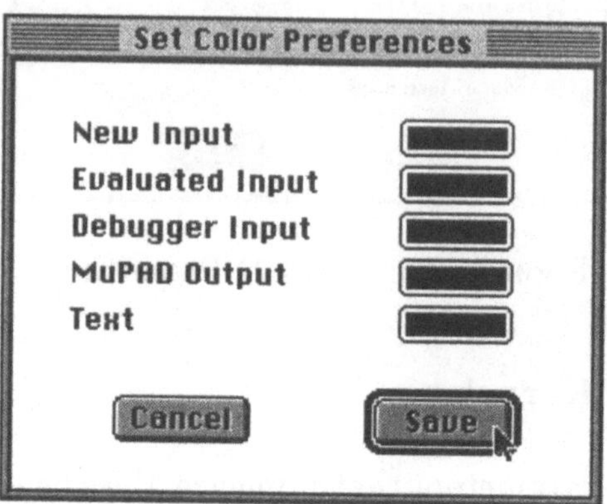

Figure 5.37: Color Presettings

5.3.11.2 Colors

These settings influence the colors in the session window. A different color can be chosen for each text type in the session window, compare section 5.3.4.3. The changes take effect immediately.

5.3.11.3 Session

These settings change the representation of the text in the session window as well as the behavior of the session window. Changes take effect immediately. The following can be changed:

Font, Size, Style: Here the font, size and style in the session window can be set. For presentations a larger font size can be advantageous.

Tabs: Here the length of the tabulator character can be set.

New Results replace old: Here it can be determined if an old result is overwritten by a new calculation or if the new result is inserted in front of the old one. (siehe Abschnitt 5.3.6.3).

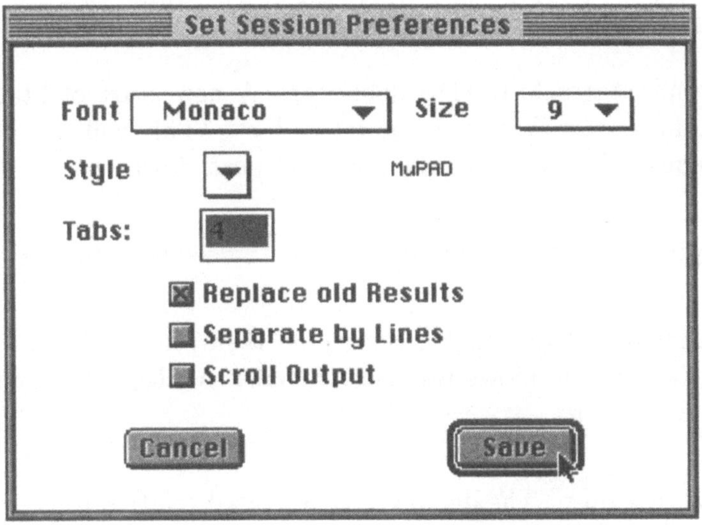

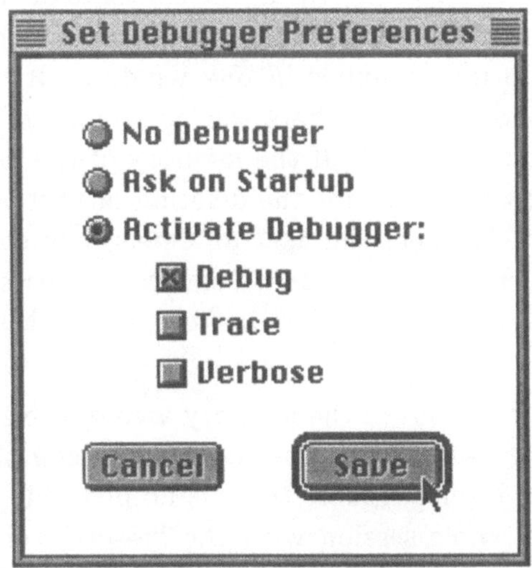

Figure 5.38: Session Presettings

Draw Separation Lines: Here it can be determined if MacMuPAD output fields are to be separated by horizontal lines or not (siehe Abschnitt 5.3.6.3).

Figure 5.39: Debugger Presettings

5.3.11.4 Debugger

This item configurates MuPAD's debugger. A new start of MacMuPAD is necessary to make the changes come into effect. (This will not be necessary in later versions.) The user has the following possibilities:

No Debugger: Instructs MacMuPAD not to use the debugger.

Ask on Startup: If this option is selected then after starting MacMuPAD a dialog appears which allows the user to make the desired debugger setting for the current session.

Activate Debugger: The debugger is activated relative to the options *Debug*, *Trace* and *Verbose* as selected by the user. he meanings of the individual options are explained in the chapter "Debugger".

5.3.12 Memory Problems

It may happen that during long or complicated calculations, MacMuPAD has not enough memory to carry out the final action correctly. This can be due to one of two different situations because MacMuPAD works in two separate memory areas: one for the user interface and one for the kernel.

In the memory area of the user interface, the contents of the different windows are stored. This memory overflows when too many windows are open simultaneously or when too much text is in one window. If this case arises then MacMuPAD warns the user and offers the possibility of freeing memory by closing a window or deleting text. If the memory of the user interface is completely full then it may happen that the deletion of text is no longer possible. In this case MacMuPAD offers the user an emergency solution. The text to be deleted is marked and <Control>-<Delete> is pressed. In this way, the text is always deleted and memory made free, but such an action cannot be undone.

All calculations are carried out in the memory area of the kernel. This memory can overflow when the user carries out too long or complicated calculations. In contrast to the user interface, here there is no possibility of freeing memory short of resetting the whole session with the "reset()" command. If there is a shortage of memory during a calculation, then MacMuPAD warns the user and tries to save all the results of the previous calculations. However, it is still possible to request variables and to save the contents of the window, but not to start further calculations. If this is tried, MacMuPAD blocks the execution of commands. After this no more commands can be evaluated and MacMuPAD answers all such attempts with a warning tone. However, the window contents can still be changed and saved because the user interface continues working, independent of the kernel.

Chapter 6

Survey about the MuPAD functions

6.1 Introduction

On the following help pages we shall describe the system functions and environment variables of MuPAD. But first an important note:

If there is nothing else noted on a help page then the parameters of the relevant function are completely evaluated and flattened.

6.1.1 The Help Mechanism in MuPAD

In MuPAD there is a brief description of every system function and environment variable. During a MuPAD session these are made available by using the **help** command.

Example 207 *With the command*

```
>> help("diff");
```

or more briefly

```
>> ?diff
```

the user can display the description of the system function diff.

In a terminal session the help text is displayed on the monitor. Under the X-Window-System output occurs with the aid of the HyTeX system. With the input of a **help** command HyTeX opens the MuPAD manual at the page given in the **help** call. In the example above this is the page describing the **diff** function.

The examples on the help pages can be directly inserted into the current MuPAD session. This is achieved by clicking the button with the prompt symbol $\boxed{>>}$ or $\underline{>>}$. If such a button is clicked then the relevant example is inserted into the current MuPAD session. The user can then change the example or directly execute it.

In many examples it is assumed that the variables do not yet have a value. This can be most simply achieved by entering **reset();** before the example is executed.

6.1.2   The Syntax of the Help Pages

On each help page the syntax of the relevant function is described under the heading "Call". Optional parameters are embedded in "⟨ ... ⟩". These parameters may be omitted when the function is called. With a non empty series of equal-type parameters the first parameter is always given followed by "...". In such a series the individual parameters must be seperated by commas.

Example 208 *The function* **append** *has the following call:*

```
append(list, expr, ...)
```

Thus, after the parameter **list** *a non empty series of* **expr** *parameters must be given. Valid calls of* **append** *are for example:*

```
append(list, expr_1);
append(list, expr_1, expr_2, expr_3);
```

In contrast, the function **return** *may be called without parameters:*

```
return(⟨ expr ... ⟩)
```

Here any number of parameters **expr** *are permitted, e.g.*

```
return();
return(expr_1, expr_2, expr_3);
```

Under the heading "Overloading" is described, which of the arguments are overloadable. That means the following: Let f be a function whose i-th argument is marked as overloadable. If f is called with an element e from a domain D as i-th argument, then $D :: f$ is called with the current parameters of f. A detailed description of the overloading mechanism can be found in section 2.3.19.2.

6.1.3 Description of Errors

Possible errors of a function are described on the help pages. However, the obvious errors, like missing parameters or parameters of the wrong type, are not included.

A wrong number of parameters returns errors like:

```
Error:  Wrong number of parameters in function call [abs]
```

Wrong parameter types return errors like the following:

```
Error:  Integer expected [igcd/ilcm]
Error:  Illegal parameter [append]
```

The latter message does not tell us very much but it is generally obvious which parameter is meant from the context.

6.2   Standard Library

■ **abs – absolute value**

Calling sequence:

abs(expr)

Parameter:

expr — expression

Overloading:

all

Summary:

abs returns the absolute value of a real or complex expression. For expressions of the type DOM_EXPR the determination of the absolute value of an expression can be defined under the function attribute "abs" in the form of a procedure or function. If the operator of expr has such an attribute then the procedure or function under this entry is called with the operands of expr and the result is returned.

If expr is a real or complex number then the kernel function abs will be used to determine its absolute value.

```
>> abs( -1.2 );

    1.2

>> abs( -8/3 );

    8/3

>> abs( 1+I );

    2^(1/2)

>> abs( PI*I );

    PI
```

```
>> abs( a+a );

    2*abs(a)
```

---

alias – defines an abbreviation or macro

Calling sequence:

```
alias(ident = expr ...)
alias(call = expr ...)
```

Parameter:

```
ident   —   identifier
call    —   function call
expr    —   expression
```

Summary:

`alias` defines an abbreviation (also called macro). Such an abbreviation is expanded into its original definition after text is read but before any evaluation takes place. This applies to any text read by the MuPAD parser, whether entered interactively or via the functions `read` or `finput`.

The abbreviation may be an identifier or a function call:

- `alias(a = b)` defines an abbreviation for the expression b. Every time a is read it is replaced by b.

- `alias(f(x1,...,xn) = b)` defines an abbreviation for b depending upon parameters. Every time a function call of the form `f(y1,...,yn)` is read this function call is replaced by `subs(b, x1=y1,...,xn=yn)`, i.e. the "formal parameters" `xi` in b are replaced by the "actual parameters" `yi`.

Multiple alias definitions may be given in a single call. The arguments of `alias` are not evaluated.

Note that an alias definition remains valid during calls of the system functions `read` and `finput`. This may cause unwanted side-effects. During the evaluation of an element of the loadproc-domain the actual alias definitions are undefined in order to avoid such side-effects. Use `unalias` to delete alias definitions.

One may not redefine keywords of the MuPAD language with an alias definition. The abbreviations must be valid expressions.

```
>> alias(d = diff): d(f(x,y), x);

   D([1], f)(x, y)


>> alias(Dx(f) = diff(f(x),x)): Dx(f+g);

   diff(f(x), x) + diff(g(x), x)
```

See also: finput, loadproc, read, unalias

■ **anames – returns the identifiers that have a value**

Calling sequence:
anames(integer)
anames(name)

Parameter:

integer — one of the numbers 0, 1, 2 or 3
name   — identifier

Summary:

anames(n) returns the set of identifiers which have a value assigned to them. The parameter n determines the identifiers to be taken into account:

n = 0: Yields all function names whose procedure body is defined by a function environment. This includes the functions of the system kernel.

n = 1: Yields all identifiers, which have been assigned procedures written in the MuPAD language.

n = 2: Yields all identifiers not included in the above criteria i.e. all variables.

n = 3: Yields all identifiers that have a value. This includes all the cases above.

anames(name) returns TRUE, if the identifier name has been assigned a value. In all other cases it returns FALSE. During this the argument name is not evaluated. Thus, e.g. the entry n := m; anames(n); returns the result TRUE, because n has a value. If, however, anames(m) is to be carried out then anames(level(n,1)) must be entered. anames is a kernel function.

```
>> f := proc() begin end_proc: anames(1);
```

```
    {f}
```

```
>> a := b: anames(a);

    TRUE
```

```
>> anames(b);

    FALSE
```

```
>> anames(sin);

    TRUE
```

See also: `load`

---

append – adds elements to a list

Calling sequence:

`append(list, expr, ...)`

Parameter:

```
list   —   list
expr   —   expression
```

Overloading:

1

Summary:

append appends the elements `expr` to the list `list` and returns the new list as
the result. `append(list, expr)` is equivalent to `[op(list), expr]`, however
append is much more efficient. **append** is a kernel function.

```
>> append([a, b], c, d);

    [a, b, c, d]
```

```
>> append([], c);
```

[c]

See also: op, concatenation operator

■ **args – accesses procedure parameters**

Calling sequence:

```
args()
args(integer)
args(integer_1..integer_2)
```

Parameter:

```
integer                 —   non-negative integer
integer_1, integer_2    —   positive integers
```

Summary:

args permits accessing of the actual parameters of a procedure and therefore can only be used in procedures and pure functions.

If **args** is called with no further parameters, then an expression sequence of all parameters is returned. The expression **args(0)** returns the number of actual parameters. By using **args(i)** the i-th parameter can be accessed. With **args(i..j)** an expression sequence containing the i-th up to the j-th parameter is returned.

With **args** the parameters are not evaluated again. However an explicit evaluation is possible by using the function **eval**.

With the expression **procname(args())** it is possible to return the function call of the current procedure with evaluated arguments. **args** is a kernel function.

```
>> f:=proc(a) begin [args()],[args(0)] end_proc: f(3, 4);
```

```
    [3, 4],[2]
```

```
>> f:=proc(a) begin args(1) end_proc: f(2, 3);
```

```
    2
```

See also: `procname`, `eval`

---

array – defines an array

Calling sequence:

```
array(range ... ⟨ , [ integer_1 ...] ⟩ )
array(range ... ⟨ , [ integer_1 ...] ⟩ , equation ...)
array(range ... ⟨ , [ integer_1 ...] ⟩ , list )
  equation  ⟶  (integer_2 ...) = expr
```

Parameter:

`range`	—	range of non-negative integers
`integer_1`	—	positive integer
`integer_2`	—	non-negative integer
`expr`	—	expression
`list`	–	list

Summary:

`array` is used to define an array. The parameters consist of a number of non-negative integer ranges that correspond to the dimension of the array. The size of the array is defined by the ranges. Optionally a list can be given at the end that contains a positive integer for each dimension of the array. This list defines the size of the subarrays into which the whole array is divided.

The other parameters are also optional and consist either of any number of equations that describe the individual initial entries of the array or a list of lists that contain the entries of the array line-by-line. If equations are used for initializing the left side of each equation must give the index of the array entry and the right side specifies the value of the corresponding entry. The initialization with lists needs a complete specification of all array elements. In the case of high dimensional arrays these lists must be correspondingly embedded so that the innermost lists contain those elements whose indicies differ only in the final dimension.

A detailed description of arrays can be found in section 2.3.12. `array` is a kernel function.

```
>> A := array(1..6, 1..4, [3, 2]);
```

```
>> A[1];A[3];
```

```
>> array(1..8, 1..8, [2, 2], (1,1) = 11, (8,8) = 88):
```

```
>> A := array(1..4, 1..3, (1,2) = 12): A[1,2];

      12
```

```
>> A := array(1..2, 1..3, [[1, 2, 3], [4, 5, 6]]): A;

      array(1..2, 1..3, (1,1) = 1, (1,2) = 2,
      (1,3) = 3, (2,1) = 4, (2,2) = 5, (2,3) = 6)
```

See also: `table`

■ **asympt − asymptotic expansions**

Calling sequence:

```
asympt(f, x ⟨ ,n ⟩)
asympt(f, x=v ⟨ ,n ⟩)
```

Parameters:

f, v	—	expressions
x	—	identifier
n	—	nonnegative integer

Summary:

`asympt(f,x=v,n)` computes an asymptotic expansion of the expression `f` with respect to `x` at the point `v`, to order `n`. The default point is `infinity`, and the default order is the value of the global variable `ORDER`, which is 6 initially.

The function `asympt` returns a domain object of the type `gseries` that can be manipulated with the standard arithmetic operations. The following methods are available too: `coeff(s, n)` gives the coefficient of the n-th monomial, `expr` converts the expansion to an expression, removing the error term, `lmonomial` gives the leading monomial, `lterm` gives the leading term, `lcoeff` gives the leading coefficient, `map` maps a function onto the coefficients, `nthterm` gives the n-th term, `nthmonomial` gives the n-th monomial, and `order` gives the error term.

```
>> s:= asympt(sin(1/x+exp(-x))-sin(1/x),x);


                                            / exp(-x) \
                  exp(-x)     exp(-x)       | ------- |
      exp(-x) -  -------  +  -------  + O|      6      |
                     2           4        \    x      /
                  2 x         24 x
```

```
>> lmonomial( s ), nthterm( s,3 );

    exp(-x), x^(-4)*exp(-x)
```

```
>> asympt( exp(sin(1/x+exp(-exp(x))))-exp(sin(1/x)),x,2);


                                        / exp(-exp(x)) \
                       exp(-exp(x))     | ------------ |
     exp(-exp(x)) +  ------------  + O|       2       |
                          x            \     x       /
```

See also: `limit`, `series`

---

atan – the inverse tangens

Calling sequence:
`atan(x)`
`atan(x,y)`

Parameter:
x — expression
y — expression

Overloading:
all

Summary:
The call **atan(x)** represents the inverse tangens of the expression **x**. If the
argument **x** is of the type DOM_FLOAT then the kernel function **atan** will be
used to compute the inverse tangens of **x**.

The two-argument function **atan(x,y)** computes the principal value of the argument of the complex number **x+I*y**, thus $-\pi < \mathrm{atan}(x, y) < \pi$.

```
>> atan(1.5);

   0.9827937232


>> atan(2,3);

   atan(3/2)


>> atan(1,infinity);

   PI * 1/2
```

■   **bernoulli – Bernoulli numbers and polynomials**

Calling sequence:

bernoulli(n ⟨ , x ⟩)

Parameter:

n — non-negative integer
x — expression

Summary:

bernoulli(n) returns the n-th Bernoulli number, bernoulli(n,x) the n-th Bernoulli polynomial in x. If the first argument is not a non-negative integer then **bernoulli** returns an error. **bernoulli** returns the function call with evaluated arguments, if the first argument cannot be evaluated to a number.

```
>> bernoulli(16);

   - 3617/510


>> bernoulli(3, x);
   subs(%, x=1/4);
   bernoulli(3, 1/4);
```

```
   x*1/2 + x^2*(-3/2) + x^3
   3/64
   3/64
```

```
>> bernoulli(4, y-1);
   expand(%);

   (y-1)^2 + (y-1)^3*(-2) + (y-1)^4-1/30
   y*(-12) + y^2*13 + y^3*(-6) + y^4 + 119/30
```

---

besselJ – Bessel function of the first kind

Calling sequence:

```
besselJ(v, x)
```

Parameters:

v, x — expressions

Overloading:

2

Summary:

besselJ(v,x) represents the Bessel function J of the first kind with order v and argument x.

```
>> besselJ(2,1+I);

   besselJ(2, 1 + I)
```

```
>> float(%);

   0.4157988694e-1 + 0.2473976415 * I
```

See also: besselY

---

besselY – Bessel function of the second kind

Calling sequence:

`besselY(v, x)`

Parameters:

`v, x` — expressions

Overloading:

2

Summary:

`besselY(v,x)` represents the Bessel function Y of the second kind with order `v` and argument `x`.

```
>> besselY(0,2);

     besselY(0, 2)

>> float(%);

     0.5103756726
```

See also: `besselJ`

■ `binomial` – Binomial coefficient

Calling sequence:

`binomial(n, k)`

Parameter:

`n, k` — expressions

Summary:

`binomial(n, k)` returns the binomial coefficient $\binom{n}{k}$. If both arguments are positive integers with $0 \le k \le n$, then

$$\text{binomial}(n, k) = \frac{n!}{k!(n-k)!}$$

The more general definition of the binomial coefficient is

$$\text{binomial}(n, k) = \frac{\text{gamma}(n+1)}{\text{gamma}(k+1)\,\text{gamma}(n-k+1)}$$

`binomial` returns the function call with evaluated arguments, if the first or second argument cannot be evaluated to a number. Exceptions to this rule are the binomial coefficients that can be simplified in a trivial manner.

```
>> binomial(46, 23);

   8233430727600

>> binomial(-235/123, 3);

   -20233265/5582601

>> binomial(163.65674, I);

   1.224281883 - 1.483129672*I

>> binomial(n, 0);

   1
```

See also: `fact`, `gamma`

`bool` – Boolean evaluation

Calling sequence:
`bool(expr)`

Parameter:
`expr` — Boolean expression

Overloading:
all

Summary:

`bool` evaluates Boolean expressions: the result is **TRUE** or **FALSE**. Evaluation is done in the same way as in the conditional part of the statements `if`, `while` or `repeat` (the so-called Boolean context).

`bool` returns the error `Can't evaluate bool` if the expression cannot be evaluated as Boolean. `bool` is a kernel function.

```
>> bool(3 < 5);

    TRUE

>> b:=a: bool(a = b and 3 < 4);

    TRUE

>> bool(4);

    Error: Can't evaluate bool [bool]
```

■ **built_in** – **defines kernel functions**

Calling sequence:

```
built_in(integer, integer_nil, string, table_nil)
built_in(integer, integer, string_nil, string)
```

Parameter:

integer	—	non-negative integer
string	—	character string
integer_nil	—	non-negative integer or **NIL**
string_nil	—	character string or **NIL**
table_nil	—	table or **NIL**

Overloading:

1, 2

Summary:

With the function **built_in** the kernel functions of the MuPAD system are defined. Kernel functions are elements of the basic type *DOM_EXEC*. In MuPAD these objects have two different tasks: they define the evaluation of an expression and they are responsible for its output. Relevant to their usage the operands have different types. Their exact significance for evaluation and output is described in section 2.3.16. Here we would like to bring to your notice once more that the definition of new *DOM_EXEC* objects should be carried out with great care.

The examples show the *DOM_EXEC* objects for evaluating and returning the system function _mult. **built_in** is a kernel function.

```
>> built_in(815, NIL, "_mult", NIL):
```

```
>> built_in(1100, 13, "*", "_mult"):
```

See also: `func_env, func`

---

bytes – occupied memory space

Calling sequence:

`bytes()`

Parameter:

none

Summary:

`bytes` returns three numbers, which give information about the occupied memory space:

- The first number gives the memory (in bytes) used logically by MuPAD.
- The second number gives the number of bytes physically allocated by the memory management.
- The third number gives the free space in the program stack (only on certain computers, e.g. the Apple Macintosh). On computers with virtual memory management and therefore with a lot of memory available the third number only returns $2^{31} - 1$.

`bytes` is a kernel function.

---

ceil – converts numbers to integer

Calling sequence:

`ceil(expr)`

Parameter:

`expr`   —   expression

Overloading:

all

Summary:

`ceil` returns the smallest integer, greater or equal to the number given by `expr`. If `expr` is a complex number then an error is returned. If the argument cannot be evaluated to a non-complex number, `ceil`returns the unevaluated function call.

If the exponents of real numbers become too large then a call of `ceil` can lead to a loss of accuracy which is not recognized as an error. `ceil` is a converting function of the kernel.

```
>> ceil(1.2);

    2


>> ceil(-8/3);

    -2


>> ceil(x);

    ceil(x)
```

See also: `round, frac, trunc, floor`

■ **coeff – returns the coefficients of a polynomial**

Calling sequence:

```
coeff(poly ⟨ ⟨ , expr_2 ⟩ , integer ⟩)
coeff(expr ⟨ , indets ⟩ ⟨ ⟨ , expr_2 ⟩ , integer ⟩)
 indets  ⟶  [ expr_1 ...]
```

Parameter:

`poly`	—	polynomial
`expr, expr_1, expr_2`	—	expressions
`integer`	—	non-negative integer

Overloading:

1

Summary:

`coeff` returns the coefficients of a polynomial. There are three different forms
of the function call of `coeff`:

- `coeff(p)` returns a sequence with all coefficients of the polynomial p.
  The coefficients are ordered according to the lexicographical ordering of
  the terms.

- `coeff(p,x,n)` returns the coefficient of the term with the factor x^n,
  where p is understood as a univariate polynomial in the variable x (x
  must naturally be a variable of the polynomial p). If a polynomial has
  more than one variable then the result is a polynomial in the remaining
  variables.

- `coeff(p,n)` returns the coefficient of the terms with the factor x^n. Here,
  x is the "main variable" (i.e. x is the first element in the list of the
  variables of the polynomial).

The first argument of `coeff` is allowed to be a polynomial or an expression
and optionally a list of indeterminates. An expression is converted into a
polynomial in the specified indeterminates (see function `poly`). `coeff` returns
`FAIL` if the expression cannot be converted into a polynomial.

If the first argument of `coeff` is a multivariate polynomial and the result a
single coefficient then this coefficient — as described above — is a polynomial
in the remaining variables. If the first argument is an expression then the
result is returned as an expression. `coeff` is a kernel function.

```
>> coeff(3*x^3+x^2*y^2+2, [x,y]);

    3, 1, 2


>> p := poly(3*x^3+x^2*y^2+2, [x,y]): coeff(p, y, 2);

    poly(x^2, [x])


>> coeff(3*x^3+x^2*y^2+2, [x,y], y, 0);

    3*x^3 + 2


>> coeff(3*x^3+x^2*y^2+2, [x,y], y, 1);
```

```
   0

>> coeff(3*x^3+x^2*y^2+2, [x,y], 2);

   y^2
```

See also: lcoeff, nterms, nthcoeff, poly, tcoeff

■ collect – collects coefficients of a polynomial

Calling sequence:

```
collect(p, [x,y,...] ⟨ , f ⟩)
```

Parameter:

p   —   polynomial expression
x,y   —   unknowns
f   —   procedure

Summary:

The call collect(p, [x,y]) collects the coefficients of p with same power in x and y. With a third argument, the procedure f is applied on the coefficients (like the function Factor in the example below).

```
>> p := x*y+z*x*y+y*x^2-z*y*x^2+x+z*x:
   collect(p,[x,y]);

   x*(z + 1) + x*y*(z + 1) + x^2*y*(-z + 1)

>> collect(p,[x]);

   x*(y + z + y*z + 1) + x^2*(y-y*z)

>> collect(p,[x],Factor);

   x*(y + 1)*(z + 1)-x^2*y*(z-1)
```

See also: factor

`combine` – combines terms of the same algebraic structure

Calling sequence:

`combine(e ⟨ , fn ⟩)`

Parameters:

e   —   expression
fn   —   optional name like sqrt, sincos

Summary:

The call `combine`(e) combines terms of the same algebraic structure. The function `combine` maps onto sets, lists and arrays, and also onto polynomials and series where `combine` maps onto the coefficients. For expressions of the type `DOM_FUNC_ENV` the function attribute `combine` will be called with the operands of expr.

With a second argument, the function `funcattr(combine,expr2text(fn))` is called. This allows the user to extend the function `combine`. The list of possible options can be obtained with `map([op(op(combine,3))],op,1)`.

```
>> combine(sin(x)+x*y*x^E);

   sin(x) + y*x^(E + 1)


>> combine(sqrt(2)*sqrt(3),sqrt);

   6^(1/2)


>> combine(sin(a)*cos(b),sincos);

   sin(a + b)*1/2 + sin(-a + b)*(-1/2)


>> map([op(op(combine,3))],op,1);
```

See also: `expand`, `factor`, `simplify`

`conjugate` – conjugate complex of an expression

Calling sequence:

```
conjugate(expr)
```

Parameter:

```
expr  —  expression
```

Overloading:

all

Summary:

conjugate calculates the complex conjugate for arbitrary given expressions. For this the method "conjugate" from the domain of expr is used. If this domain does not contain this method then this function returns the expression conjugate(expr).

Exceptions to this are expressions of the types DOM_INT, DOM_RAT, DOM_FLOAT, DOM_COMPLEX and DOM_EXPR, which are directly processed by the function conjugate.

For expressions of the type DOM_EXPR the conjugate can be defined in form of a procedure or function in the function attribute "conjugate". If the operator of expr has such an attribute then the procedure or function in the entry of the attribute "conjugate" is called with the operands of the argument expr and the result is returned.

If such a function attribute is not defined, but the operator of expr is one of the function environments _plus, _mult or _power, then this is directly processed by the function conjugate and the result is returned. In all other cases the expression conjugate(expr) is returned.

```
>> conjugate( (1+I)*exp(2-3*I) );

   (1 - I)*exp(2 +3*I)


>> conjugate( x+2*sin(3-5*I) );

   conjugate(x) + 2*sin(3 + 5*I)


>> f := fun(sin(2*args(1))*cos(2*args(1))):
   f := funcattr( f,"conjugate",f@conjugate ):
   conjugate( f(x) );
```

```
    cos(2*conjugate(x))*sin(2*conjugate(x))
```

See also: `funcattr, Re, rectform, Im`

`contains` – tests if an element or index exists

Calling sequence:

```
contains(container, expr)
contains(list, expr ⟨ , integer) ⟩
```

Parameter:

`container`	—	set, list, array, table, or domain
`list`	—	list
`expr`	—	expression
`integer`	—	integer

Overloading:

1

Summary:

`contains` tests if the expression `expr` is contained as an element or respectively as an index in the container `container`. The container can be a set, a list, an array, a table or a domain.

If a set is tested then one of the Boolean values **TRUE** or **FALSE** is returned, according to if `expr` is an element of the set `container` or not.

If a list is tested, then `contains` returns the position of the searched for element. If the element exists more than once then only the first occurrence is taken into account. The position is returned as a positive integer. If the element does not exist, `contains` returns the value 0.

The position in the list from where the search is to begin can be specified by giving a third argument. Entries occurring before this position are not taken into account.

If the container is an array, a table or a domain then `contains` returns **TRUE** when a value is stored under the index `expr`, otherwise **FALSE** is returned. `contains` is a kernel function.

```
>> contains({a, b, c}, a);
```

```
    TRUE

>> contains({a, b, c+d}, c);

    FALSE

>> contains([a, b, c], b);

    2

>> contains([a, b, c], d);

    0

>> contains([a, b, a, b], b, 3);

    4

>> t := table(13="val"): contains(t, 13), contains(t, "val");

    TRUE, FALSE
```

See also: has

■ content – content of a polynomial

Calling sequence:
```
content(poly)
content(expr ⟨ , indets ⟩)
 indets  ⟶   [ expr_1 ...]
```

Parameter:
```
poly          —  polynomial
expr, expr_1  —  expressions
```

Overloading:
1

Summary:

content calculates the content of a polynomial, i.e. the greatest common divisor of the coefficients of the polynomial. The greatest common divisor of the coefficients must be able to be calculated using the function gcd.

The first argument of content is allowed to be a polynomial or an expression and optionally a list of indeterminates. An expression is converted into a polynomial in the specified indeterminates (see function poly). content returns FAIL if the expression cannot be converted into a polynomial.

```
>> content(poly(6*x^3*y + 3*x*y + 9*y, [x,y]));

   3

>> content(poly(6*x^3*y + 3*x*y + 9*y, [x]));

   3*y

>> content(4*x*y + 6*x^3 + 6*x*y^2 + 9*x^3*y, [x]);

   3*y + 2
```

See also: gcd, poly, primpart

---

context – evaluates the expression in a different context

Calling sequence:
context(expr)

Parameter:
expr — any expression

Summary:

context firstly evaluates expr as usual. The result is then re-calculated in the context that was valid before the present procedure was called. This function is necessary for procedures that use the option hold, as well as for certain cases in the evaluation of domain methods.

The function must not be called on the interactive level or directly in a parallel statement or in the function context itself. Therefore, it is *not* possible to carry out evaluations on more than one context-level higher by using nested context calls. context is a kernel function.

```
>> i:=2:
   f := proc()
   local  i;
   begin
      print(i,context(i));
   end_proc(i);

     i,2
```

■ **contfrac – evaluates to a continued fraction**

Calling sequence:

```
contfrac(e ⟨ , n ⟩)
```

Parameters:

```
e  —    constant expression
n  —    positive integer
```

Summary:

contfrac(e,n) computes a continued fraction approximation for e, using the
first n digits of its floating-point evaluation. When n is not given, the value
of the variable DIGITS is used. In the output, the sign ... represents a real
number between one and infinity.

Continued fractions belong to the special domain CF, and the usual arithmetic
operations can be applied to them.

```
>> a:=contfrac(PI,5): b:=contfrac(sqrt(7),2): a,b;
```

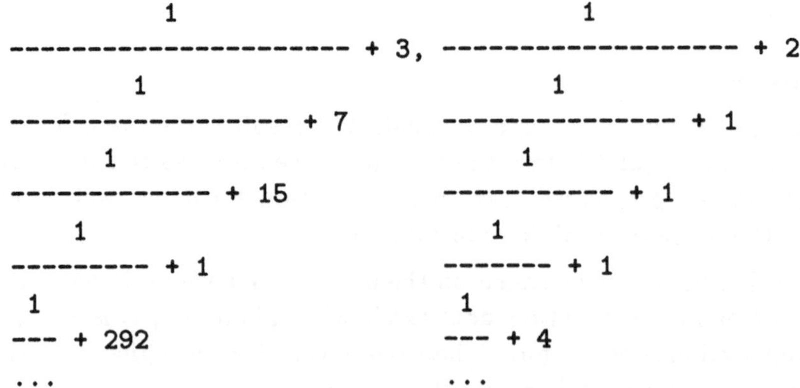

```
>> a+b, a*b, a^3;

          1                  1                  1
------------------ + 5,  -------- + 8,  -------------- + 31
          1                  1                  1
----------- + 1          --- + 3          ------- + 159
     1                   . . .                1
------- + 3                               --- + 3
  1                                       . . .
--- + 1
. . .
```

See also: DIGITS, float

D − differential operator

Calling sequence:

```
D(expr)
D([integer ...], expr)
```

Parameter:

```
expr      —   expression
integer   —   non-negative integer
```

Overloading:

1

Summary:

D as a differential operator calculates the derivative of a function.

- If f is a function with one argument then D(f) returns the derivative of f.
- If f is a function with more than one argument then D([i], f) returns the i-th partial derivative of f.

An expression with the form D([i,j], f) is equivalent to D([i], D([j], f)). It is assumed that the partial derivatives commute. Furthermore D([], f) = f.

The argument expr must be an expression that can be used as a function. Such an expression can contain identifiers, functions, procedures and constants.

Furthermore the operators +, *, ^ and @ can be used. When differentiating expressions the rules for sums and products and the chain rule are used.

For a function environment a user-defined differential procedure can be given. This must be given as an attribute under the index "D" in the function environment: compare to the function **funcattr**. The procedure is called exactly like D with the index list and the function environment as arguments.

```
>> D(sin);

   cos
```

```
>> D(exp + ln);

   1/id + exp
```

```
>> D(f @ (g+h));

   (D(g)+D(h)) * (D(f) @ (g+h))
```

```
>> D([1,2], func(sin(x*y), x, y));

   D([1, 2], _mult)*cos@_mult +
   D([1], _mult)*D([2], _mult)*(-sin)@_mult
```

See also: diff, funcattr

■ **Dpoly – differential operator for polynomials**

Calling sequence:
```
Dpoly(poly)
Dpoly([integer ...], poly)
```

Parameter:
```
poly      —   polynomial
integer   —   positive integer
```

Overloading:

2

Summary:

Dpoly, as the differential operator D, calculates the derivative of a polynomial.

- If p is a univariate polynomial then Dpoly(p) returns the derivative of p with respect to its variable.
- If p is a multivariate polynomial then Dpoly([i], p) returns the partial derivative of p with respect to the i-th variable. (This is the i-th element in the list of variables of p.)

  If i is greater than the number of variables then Dpoly returns the zero polynomial.

An expression of the form Dpoly([i,...j], p) is equivalent to the expression Dpoly([i],... Dpoly([j], p)...). It is assumed that the partial derivatives commute. Furthermore Dpoly([], p) returns p.

If the coefficients of the polynomial are elements of a domain Dom, then this domain must have the method "intmult": Dom::intmult(e,i) must calculate the integer multiple i * e of a domain element e and a positive integer i. Dpoly is a kernel function.

```
>> Dpoly(poly(2*x^2 + x + 1));

   poly(4*x + 1, [x])


>> Dpoly([1], poly(x^2*y + 3*x + y, [x,y]));

   poly(2*x*y + 3, [x,y])


>> Dpoly([1,2], poly(x^2*y + 3*x + y, [x,y]));

   poly(2*x, [x,y])
```

See also: D, diff

---

debug – **user controlled execution of procedures**

Calling sequence:

```
debug(statement ...)
```

Parameter:

```
statement  —  statement
```

Summary:

debug is used in the debug mode of MuPAD for the user controlled execution of the statement **statement**. If this statement contains one or more procedure calls then the interactive mode of the debugger is activated. In this mode the commands for controlling the program run or commands that give information about the program status can be executed. A detailed description can be found in chapter 3. Please note:

- MuPAD must be in the debug mode.
- Only user-defined procedures can be processed.
- Interactively entered procedures cannot be processed.

debug is a kernel function.

See also: `trace`, `PRINTLEVEL`

■ decompose – decomposes polynomials

Calling sequence:

```
decompose(p ⟨ , x ⟩)
```

Parameters:

```
p  —  polynomial
x  —  variable
```

Summary:

decompose(p,x) decomposes the polynomial p with respect to the variable x. decompose returns a sequence of polynomials $q_1 \ldots q_n$ such that $p(x) = q_1(q_2(\ldots(q_n(x))\ldots))$. This decomposition is not unique but one is always found when one exists. If no decomposition exists, the input polynomial is returned.

When the polynomial p contains two or more unknowns, the second argument must be given to indicate with respect to which variable the decomposition has to be done.

```
>> decompose(x^6+6*x^4+x^3+9*x^2+3*x-5);

   x + x^2-5, x*3 + x^3

>> decompose(x^6-9*x^5+27*x^4-27*x^3-2*x^2*y+6*x*y+1,x);

   x*y*(-2) + x^3 + 1, x*(-3) + x^2

>> decompose(x^4-3*x^3-x+5);

   -x + x^3*(-3) + x^4 + 5

>> decompose(poly(x^4-3*x^3-x+5,[x],IntMod(5)));

   poly(-x + x^2,[x],IntMod(5)), poly(x + x^2,[x],IntMod(5))
```

See also: `factor, solve`

---

degree – degree of a polynomial

Calling sequence:

degree(poly ⟨ , *expr_2* ⟩)
degree(expr ⟨ , *indets* ⟩ ⟨ , *expr_2* ⟩)
 indets ⟶ [expr_1 ...]

Parameter:

poly   —   polynomial
expr, expr_1, expr_2   —   expressions

Overloading:

1

Summary:

degree(p) calculates the total degree of the polynomial p. The degree of a
zero polynomial is defined as 0. degree(p,x) calculates the degree of the
polynomial p with respect to the variable x.

The first argument of degree is allowed to be a polynomial or an expression
and optionally a list of indeterminates. An expression is converted into a

polynomial in the specified indeterminates (see function `poly`). `degree` returns
`FAIL` if the expression cannot be converted into a polynomial. `degree` is a
kernel function.

```
>> degree(x^3+x^2*y^2+2);

    4


>> p := poly(x^2*z+x*z^3+1, [x,z]): degree(p);

    4


>> degree(x^2*z+x*z^3+1, [x,z], z);

    3


>> degree(0, [x,y]);

    0
```

See also: `degreevec, ldegree, poly`

■ **degreevec — exponents of the leading term of a polynomial**

Calling sequence:

```
degreevec(poly ⟨ , order ⟩)
degreevec(expr ⟨ , indets ⟩ ⟨ , order ⟩)
 indets  ⟶  [ expr_1 ...]
```

Parameter:

`poly`	—	polynomial
`expr, expr_1`	—	expressions
order	—	`LexOrder`, `DegreeOrder` or `DegInvLexOrder`

Overloading:

1

Summary:

degreevec returns a list containing the exponents of the leading term of a polynomial. If **x1^e1 *...* xm^em** is the leading term of the polynomial then the list **[e1,...,em]** is returned. For a zero polynomial a list containing 0s is returned.

With the option *order* the desired term ordering can be given, see section 2.3.13. If no order is given then the lexicographical one is used.

The first argument of **degreevec** is allowed to be a polynomial or an expression and optionally a list of indeterminates. An expression is converted into a polynomial in the specified indeterminates (see function **poly**). **degreevec** returns **FAIL** if the expression cannot be converted into a polynomial. **degreevec** is a kernel function.

```
>> degreevec(x^3+x^2*y^2+2, [x,y]);

   [ 3, 0 ]


>> degreevec(x^3+x^2*y^2+2, [x,y], DegreeOrder);

   [ 2, 2 ]


>> p := poly(x^2*z+x*z^3+1, [x]): degreevec(p);

   [ 2 ]


>> degreevec(0, [x,y,z]);

   [ 0, 0, 0 ]
```

See also: **degree, ldegree, lterm, nthterm, poly**

---

denom − denominator of a rational expression

Calling sequence:

denom(expr)

Parameter:

expr — expression

Summary:

denom returns the denominator of the rational expression **expr**. **expr** is converted to normal form with **normal**. The denominator is an expanded polynomial expression. See function **normal**.

```
>> denom(-3/4);

   4

>> denom(x + 1/(2/3*x - 2/x));

   2*x^2 - 6

>> denom((cos(x)^2 - 1) / (cos(x) - 1));

   1
```

See also: gcd, normal, numer

■ **diff** – differentiates an expression or a polynomial

Calling sequence:

diff(expr ⟨ , ident ... ⟩)

Parameter:

expr   —   expression or polynomial
ident   —   identifier

Overloading:

1

Summary:

diff differentiates **expr** with respect to the identifier given by **ident**. If more than one identifier is given then the expression is differentiated corresponding to the first identifier, this result then according to the second indentifier and so on until all identifiers have been processed. If the identifiers are given in

the form `ident $ n` through the sequence operator then the n-th derivative
corresponding to the identifier `ident` is calculated. If no identifier is given
then the expression is evaluated and returned. In this case a differentiation
does not take place.

For functions whose derivative is not known by `diff` the function call with
evaluated arguments is returned.

With the function `funcattr` the effect of differentiation on any system and
user-defined function can be defined. For this purpose a procedure must be
entered in the corresponding function environment under the index "diff".
This procedure is called with the expressions to be evaluated and the differentiation parameter as arguments. The result of the procedure is then used
as the result of the differentiation.

If the argument `expr` is a polynomial whose coefficients are elements of a
domain Dom then this domain must contain the methods `diff` and `intmult`:

- `Dom::diff(e,x)` must return the derivative of the domain element `e`
  corresponding to `x`.

- `Dom::intmult(e,i)` must calculate the multiple `i * e` of the domain
  element `e` for a positive integer `i`. For calculating `diff(poly(e*x^3),x)`
  `= poly(3*e*x^2)` correctly, `3 * e` must be calculated in the domain Dom.

`diff` is a kernel function.

```
>> diff(x^2,x);

   2*x


>> diff(x^2*sin(y),x,y);

   x*cos(y)*2


>> diff(sin(x)*cos(x),x$3);

   cos(x)^2*(-4)+sin(x)^2*4


>> diff(poly(sin(a)*x^3 + 2*x, [x]), a, x);

   poly(x^2*(cos(a)*3), [x])
```

```
>> diff(f(x),x);

   diff(f(x),x)


>> f(2) := 4:
   diff_proc := proc(expr,x)
                begin
                     g(op(expr,1))*diff(op(expr,1),x)
                end_proc:
   f := funcattr(f, "diff", diff_proc):
   diff(f(sin(x)),x);

   cos(x)*g(sin(x))
```

See also: D

■ **DIGITS – significant digits of floating-point numbers**

Calling sequence:

```
DIGITS := integer
```

Parameter:

```
integer   —   positive integer
```

Summary:

DIGITS gives the number of significant digits when dealing with floating-point numbers. DIGITS can be set to any integer between 1 and $2^{31} - 1$. If a floating-point number is given with more than DIGITS places then all extra decimal places are removed.

Through the assignment DIGITS := NIL;, DIGITS is set to the default value 10. DIGITS is an environment variable.

```
>> 1.756746574666;

   1.756746574


>> float(PI);
```

```
    3.141592653
```

```
>> DIGITS := 40: float(PI);
```

```
    3.141592653589793238462643383279502884197
```

```
>> DIGITS := 1: float(PI), 3.9, -3.2;
```

```
    3., 3., -3.
```

See also: float

---

divide − divides polynomials

Calling sequence:

```
divide(poly_1, poly_2 ⟨ , opt ⟩)
divide(expr_1, expr_2 ⟨ , indets ⟩ ⟨ , opt ⟩)
 indets   ⟶   [ expr ...]
```

Parameter:

poly_1, poly_2	—	polynomials
expr, expr_1, expr_2	—	expressions
opt	—	one of the options Quo, Rem or Exact

Overloading:

1,2

Summary:

divide(p,q) divides the univariate polynomial p by the polynomial q. If no option is given then the quotient s and the remainder r are calculated such that $p = sq + r$. The sequence consisting of the quotient s and the remainder r is returned. The function returns **FAIL** if division is not possible.

- If the identifier **Quo** is given as the option *opt* then only the quotient s is calculated.

- If the identifier **Rem** is given as option then only the remainder r is returned.

divide(p,q,Exact) divides the multivariate polynomial p by q. The quotient p/q is returned as the result. If p cannot be divided by q then the function returns **FAIL**.

If the arguments are polynomials with coefficients from a domain then this domain must contain the method **divex** ("divide exact"). With this method the coefficients are then divided. If the domain elements cannot be divided then the method must return **FAIL**.

The first two arguments can be either polynomials or expressions and optionally a list of variables. If the arguments are expressions, these are converted into polynomials (see function **poly**). **divide** returns **FAIL** if the expressions cannot be converted. Both polynomials must be of the same type, i.e. their variables and coefficient rings must be identical. The result has the same type as the first two arguments. **divide** is a kernel function.

```
>> divide(x^3+x+1, x^2+x+1);

   x - 1, x + 2


>> divide(a*x^3+x+1, x^2+x+1, [x], Quo);

   a*x - a


>> divide(a*x^3+x+1, x^2+x+1, [x], Rem);

   x + a + 1


>> p := poly(x^2-x*y-x+y, [x,y]): q := poly(x-1, [x,y]):
   divide(p, q, Exact);

   poly(x-y, [x,y])
```

See also: `content, multcoeffs, pdivide, poly`

■ **domain – creates a domain**

Calling sequence:

```
domain()
domain(dom)
domain(table)
```

Parameter:

```
dom      —  any domain
table    —  table
```

Summary:

`domain()` creates a domain with no entries. Entries can be added to this domain later with `domattr`.

`domain(dom)` creates a copy of the domain `dom`. This is necessary because changes to a domain are always carried out directly. If a domain is changed then the value of this domain is changed in all data in which it occurs. However, if the domain is copied (with the function `domain`) before the changes are made, then the changes do not occur in the other domain because the newly created domain is not included in any of them (see also the examples below).

`domain(table)` creates a new domain which includes those entries also found in the table `table`. `domain` is a kernel function.

```
>> domain();

   domain( )


>> d1:=domain(): d2:=d1: d2::x:=y: d1::x;

   y


>> d1:=domain(): d2:=domain(d1): d2::x:=y: d1::x;

   FAIL


>> d:=domain(table("x"=y)): d::x;

   y
```

See also: `domattr, new`

■ domattr – accesses the contents of a domain

Calling sequence:

```
domattr(dom, expr)
```

Parameter:

```
dom    —   domain or domain element
expr   —   expression
```

Summary:

Similar to a table, a domain contains entries that can be accessed with indices.

`domattr(dom, expr)` returns the entry of `dom` with the index `expr`. If `dom` has no entry with this index then the function returns `FAIL`. If the parameter `dom` is not a domain but a domain element then `dom` is replaced by the corresponding domain and the operation is carried out on this.

If this function occurs on the left side of an assignment the right side of the assignment is inserted in `dom` under the index `expr`. Here the function does not copy the domain and side-effects can occur: i.e. the domain is changed in every datum that it contains. Through this characteristic it is possible for a domain to contain itself as a sub-domain.

The operator `::` exists for this function. The parser automatically converts an entry of the form `x::y` into the function call `domattr(x,"y")`. The value after the colons must be either a variable or a keyword. Please note that the variable after the colons is converted into a string! `domattr` is a kernel function.

```
>> d1:=domain(): d2:=d1: d2::x:=y: d1::x;
```

```
     y
```

```
>> d:=domain(table("x"=y)): d::x;
```

```
     y
```

```
>> d:=domain(table("x"=y)): e:=new(d): e::x;
```

        y

See also: domain

---

domtype – determines types

Calling sequence:

domtype(expr)

Parameter:

expr — expression

Overloading:

1

Summary:

domtype returns the domain to which **expr** belongs, i.e. the type of the argument. The pre-defined basic types and their significance are listed in the table A.1 in the appendix. In contrast to most other functions, arguments which are expression sequences, are not flattened. **domtype** is a kernel function.

```
>> domtype(2.345434345);

    DOM_FLOAT


>> domtype(x - y);

    DOM_EXPR


>> domtype((a:=5));

    DOM_INT


>> a:=3,4: domtype(a);
```

`DOM_EXPR`

See also: `testtype, type`

■ **editing of lines in the terminal version**

Summary:

The line editor is only available in the terminal version of MuPAD, not under the X-Window-System or on the Macintosh.

The current text line can be edited with the line editor during interactive input. The commands of the line editor are entered as control characters i.e. by simultaneous pressing of the control key and a second one, e.g. <Ctrl-A> indicates the simultaneous pressing of the control and A keys. The available commands are:

<Ctrl-A>	cursor to line beginning.
<Ctrl-Y>	cursor to the beginning of the previous word.
<Ctrl-B>	cursor one character to the left.
<Ctrl-F>	cursor one character to the right.
<Ctrl-Z>	cursor to the beginning of the next word.
<Ctrl-E>	cursor to the line end.
<Ctrl-U>	delete the complete input line.
<Ctrl-W>	delete from the cursor position to the beginning of the previous word.
<Ctrl-H>	delete the character left of cursor.
<Ctrl-D>	delete the character under the cursor.
<Ctrl-T>	delete the next word.
<Ctrl-K>	delete to end of line.
<Ctrl-L>	insert the last input line before the current cursor position.
<Ctrl-P>	reproduction of the last input line. Repeated pressing of <Ctrl-P> successively reproduces the previous input line. If the cursor is not at the beginning of the line then the previous lines are searched for an entry that corresponds to the characters of the current input.
<Ctrl-N>	analogue to <Ctrl-P>, but the previous input is run through in reverse order.

---

eint − the exponential integral

Calling sequence:

eint(x)

Parameter:

x — expression

Overloading:

all

Summary:

The exponential integral is defined as

$$\text{eint}(\mathbf{x}) = \int_{x}^{\infty} \exp(-t)\, t^{-1}\, dt$$

At the moment, a numerical evaluation is only possible for positive real arguments. eint(x) is equivalent to igamma(0, x). eint returns the function call with evaluated arguments, if the argument cannot be evaluated to a number. eint is a kernel function.

```
>> eint(2.4);

     0.2844026093e-1
```

See also: igamma

---

erf, erfc − error function and complementary error function

Calling sequence:

erf(x)
erfc(x)

Parameter:

x — expression

Overloading:

all

Summary:

The error function is defined as

$$\mathtt{erf(x)} = \frac{2}{\sqrt{\pi}} \int_0^x \exp(-t^2)\, dt$$

The complementary error function is defined as

$$1 - \mathtt{erf(x)}$$

At the moment, evaluation is only possible for real arguments. **erf** and **erfc** return the function call with the evaluated argument when the argument cannot be evaluated to a number. **erfc** is a function of the system kernel.

```
>> erfc(-0.2);

   1.222702589


>> erfc(2);
   float(%);

      erfc(2)
      0.4677734981e-2
```

■ **error – user-specified error termination**

Calling sequence:

error(string)

Parameter:

string   —   character string

Summary:

error returns the error message given through **string** and effects an return to the interactive level. The statements in a procedure following the **error** call are not executed.

Errors can be intercepted with the function **traperror**. If **error** is called from a **traperror** environment then execution continues in the **traperror** environment and not on the interactive level. **error** is a kernel function.

```
>> error("Fault");
```

```
    Error: Fault

>> f := proc(s) begin error(s) end_proc: sin(f("Fault"));

    Error: Fault [f]
```

See also: traperror

---

eval – evaluates special functions subsequently

Calling sequence:

eval(expr)

Parameter:

expr — expression

Summary:

eval works only on certain system functions that in some way return uneval-uated results. If the call of such a system function occurs in the argument of eval then the result of the function is subsequently re-evaluated. For the evaluation the current substitution depth is used.

The referred system functions are the functions args, coeff, evalp, expr, hold, input, last, lcoeff, nthcoeff, subs, subsex, subsop, tcoeff and text2expr. They all return a result that is generally not completely evaluated. Polynomial functions return a not completely evaluated result, if the argument is a polynomial and the result an expression. (See section 2.3.13.)

A further special case of the function eval is when the environment variable EVAL_STMT has the value FALSE. In this case eval forces the execution of statements whose evaluation was prevented by the value of EVAL_STMT. eval is a kernel function.

```
>> a:=b: b:=c: text2expr("a+a"), eval(text2expr("a+a"));

    a+a, 2*c

>> EVAL_STMT := FALSE: d:=(e:=f): d, e, eval(d), e;
```

```
    (e := f), e, f, f
```

See also: `args`, `coeff`, `evalp`, `expr`, `hold`, `input`, `last`, `lcoeff`, `nthcoeff`, `subs`, `subsex`, `subsop`, `tcoeff`, `text2expr`, `EVAL_STMT`

■ **evalassign – assignment with evaluation of the left side**

Calling sequence:

```
evalassign(expr_1, expr_2, integer)
```

Parameter:

`expr_1`, `expr_2`	—	expressions
`integer`	—	non-negative integer

Summary:

`evalassign` makes it possible to firstly evaluate an expression with a given substitution depth and then to assign the result a new value. For this, `expr_1` is firstly evaluated with the depth `integer`. The result must be an expression permissible for the left side of an assignment. Finally the value `expr_2` is completely evaluated and the result assigned to this left side. The call `evalassign(expr_1, expr_2, 0)` is equivalent to the statement `expr_1 := expr_2`.

```
>> a:=b: evalassign(a, 100, 1): level(a, 1), b;

   b, 100
```

■ **evalp – evaluates a polynomial**

Calling sequence:

```
evalp(poly, expr_2 = expr_3, ...)
evalp(expr ⟨ , indets ⟩, expr_2 = expr_3, ...)
 indets  ⟶  [ expr_1 ...]
```

Parameter:

`poly`	—	polynomial
`expr`, `expr_1`, `expr_2`, `expr_3`	—	expressions

Overloading:

1

Summary:

evalp(p, x1=y1,..., xn=yn) evaluates the polynomial p by substituting the variables **xi** of p by **yi**, i.e. p is evaluated with **x1=y1, ..., xn=yn**. The values **yi** may be expressions that could also be used as coefficients.

The variables are evaluated in the sequence given by the equations using Horners rule; polynomials in the remaining variables occur as intermediate results. The evaluation of variables with the value 0 is most effective and should take place first. After that the remaining main variable should be evaluated first.

The first argument of **evalp** is allowed to be a polynomial or an expression and optionally a list of indeterminates. An expression is converted into a polynomial in the specified indeterminates (see function **poly**). **evalp** returns **FAIL** if the expression cannot be converted into a polynomial.

If the polynomial still has free variables after evaluation, the result is a polynomial in these variables. In this case the type of the result depends on the first argument. If the argument is a polynomial then a polynomial is returned as result.

If all variables are evaluated then the result is an expression from the coefficient ring of the polynomial. **evalp** is a kernel function.

```
>> evalp(x^2+2*x+3, x=a+2);

   (a+2)*(a+4)+3


>> p := poly(x^2+x*y+2, [x,y]): evalp(p, x=3);

   poly(3*y + 11, [y])


>> p := poly(x^2+x*y+2, [x,y]): evalp(p, x=3, y=2);

   17


>> evalp(x*y*z+x^2+y^2+z^2, x=1, y=1);
```

```
    z^2 + z + 2
```

See also: `coeff, poly`

■ **EVAL_STMT** − controls the execution of statements in expressions

Calling sequence:

```
EVAL_STMT := bool
```

Parameter:

`bool` — Boolean value

Summary:

`EVAL_STMT` controls the execution of statements that occur in expressions; such as `f((a:=b))`. If `EVAL_STMT` is set to `TRUE`, then such statements are executed. In the example above, firstly `b` is assigned to `a` and then `f` is called with `b` as argument. If, however, `EVAL_STMT` is set to `FALSE` then the statements in expressions are not executed.

Through the assignment `EVAL_STMT := NIL;`, `EVAL_STMT` is set to its default value `TRUE`. `EVAL_STMT` is an environment variable.

```
>> a + (b := 3), b;

   a+3, 3


>> EVAL_STMT := FALSE: c + (d := 3), d;

   c + (d := 3), d
```

■ **expand** − expands an expression

Calling sequence:

```
expand(expr)
expand(expr, expr_1 ...)
```

Parameter:

```
expr    —  expression
expr_1  —  expression
```

Overloading:

1

Summary:

expand expands the algebraic expression **expr**. The most important use of **expand** is the multiplication of products in sums. Furthermore the powers of sums are multiplied as long as they are positive integers. Products in the divisors of fractions are not multiplied.

With the optional arguments **expr_1, ...** etc. the expansion of certain sub-expressions of **expr** can be prevented: if these expressions are given as arguments then the sub-expressions of **expr** that correspond to these expressions are not expanded. More exactly the following occurs: in **expr** the sub-expressions **expr_1 ...** are parallelly substituted with auxiliary variables (see **subs**), then **expr** is expanded. In the expanded expression the auxiliary variables are then replaced by the original expressions.

The expansion of functions can be controlled by the user's own expansion routines. For this the function to be expanded must have a function environment in which a procedure is entered as an attribute under the index "**expand**" (compare to the function **funcattr**). This procedure is called by **expand** as soon as this function environment occurs as the zero operand in a sub-expression of **expr**. The procedure is called with the operands of the sub-expression as arguments. The operands are not expanded before the call of the expansion routine. The result of the expansion routine is also not expanded.

Furthermore for domains that are not basic types user expansion routines can also be given. These must be entered under the index "**expand**" in the domain. The routine is called if an element of the domain is contained in the expression to be expanded. The domain element is entered as argument, its result is not expanded. If no expansion routine is given for a domain then the domain elements are processed as constants on expansion.

expand is a kernel function.

```
>> expand((x+1)*(y+z));

   y+z+x*y+x*z


>> expand((x+1)/(y+z));
```

```
   1/(y+z)+x/(y+z)
```

```
>> expand((x+1)*(y+z), x+1);

   y*(x+1)+z*(x+1)
```

```
>> expand((x+y)^(z+2));

   x*y*(x+y)^z*2+x^2*(x+y)^z+y^2*(x+y)^z
```

```
>> expand((x+y)^(z-2));

   (x+y)^(-2)*(x+y)^z
```

```
>> sin:= funcattr(sin, "expand", proc(x)
       local a,b;
   begin
       x:= expand(x);
       if type(x) = "_plus" then
          a:= op(x,1); b:= x - a;
          expand(sin(a)*cos(b)+cos(a)*sin(b))
       else
          sin(x)
       end_if
   end_proc):
   expand(sin(a*(b+c)));

    cos(a*b)*sin(a*c)+cos(a*c)*sin(a*b)
```

```
>> expand(sin(a+b)*c);

   c*cos(a)*sin(b)+c*cos(b)*sin(a)
```

See also: funcattr, subs

■ **export – exports library files**

Calling sequence:

```
export(library)
export(library, function)
```

Parameter:

```
library    —    library name
function   —    function name
```

Summary:

export exports functions of the specified library. That is the functions of the library are made globally available. Before exporting the functions of a library, the library has to be loaded via **loadlib**. This is done automatically for some standard libraries. **export(library)** exports all functions of **library**. After **export(library, function)** the function **function** is globally available, i.e **function(...)** can be used instead of **library::function(...)**.

```
>> loadlib("combinat"):


>> powerset(2);

    powerset(2)


>> combinat::powerset(2);

    {{}, {2}, {1}, {1, 2}}


>> export(combinat, powerset):


>> powerset(2);

    {{}, {2}, {1}, {1, 2}}
```

See also: `loadlib, info`

---

expose – expose definition

Calling sequence:

```
expose(expr)
```

Parameter:

`expr` — expression

Overloading:

1

Summary:

`expose` exposes the definition of `expr` by returning certain operands. Currently the following cases are handled:

- If `expr` is a functional environment then its first operand, the executing function or procedure, is shown.
- If `expr` is a domain then the operands of the domain are shown.

Other arguments are simply returned. Note that one always may explore objects by using the functions `op` and `extop`.

```
>> expose(sin);

   proc(x)
     name sin;
     local f;
   begin
     if args(0) <> 1 then
       error("wrong no of args")
     end_if;
     ...
```

See also: `extop`, `op`

■ **expr – converts a polynomial into an expression**

Calling sequence:

```
expr(pol)
expr(exp)
```

Parameter:

```
pol  —  polynomial
exp  —  expression
```

Overloading:

all

Summary:

expr converts the polynomial **pol** (basic type *DOM_POLY*) into an expression. If **pol** is a polynomial with domain elements as coefficients then these are also converted into expressions using the method **expr** of the domain.

If the argument is not a polynomial, then it is evaluated and returned.

expr is a kernel function.

```
>> p := poly(x^2+y, [x]): expr(p);

   y + x^2
```

See also: poly

---

expr2text – converts MuPAD expressions into character strings

Calling sequence:

```
expr2text(⟨ expr ... ⟩)
```

Parameter:

```
expr  —  expression
```

Summary:

expr2text converts the expression **expr** into a character string that corresponds to the output of the expression **expr** with the Pretty-Printer turned off.

If the function is called without arguments then an empty character string is created. Several arguments are interpreted as an expression sequence and are correspondingly converted into one character string. The arguments are evaluated before conversion.

expr2text is a kernel function.

```
>> expr2text(a+b);

   "a+b"

>> expr2text(a,b,c);

   "a, b, c"

>> expr2text(hold((a:=b; c:=d)));

   "(a:=b; \nc:=d)"
```

See also: `text2expr, textinput`

■ **extnops – number of operands in a domain element**

Calling sequence:
`extnops(domelem)`

Parameter:
`domelem` — domain element

Summary:

A domain element consists of a reference to the corresponding domain and a list of values that represent its contents. **extnops** returns the number of operands in this list. **extnops** is a kernel function.

```
>> d:=domain(): e:=new(d,1,2,3,4): extnops(e);

   4
```

See also: `extop, extsubsop, new, nops`

■ **extop – accesses the operands of a domain element**

Calling sequence:

```
extop(domelem)
extop(domelem, integer)
extop(domelem, integer1, integer2 〈 , integer ... 〉)
```

Parameter:

`domelem`	—	domain element
`integer1, integer2`	—	positive integers
`integer`	—	positive integer

Summary:

A domain element consists of a reference to the corresponding domain and a list of values which represent its contents. The function **extop** allows access to the domain and the operands of this list.

extop(domelem) returns an expression sequence of all the operands in the list of the domain element **domelem**.

extop(domelem, integer1) returns the **integer1**-th entry of **domelem**. If **integer1** is 0, the function returns the domain to which the domain element belongs.

extop(domelem, integer1, integer2 〈 , integer... 〉) gives an expression sequence of those entries in **domelem**, whose indices have been passed to the function. **extop** is a kernel function.

```
>> d:=domain(): e:=new(d,1,2,3,4): extop(e);

    1, 2, 3, 4


>> d:=domain(): e:=new(d,1,2,3,4): extop(e,2);

    2


>> d:=domain(): e:=new(d,1,2,3,4): extop(e,2,4);

    2, 4


>> d:=domain(): e:=new(d,1,2,3,4): extop(e,0);
```

```
domain()
```

See also: `extnops, extsubsop, new, op`

■ **extsubsop – substitutes operands of a domain element**

Calling sequence:

```
extsubsop(domelem ⟨ , equal ... ⟩)
```

Parameter:

```
domelem   —   domain element
equal     —   equation with a positive integer on the left side
```

Summary:

A domain element consists of a reference to the corresponding domain and a list of values that represent its contents.

`extsubsop` replaces the operands of the list of the domain element `domelem` that correspond to the indices of the left side of the equation with the relevant right side of the equation. If the left side of the equation has the value 0 then the right side of the equation must be a domain. The function then replaces the domain to which the domain element belongs with the given one. `extsubsop` is a kernel function.

```
>> d := domain(): e := new(d,1,2,3,4):
   extsubsop(e,1=5): extop(e,1);

   1


>> d := domain(): e := new(d,1,2,3,4):
   e := extsubsop(e,1=5): extop(e,1);

   5


>> d := domain(): e := new(d,1,2,3,4):
   e := extsubsop(e,6=8): e;
```

```
    new(domain( ), 1, 2, 3, 4, NIL, 8)
```

See also: `extnops`, `extop`, `new`, `subsop`

---

fact – factorial function

Calling sequence:

`fact(integer)`

Parameter:

`integer` — non-negative integer

Overloading:

all

Summary:

`fact` is the factorial function that, when `fact(n)` is called, returns the product of the integers from 1 to n. `fact` returns an error when the argument `integer` is a number but not a non-negative integer. `fact` returns the function call with evaluated arguments, if the argument cannot be evaluated to a number. `fact` is a kernel function.

```
>> fact(20);

   2432902008176640000
```

---

factor, Factor – factorizes polynomials

Calling sequence:

`factor(p)`
`Factor(p)`

Parameter:

`p` — polynomial or expression

Overloading:

1

Summary:

factor calculates the factorization of a polynomial. The polynomial may be either a multivariate polynomial over the rationals or an univariate polynomial over a residue class ring `IntMod(p)` with a prime number p.

The argument of **factor** may be either a polynomial or an expression. An expression is converted into a polynomial (see function `poly`). **factor** returns FAIL if the expression cannot be converted.

`factor(a)` returns the list `[u, f1, e1, ..., fn, en]` with `a = u * f1^e1 * ...* fn^en`, where the `fi` are the irreducible factors and u is the content of a. The `fi` are pairwise different and primitive.

The function Factor returns an expression instead of a list. It is intended for interactive use, whereas **factor** is more useful while programming.

```
>> factor(3*x^4+5*x^3-3*x-1);

   [1, -x + x^2*3-1, 1, x + 1, 2]


>> factor(x^3+3*x^2+5*x+9);

   [1, x*5 + x^2*3 + x^3 + 9, 1]


>> Factor(poly(x^3+3*x^2+5*x+9, IntMod(13)));

   poly(x-3, [x], IntMod(13)) * poly(x-5, [x], IntMod(13)) *
   poly(x-2, [x], IntMod(13))


>> Factor(4*x^3+4*x^2*y+3*x*y+3*y^2);

   (x + y)*(y*3 + x^2*4)


>> Factor(2/3*x^2+1/3*x-1/3);

   1/3*(x + 1)*(x*2-1)
```

See also: poly, ifactor, sqrfree

■ **fclose** – closes a file

Calling sequence:

`fclose(integer)`

Parameter:

`integer` — positive integer

Overloading:

all

Summary:

`fclose` closes the file indicated by the file descriptor `integer`. The corresponding file must have already been opened with `fopen`. `fclose` is a kernel function.

```
>> fclose(123);
```

See also: `fopen`

---

fft – fast-Fourier transformation

Calling sequence:

`fft(list, m)`

Parameter:

`list` — liste of numbers with 2^m entries
`m` — non-negative integer

Summary:

`fft` calculates the fast-Fourier transformation of the passed list `list`. The list must contain 2^m numerical elements. The result is also a list with 2^m numbers.

```
>> A := [1.0, 2.0, 3.0, 4.0]: fft(A, 2);

    [1.0e1, - 2.0 + 2.0*I, - 2.0, - 2.0 - 2.0*I]


>> A := [2.0, 3.0, -1.0, 4.0, 2.3, 1.3, 4.3, 3.7]: fft(A, 3);
```

```
[1.960000000e1,   0.6899494936 + 3.885786437 I,
1.0 + 3.400000000 I,   - 1.289949493 - 6.714213562 I,
- 4.400000000,   - 1.289949493 + 6.714213562 I,
1.0 - 3.400000000 I,   0.6899494936 - 3.885786437 I]
```

See also: `ifft`

◼ `finput` – reads expressions from a file

Calling sequence:

```
finput(integer ⟨ , identifier ... ⟩)
finput(string ⟨ , identifier ... ⟩)
```

Parameter:

`integer`	—	positive integer
`identifier`	—	identifier
`string`	—	character string

Summary:

`finput` reads expressions from a file. The first argument specifies the file. The other arguments of `finput` must be identifiers. The expressions are read from the file in their sequence and assigned to these identifiers without being evaluated. When called, the identifiers are also not evaluated. The result of `finput` is the final read expression.

If one of the expressions cannot be read then its corresponding identifier and all those following are assigned no values. In this case the return value of `finput` is the element of the type *DOM_NULL*.

If `finput` is called without identifiers as arguments then only one expression is read. This is the return value of the function. If no expression can be read then the return value is the element of the type *DOM_NULL*.

An expression read from a text file can be more than one line long. It is only finished when the expression is syntactically complete. Additionally, the expression must be closed with either a semicolon or a colon.

The file is specified as follows:

- If a positive integer `integer` is given as the first argument then this is considered the file descriptor. The corresponding file must have previously been opened with `fopen`.

After the data have been read the file remains open. When next reading from this file, reading continues at the current position.

- If a character string is given with the argument **string** then the file with this name is opened. The file can be a text or a binary file. It is searched for in the current directory.

  The data are read starting at the beginning of the file. After the data have been read, the file is closed.

- In MacMuPAD, if an empty character string is given in the argument **string** then a dialog is opened in which the user can choose a file.

If data are to be read subsequently from a file through several calls of **finput** then a file descriptor must be used.

In contrast to most other functions, arguments which are expression sequences, are not flattened. **finput** is a kernel function.

```
>> fprint("test", 1, 2, 3, 4):
   finput("test", d1, d2): d1, d2;

   1, 2


>> fp:=fopen("test"): finput(fp, d1, d2):
   finput(fp, d1, d2): d1, d2;

   3, 4
```

See also: `fclose, fopen, fread, ftextinput, input, read`

---

float – evaluates to a floating-point number

Calling sequence:

`float(expr)`

Parameter:

`expr` — expression

Overloading:

all

Summary:

`float` evaluates sub-expressions of `expr` as far as possible to floating-point numbers. Arithmetic expressions consisting of integers, rational and complex numbers, floating-point numbers, the constants `EULER`, `PI` and `E` or functions such as `exp` and `sin` are evaluated.

If the argument is a table or an array then the elements are not evaluated to floating-point numbers. Elements of tables and arrays are only evaluated when they are accessed with their indices (see section 2.3.11).

The accuracy of the evaluation is determined by the global variable `DIGITS`. `DIGITS` has the default value 10, which means that floating-point numbers are printed with ten digits. The user can set the value of `DIGITS` to between 0 and $2^{31} - 1$. `float` is a converting function of the kernel.

```
>> float(E+sin(PI/4));

   3.425388609


>> float(sin(2*a)+5*c/3);

   c*1.666666666 + sin(a*2.0)


>> float(2/3*I+3);

   3.0 + 0.6666666666*I
```

See also: `DIGITS`

◼ floor – rounding-off

Calling sequence:

`floor(expr)`

Parameter:

`expr` — expression

Summary:

`floor` returns the largest integer that is smaller than or equal to the one given by **expr**. If **expr** is a complex number then an error is returned. if the argument cannot be evaluated to a number, `floor` returns the unevaluated function call.

If the exponents of real numbers are too large then a call of `floor` can lead to a loss of accuracy which is not identified as an error. `floor` is a converting function of the kernel.

```
>> floor(1.2);

    1
```

```
>> floor(-8/3);

    -3
```

```
>> floor(x+3/2);

    floor(x + 3/2)
```

See also: ceil, round, frac, trunc

---

fopen − opens a file

Calling sequence:

```
fopen(string)
fopen( ⟨ type, ⟩ string, mode )
```

Parameter:

type   —   one of the options **Bin** or **Text**
string —   character string
mode   —   one of the options **Write** or **Append**

Summary:

`fopen` opens the file with the name **string**. If, in MacMuPAD, an empty character string is given by the argument **string** then a dialog is opened in which the user can choose a file.

fopen returns a positive integer as the return value; the so-called *file descriptor*. This number is used to identify the file in later operations. **FAIL** is returned if the file cannot be opened. The file can be closed with the function **fclose**.

- If no further options are given then the file is opened for reading. **fopen** automatically identifies if the file is text or binary.

- If one of the options **Write** or **Append** is given then the file is opened for writing. If necessary, a non-existing file is newly created. With **Write** the old contents of the file are overwritten. With **Append** the new data is appended at the end of the old data, i.e. the old data is not lost.

  If the variable **WRITE_PATH** has a value then the file is created in the given directory, otherwise it is created in the current directory.

  In MacMuPAD, if **fopen** is called with the option **Write** while in the interactive level then the user is warned that a file is about to be overwritten. He can then cancel the operation.

While opening the file with **Write** or **Append** optionally one of the options **Bin** or **Text** can be given. Then the file is opened as a text, or respectively, a binary file. If no option is given and a new file is created, it is opened as a binary file.

If an existing text file is opened with **Append** and **Bin** then this leads to an error. Analogue when an existing binary file is opened with **Append** and **Text** an error is also returned. **fopen** is a kernel function.

```
>> f := fopen(Bin, "test.mb", Append);
```

```
        123
```

See also: `fclose, finput, fprint, ftextinput, protocol, read, write`

■ **fprint − writes to a file**

Calling sequence:

```
fprint(⟨ Unquoted, ⟩ integer, expr ...)
fprint(⟨ Unquoted, ⟩ ⟨ type, ⟩ string, expr ...)
```

Parameter:

```
Unquoted   —   option
integer    —   non-negative integer
type       —   one of the options Bin or Text
string     —   character string
expr       —   expression
```

Summary:

fprint is used to write text or expressions to a file. The argument expr is evaluated and the result written. If several arguments are specified they are evaluated sequentially and written. Text output occurs without the Pretty-Printer. Usually there is a colon between each value so that the output can be read-in again using finput. After every output a new line is started.

If the option Unquoted is given as the first argument then character strings without quotation marks are written. Furthermore, at the end of a line no colon occurs. Generally data output in this way *cannot* be read-in again by using finput.

Not only expressions can be used as arguments, but also any statement can be used, e.g. also assignments or procedure definitions. Statements must, however, be additionally bracketed, e.g. as in fprint("test.mb",(a:=2)).

fprint returns the element of the type *DOM_NULL* as the result.

The file can be specified by its name or with a file descriptor:

- If a positive integer integer is given then this is considered the file descriptor. The corresponding file must have already been opened with fopen.

- If the value 0 is given as a non-negative integer then output takes place on the screen.

- If a character string is given with the argument string then the file with this name is opened and data already in the file is overwritten. If this file does not exist then it is newly created.

  In MacMuPAD, if fprint is called on the interactive level then the user is warned that a file is about to be overwritten. The operation can then be cancelled.

  If, additionally, one of the options Bin or Text is given then the file is opened as a binary or, respectively, a text file. Without one of these options the file is opened as a binary file.

  After data have been written the file is closed.

- In MacMuPAD, if an empty character string is given by the argument string then a dialog is opened in which the user can choose a file.

`fprint` is a kernel function.

```
>> fprint(Bin, "test1.mb", (d:=5), d*3):
   finput("test1.mb", e, f): e, f;

     5, 15


>> fprint(Unquoted, Text, "test2.mu", "Hello ", " World !"):
   ftextinput("test2.mu");

     "Hello : World !"
```

See also: `fclose`, `fopen`, `print`, `protocol`, `write`

■ frac – fractional part of a number

Calling sequence:

`frac(expr)`

Parameter:

`expr` — expression

Summary:

`frac(x)` returns the value $x - \lfloor x \rfloor$, i.e. the fractional part of **x**. This is a number from the interval $[0, 1[$. With complex arguments `frac` returns an error. if the argument cannot be evaluated to a number, `frac` returns the unevaluated function call. `frac` is a kernel function.

```
>> frac(8/3);

     2/3


>> frac(-2.3);

       0.7000000000


>> frac(x);
```

```
frac(x)
```

See also: `ceil, floor, round, trunc`

`fread` – **reads and executes a file**

Calling sequence:

```
fread(integer)
fread(string)
```

Parameter:

integer   —   positive integer
string   —   character string

Summary:

`fread` reads-in the contents of a file and executes the statements in the file.
`fread` is similar to `read`, the only difference is that `fread` does not search for
files in the directories given by `READ_PATH` and `LIB_PATH`. `fread` is a kernel
function.

```
>> a:=3: b:=5: write("testfile.mb",a,b): reset(): a, b;
   fread("testfile.mb"): a, b;

   a, b
   3, 5
```

See also: `LIB_PATH, read, READ_PATH`

`ftextinput` – **reads text line-by-line from a file**

Calling sequence:

```
ftextinput(integer 〈 , identifier ... 〉)
ftextinput(string 〈 , identifier ... 〉)
```

Parameter:

integer   —   positive integer
identifier   —   identifier
string   —   character string

Summary:

`ftextinput` reads-in character strings line-by-line from a text file. The first argument specifies the file. Any other arguments of `ftextinput` must be identifiers. Individual text lines are read-in from the file in sequence and assigned to these identifiers. The result of `ftextinput` is the final read-in text line.

If a line cannot be read-in then the corresponding identifier and those following are not assigned any lines. In this case the result of `ftextinput` is the element of the type *DOM_NULL*.

If `ftextinput` is called without identifiers as arguments then only one line is read-in which is then the return value of the function. If the line cannot be read-in then the element of the type *DOM_NULL* is returned.

The file is specified as follows:

- If the first argument is a positive integer `integer` then this is considered a file descriptor. The corresponding text file must have already been opened with `fopen`.

  After the text lines have been read-in the file remains open. The next reading from the file with this descriptor begins at the current position.

- If a character string is given as the argument `string` then the file with this name is newly opened. The file is searched for in the current directory.

  The text lines are read-in starting at the beginning of the file. After the data have been read the file is closed.

- In MacMuPAD, if an empty character string is given as the argument `string` then a dialog is opened in which the user can select a file.

When a file name is given then only the lines at the beginning of a file can be read. If successive text lines are to be read from a file with several calls of `ftextinput` then a file descriptor must be used.

In contrast to most other functions, arguments which are expression sequences, are not flattened. `ftextinput` is a kernel function.

```
>> fprint(Unquoted, Text, "test", "AA\nBB\nCC\nDD\n"):
   ftextinput("test", z1, z2): z1, z2;

     "AA", "BB"


>> fp:= fopen("test"): ftextinput(fp, z1, z2):
   ftextinput(fp, z1, z2): z1, z2;
```

```
    "CC", "DD"
```

See also: `fclose, fopen, finput, fread, read, textinput`

---

fun – generates pure functions

Calling sequence:

```
fun(expr)
```

Parameter:

`expr` — expression

Summary:

`fun(expr)` generates a function from the expression `expr` with the variables `args(1) ... args(args(0))` and returns it as a so called pure function. The expression `expr` is not evaluated.

A pure function is executed in a similar manner to a procedure. However, it has no formal parameters or local variables. (This means that it also does not have any local environment-variables.) Any arguments passed to a pure function can only be accessed by using `args`.

If a pure function is called with too few arguments then an error is returned. Any extra arguments are ignored.

Compared to the usual functions, pure functions are much faster.

```
>> f:= fun(sin(args(1)^2 + args(2)^2)): f(x,y);

    sin(x^2 + y^2)
```

See also: `args, func`

---

func – generates functions from expressions

Calling sequence:

```
func(expr ⟨ , ident ... ⟩)
```

Parameter:

```
expr    —    expression
ident   —    identifier
```

Summary:

`func(expr, x1,...,xn)` generates a function with the variables `x1` ... `xn` from the expression `expr` and returns this as a functional expression or a pure function (see `fun` and section 2.7). `func` allows an easier defining of pure functions as `fun`.

Firstly, `func` tries to generate a functional expression from `expr`. If this is not possible then a pure function is created from `expr`. With this all identifiers `x1` ... `xn` in `expr` are substituted by arguments of pure functions.

If no variables are given then they are determined with the function `indets`. In this case the order of the variables is random. `func` does not evaluate its arguments.

```
>> func(sin(x), x);

    sin


>> func(x + 1);

    id + 1


>> F:= func(f(x, y), x, y);
   F(2,a);

    f
    f(2, a)


>> G:= func(f(x, y), x);
   G(2);

    func(f(x, y), x)
    f(2, y)
```

See also: `fun`, `indets`

■ `funcattr` – **accesses the attributes of a function environment**

Calling sequence:

```
funcattr(func_env, expr1)
funcattr(func_env, expr1, expr2)
```

Parameter:

```
func_env        —   function environment
expr1, expr2    —   expressions
```

Summary:

`funcattr(func_env, expr1)` returns the attribute of the function environment **func_env** with the index **expr1**. If the function does not have an attribute with this index then it returns the value **FAIL**.

`funcattr(func_env, expr1, expr2)` enters the attribute **expr2** under the index **expr1** in a copy of the function environment **func_env**. The so-created new function environment is then returned. If **func_env** does not have the type *DOM_FUNC_ENV*, then the argument is converted into a function environment.

The attributes of a function environment can be used to store special information about the represented function. E.g. if the function **diff** does not know how to differentiate another function, it searches under the index **"diff"** for an attribute that can calculate the derivative. Similarily, the function **expand** searches under the index **"expand"** for an attribute that can perform an expansion.

The attributes of a function are stored in a table in the 3rd entry of a function environment. **funcattr** is a kernel function.

```
>> funcattr(poly(x^2,[x]), polynom, TRUE);

   poly(x^2, [x])


>> tan := funcattr(tan, "diff", proc(x, var) begin
      diff(op(x, 1), var)*(1+tan(op(x,1))^2)
   end_proc):
   diff(tan(2*x),x);

   2*tan(2*x)^2 + 2


>> tan := funcattr(tan, "expand", proc(x) begin
      sin(x)/cos(x)
```

```
  end_proc):
  expand(tan(x));

  sin(x) / cos(x)
```

■ func_env – defines function environments

Calling sequence:

```
func_env(func_1, func_2, ⟨ table ⟩)
```

Parameter:

func_1, func_2	—	any expressions
table	—	table

Summary:

func_env allows the definition of function environments. These are elements of the domain *DOM_FUNC_ENV*. Function environments appear as the zero operand of expressions. The first argument of func_env determines the evaluation of the expression, the second argument determines the output of the expression. With the 3rd argument the created function environment can be given attributes to which functions such as **diff** and **expand** react.

Function environments are described in more detail in section 2.3.16.

Only advanced users should define or modify function environments.

The examples below show the function environments of the system functions _mult and _assign. func_env is a kernel function.

```
>> func_env(built_in(815, NIL, "_mult", NIL),
   built_in(1100, 13, "*", "_mult")):


>> func_env(built_in(900, NIL, "_assign", NIL),
   built_in (1102, 17, NIL, "_assign")):
```

See also: built_in, funcattr

■ gamma – Gamma function

Calling sequence:

`gamma(x)`

Parameter:

x   —   expression

Summary:

The Gamma function is defined as

$$\text{gamma}(\mathbf{x}) = \Gamma(x) = \int_0^{\infty} \exp(-t)\, t^{x-1}\, dt$$

If the argument x is a non-negative integer then is $\Gamma(x) = (x - 1)!$. The Gamma function can be seen as a generalization of the factorial function. The Gamma function is also defined for complex arguments. **gamma** returns the function call with evaluated arguments, if the argument cannot be evaluated to a number. **gamma** is a kernel function.

```
>> gamma(3.4);

   2.981206426
```

```
>> gamma(23);

   1124000727777607680000
```

```
>> gamma(I);
   float(%);

   gamma(I)
   - 0.1549498283 - 0.4980156681*I
```

See also: `igamma, fact`

■ **gcd − greatest common divisor of polynomials**

Calling sequence:

```
gcd(poly ...)
gcd(expr ...)
```

Parameter:

poly — polynomial
expr — expression

Overloading:

all

Summary:

gcd calculates the greatest common divisor of any number of polynomials. The polynomials may either have the rationals, a residue class ring IntMod(p) with a prime number p or a domain as their coefficient ring.

If the arguments are polynomials with coefficients from a domain then the domain must have the methods gcd and divex ("divide exact"). The method gcd must return the greatest common divisor for any number of domain elements. The method divex must divide an domain element by another; if the domain elements cannot be divided then the method must return FAIL.

The arguments of gcd may be either polynomials or expressions. Expressions are converted into polynomials (see function poly). gcd returns FAIL if an expression cannot be converted.

```
>> gcd(6*x^3 + 9*x^2*y^2, 2*x + 2*x*y + 3*y^2 + 3*y^3);

   2*x + 3*y^2
```

```
>> a:= poly(2*x^2  - 4*x*y - 2*x + 4*y, [x,y], IntMod(17)):
   b:= poly(x^2*y  - 2*x*y^2, [x,y], IntMod(17)):
   gcd(a, b);

   poly(x - 2*y, [x,y], IntMod(17))
```

See also: content, igcd, lcm, poly

■ **gcdex – extended gcd algorithm for polynomials**

Calling sequence:

```
gcdex(a, b, x)
gcdex(p, q)
```

Parameters:

a, b — expressions
x — unknown
p, q — univariate polynomials

Summary:

gcdex(a,b,x) returns a tuple g,s,t where g is the gcd of a and b with respect to the variable x, and s,t are polynomials such that $g = sa + tb$, with degree(s,x)<degree(b,x), and degree(t,x)<degree(a,x).

```
>> gcdex(poly(x^3+1),poly(x^2+2*x+1));

    poly(x + 1, [x]), poly(1/3, [x]), poly(x*(-1/3) + 2/3, [x])
```

```
>> gcdex(x^3+a,x^2+1,x);

    1, (a + x)/(a^2 + 1), 1/(a^2 + 1)*(-a*x-x^2 + 1)
```

See also: gcd, igcdex, pdioe, poly

---

genident – creates a free identifier

Calling sequence:

genident(⟨ string ⟩)

Parameter:

string — string

Summary:

genident creates a new free identifier of the form X.i with a prefix X and a non-negative integer i. The identifier will not be contained in any expression existing at the time when genident is called.

If a string is given as argument then it is used as prefix. The default prefix is X. genident is a kernel function.

```
>> x := genident(), genident("Y"), genident();
```

```
X1, Y1, X2
```

See also: `anames`

■ **genpoly** – creates a polynomial using a b-adic expansion

Calling sequence:

```
genpoly(poly, integer_1, expr_2)
genpoly(expr ⟨ , indets ⟩, integer_1, expr_2)
genpoly(integer, integer_1, expr_2)
 indets  ⟶  [ expr_1 ...]
```

Parameter:

`integer`	—	integer
`integer_1`	—	integer larger than 1
`poly`	—	polynomial
`expr, expr_1, expr_2`	—	expressions

Overloading:

1

Summary:

`genpoly(n,b,x)` creates a polynomial in the variable x from an integer n by using a b-adic expansion. b must be an integer larger than 1.

The b-adic expansion of n is defined as $n = \sum_{i=0}^{m} c_i b^i$, such that the c_i are integers modulo b and the modulus is calculated symmetrically. Through this expansion the polynomial $\sum_{i=0}^{m} c_i x^i$ is created. The polynomial is defined over the coefficient ring `Expr`.

If the first argument of `genpoly` is a polynomial then this must have the coefficient ring `Expr` and may only have integer coefficients. Furthermore, the third argument must not be a variable of the polynomial. In this case each coefficient is b-adic expanded to a polynomial in x. The result is a polynomial with the variable x, followed by the variables of the given polynomial (x is the main variable of the new polynomial).

The first argument may also be an expression with an optional list of variables. If the argument is an expression then this is converted into a polynomial (see function `poly`).

If the first argument is an integer then the result is a polynomial. If it is a
polynomial or an expression then the result has the same type. **genpoly** is a
kernel function.

```
>> genpoly(15, 7, x);

   poly(x*2 + 1, [x])

>> genpoly(15*y^2 - 6*y + 3, 7, x);

   x*y^2*2 - x*y + y^2 + y + 3

>> genpoly(poly(15*y^2 + 8*z, [y, z]), 7, x);

   poly(x*y^2*2 + x*z + y^2 + z, [x, y, z])
```

See also: `evalp, poly`

---

global – accesses a net variable

Calling sequence:

```
global(identifier)
global(identifier, expr)
```

Parameter:

```
identifier  —  identifier
expr        —  expression
```

Summary:

`global(identifier)` returns the value of the net variable with the name
`identifier`. If the net variable with this name is undefined then the function
returns the unevaluated function call. `identifier` is not evaluated.

`global(identifier, expr)` assigns the value `expr` to the net variable with
the name `identifier`. `expr` is evaluated before the assignment. `identi-
fier` is not evaluated and is considered as the name of the net variable. This
function is described in more detail in section 2.8.2.1. `global` is a kernel
function.

■ **has − tests for the existence of a given subexpression**

Calling sequence:

```
has(expr, expr_1)
has(expr, list)
has(expr, set)
```

Parameter:

```
expr, expr_1   —   expressions
list           —   list
set            —   set
```

Summary:

has tests if the expression **expr** contains the subexpression **expr_1**. Only complete subexpressions that can be replaced with **subs** are found.

If the second argument of **has** is a list or set then the elements are tested to see if one of them is a subexpression of **expr**.

```
>> has(a+b+c, b);

    TRUE
```

```
>> has(a+b+c, a+b);

    FALSE
```

```
>> has((a+b)*c, [a+b, d]);

    TRUE
```

```
>> has(cos(x + sin(y)^2), sin);

    TRUE
```

```
>> has([f(a) + 2, sin(x), 3], {f, g});
```

    TRUE

See also: contains, subs

---

help, ? – displays a help text

Calling sequence:

help("keyword")
?keyword

Parameter:

keyword — keyword

Summary:

help("keyword") displays a help text with information about the keyword keyword. The exact form of the output depends on the computer. In the terminal version information is displayed as ASCII text. In xmupad, the Open-Windows interface to MuPAD, the user can obtain information from the hypertext on-line documentation. See also section 2.1.1.

Instead of help("keyword"), the user can also simply enter ?keyword. However, the ? command is not a MuPAD function and therefore cannot be used in expressions. It can only be entered interactively and in a separate line. Note that neither the keyword need be set in quotation marks nor the command closed with a semicolon.

The keyword can contain the "wildcards" ? and * : ? stands for any or a missing character, * stands for any (also an empty) series of characters. ?*normal* returns normal groebner::normalf linalg::normalize stats::normal for example.

In the terminal version the user only has access to a short description of system functions and system variables. The hypertext on-line documentation also includes information about types and control structures, e.g. if or for. The command anames(0); gives an overview of the system functions and ?* gives an overview of the help pages available in MuPAD.

Commands must be closed with a semicolon or colon. MuPAD executes this command when the user presses the <Return> or <Enter> key. The command can also be more than one line long. More than one command can be entered in one line if each command is ended with a semicolon. Two exceptions to this are the ? and ! commands. quit ends MuPAD.

For any error found in the system send an e-mail to

MuPAD-bugs@mathematik.uni-paderborn.de

General information about MuPAD, e.g. FAQ service, new developments etc., can be found on the MuPAD WWW home page

http://math-www.uni-paderborn.de/MuPAD/

See also: `anames`

■   **`history` – displays the entries of the history table**

Calling sequence:

`history()`

Parameter:

—   none

Summary:

`history` displays the current contents of the history table and returns the "empty" datum of the type *DOM_NULL* as the return value. The output of `history` is series of pairs consisting of the number in the history table and the corresponding entry. `history` is a kernel function.

```
>> a:= 7: b: history();

    1  7
    2  b
```

See also: `last, HISTORY`

■   **`HISTORY` – length of the history table**

Calling sequence:

```
HISTORY := [ integer_1, integer_2 ]
HISTORY := integer
```

Parameter:

`integer, integer_1, integer_2`   —   non-negative integer

Summary:

HISTORY defines the length of the history table. HISTORY also defines the number of results that can subsequently be accessed with `last`:

- If the value of HISTORY is given as a list of two integers then the first integer gives the length of the history table on the interactive level and the second gives the length of the table used in procedures.

- If only one integer is given as the value of HISTORY then its meaning depends on the current context. If the assignment is given on the interactive level then the integer gives the length of the history table on the interactive level. However, if the assignment is given in a procedure then this sets the length of the history table used for procedures.

With the assignment `HISTORY := NIL;`, HISTORY is set to its default value of [20, 3]. The highest possible value for both components of HISTORY is $2^{31} - 1$. HISTORY is an environment variable.

```
>> f := proc() local HISTORY; begin
       HISTORY := 5; a; b; c; d; e; last(5)
   end_proc:
   f();


      a
```

See also: `history, last`

■ **hold** − **prevents evaluation**

Calling sequence:

`hold(expr)`

Parameter:

`expr` — expression

Summary:

`hold` prevents evaluation of the expression `expr`. More exactly, the expression `hold(expr)` is evaluated to the unchanged expression `expr`. `hold` is a kernel function.

```
>> 2+3*0-7+0, hold(2+3*0-7+0);

   -5, 2+3*0-7+0

>> a:=b: b+a+b, hold(b+a+b);

   b*3, b+a+b
```

See also: `eval, val`

■ `icontent` – integer content of a polynomial

Calling sequence:

```
icontent(poly)
icontent(expr 〈 , indets 〉)
 indets  ⟶  [ expr_1 ...]
```

Parameter:

```
poly           —   polynomial
expr, expr_1   —   expressions
```

Overloading:

1

Summary:

`icontent` calculates the content of a polynomial with integer or rational coefficients.

- A polynomial with integer coefficients has the greatest common divisor of the coefficients as its content.
- A polynomial with rational coefficients has the greatest common divisor of the numerators of the coefficients divided by the least common multiple of the denominators as its content.

The first argument of `icontent` is allowed to be a polynomial or an expression and optionally a list of indeterminates. An expression is converted into a polynomial in the specified indeterminates (see function `poly`). `icontent` returns `FAIL` if the expression cannot be converted into a polynomial. `icontent` returns an error if the polynomial has no rational coefficients. `icontent` is a kernel function.

```
>> icontent(6*x*y+9*y^2+21, [x,y]);

    3
```

```
>> p := poly(6/7*x*y+9/4*y+12/14): icontent(p);

    3/28
```

```
>> icontent(0, [x,y]);

    0
```

See also: poly

---

id – the identical function

Calling sequence:

```
id(expr ...)
id()
```

Parameter:

expr — expression

Summary:

id is the identical function. id evaluates its arguments and returns them as an expression sequence. id () returns the element of the type *DOM_NULL*. id is a kernel function.

```
>> a:= 2: id(a, b, 4+2);

    2, b, 6
```

```
>> domtype(id());

    DOM_NULL
```

---

ifactor – factorization of integers

Calling sequence:

```
ifactor(integer)
```

Parameter:

```
integer   —   integer
```

Summary:

`ifactor` returns the factorization of `integer`. The result is the list `[s, p1, e1, ..., pn, en]`. Whereby `pi` is the i-th prime factor and `ei` the corresponding exponent; `s` is the sign of `integer` (1 or -1). The prime factors are sorted in ascending order. `ifactor(0)` returns 0.

`ifactor` returns an error when the argument is a number but not an integer. `ifactor` returns the function call with an evaluated argument if the argument is not a number. `ifactor` is a kernel function.

```
>> ifactor(11);

    [1, 11, 1]


>> ifactor(-8434536348);

    [- 1, 2, 2, 3, 1, 7, 1, 100411147, 1]
```

See also: `isprime`

■ ifft – inverse fast Fourier transformation

Calling sequence:

```
ifft(list, m)
```

Parameter:

```
list   —   list of numbers with 2^m entries
m      —   non-negative integer
```

Summary:

`ifft` calculates the inverse fast Fourier transformation of the given list `list`. The list must contain 2^m numerical elements. The result is also a list with 2^m numbers.

```
>> A := [1.0, 2.0, 3.0, 4.0]:
   fft(A, 2):
   ifft(%, 2);

     [1.0, 2.0, 3.0, 4.0]
```

See also: fft

■ **igamma** – incomplete Gamma function

Calling sequence:

igamma(x, y)

Parameter:

x, y   —   expressions

Summary:

The incomplete Gamma function is defined as

$$\text{igamma}(\mathbf{x}, \mathbf{y}) = \int_{y}^{\infty} \exp(-t)\, t^{x-1}\, dt$$

At the moment, numerical calculation is only available for positive real numbers.

igamma(x, 0) is equivalent to gamma(x). igamma returns the function call with evaluated arguments, if the argument cannot be evaluated to a number. igamma is a kernel function.

```
>> igamma(2,4);

     exp(- 4)*5
```

```
>> igamma(2.0,4);

     0.9157819444e-1
```

See also: gamma

---

igcd – greatest common divisor of integers

Calling sequence:

```
igcd(integer, integer ...)
```

Parameter:

```
integer  —  integer
```

Summary:

igcd returns the greatest common positive divisor of a sequence of integers. igcd returns 0 when all arguments are 0.

igcd returns an error when one of the arguments is a number but not an integer. If any argument is not a number, igcd returns the function call with evaluated arguments. igcd is a kernel function.

```
>> igcd(-10, 6);

     2


>> a:=4420,128,8984,488: igcd(a);

     4
```

See also: gcd, igcdex, ilcm

■ **igcdex – extended Euclidean algorithm for integers**

Calling sequence:

```
igcdex(integer_1, integer_2)
```

Parameter:

```
integer_1, integer_2  —  integers
```

Summary:

igcdex(a, b) calculates the greatest common positive divisor d of the numbers a and b as well as the integers x and y, for which d = a*x + b*y is valid, using the extended Euclidean algorithm. The expression sequence d, x, y is returned. igcdex(0, 0) returns the sequence 0, 1, 0.

If one of the arguments is a number but not an integer then igcdex returns an error. If any argument is not a number, igcdex returns the function call with evaluated arguments. igcdex is a kernel function.

```
>> igcdex(-10, 6);

    2, 1, 2

>> igcdex(3839882200, 654365735423132432848652680);

    109710920, -681651885490791809, 4
```

See also: `igcd, ilcm`

`ilcm` − least common multiple of integers

Calling sequence:

`ilcm(integer, integer ...)`

Parameter:

`integer` — integer

Summary:

`ilcm` calculates the least common positive multiple of a sequence of integers. `ilcm` returns 0 if one of the arguments is 0.

`ilcm` returns an error when one of the arguments is a number but not an integer. If any argument is not a number, `ilcm` returns the function call with evaluated arguments. `ilcm` is a kernel function.

```
>> ilcm(9, 6, 3);

    18

>> ilcm(-10, 6);

    30
```

See also: `igcd, igcdex`

`Im` − imaginary part of an expression

Calling sequence:

`Im(expr)`

Parameter:

`expr` — expression

Overloading:

All

Summary:

Im returns the imaginary part of any expression. For this the method `"Im"` from the domain of `expr` is used. If this domain does not have such a method then the function returns the expression `Im(expr)`.

Exceptions to this are expressions of the type `DOM_INT`, `DOM_RAT`, `DOM_FLOAT`, `DOM_COMPLEX` and `DOM_EXPR`, which are processed directly by the function Im.

For expressions of the type `DOM_EXPR` the determination of the imaginary part of an expression can be defined under the function attribute `"Im"` in the form of a procedure or a function. If the operator of `expr` has such an attribute then the procedure or function under the entry of the attribute `"Im"` is called with the operands of the argument `expr` and the result is returned.

If such a function attribute is not defined and the operator of `expr` is one of the function environments `_plus`, `_mult` or `_power`, then this is processed directly by the function Im and the result is returned. In all other cases the expression `Im(expr)` is returned.

Use the function `rectform` if you have unknowns which represent real or complex quantifiers (see examples below).

```
>> Im( 2*exp(1+I) );

    E*sin(1)*2


>> Im( u+v*I );

    Im(u) + Re(v)


>> Im( x+Re(2+3*I) );

    Im(x)
```

```
>> rectform( tan(x+I*y),{x,y} );

        sin(2 x)          /        sinh(2 y)        \
    -------------------- + | -------------------- | I
    cos(2 x) + cosh(2 y)   \ cos(2 x) + cosh(2 y) /


>> Im( % );

    sinh(y*2)/(cos(x*2) + cosh(y*2))
```

See also: conjugate, funcattr, Re, rectform

---

info – prints short information

Calling sequence:

info(expr)

Parameter:

expr   —   expression

Summary:

info prints a short descriptive information about expr onto the screen, if available.

One may add ones own informations to functions and domains: If expr is a function then info executes any attribute stored under the index "info" in the function environment of expr. If expr is a domain then info executes any domain entry stored under the index "info".

```
>> info(groebner);

    Library 'groebner': Calculation of Groebner-bases for
    polynomial ideals
    Interface:
    groebner::gbasis
    groebner::spoly
    groebner::normalf
```

See also: `help, userinfo`

■ `indets` – **determines the indeterminates of an expression**

Calling sequence:

`indets(expr ⟨ , `*opt*` ⟩)`

Parameter:

`expr`   —   expression
opt   —   one of the options `RatExpr` or `PolyExpr`

Overloading:

all

Summary:

`indets` returns the indeterminates of an expression as a set:

- A call in the form `indets(expr)` returns all indeterminates of the expression `expr`. Indeterminates are those identifiers with no values. Exceptions are the zero operands of sub-expressions.
- With a call in the form `indets(expr, RatExpr)` the expression `expr` is considered a rational expression. Non-rational sub-expressions like `sin(x)` or `f(a, b)` are considered indeterminates and are returned. In contrast sub-expressions such as `f(2, PI)` are considered constants even when the identifier `f` has no value. (The expression is constant because the operands of `f` are constant.)
- With a call in the form `indets(expr, PolyExpr)` the expression `expr` is considered a polynomial expression. Similar to `RatExpr` non-polynomial sub-expressions like `exp(t+1)` or `x^(-3)` are considered indeterminates.
- If `expr` is a list, set, table or an array then `indets` is used recursively on the individual elements.

`indets` is a kernel function.

```
>> e := 1/x[2] + f(PI/3) + sin(y)*y^3 + z^(-3):
   indets(e);
   indets(e, RatExpr);
   indets(e, PolyExpr);
```

```
{x, y, z}
{x[2], x, y, z, sin(y)}
{1/x[2], z^(-3), x, y, z, sin(y)}
```

```
>> indets( [ a, 2, {f(3), g(c)}, b+4 ] );

   {a, b, c}
```

See also: op

`index_val` – indexed access without evaluation

Calling sequence:

`index_val(expr, expr_1 ...)`

Parameter:

`expr, expr_1` — expressions

Overloading:

1

Summary:

`index_val` permits indexed access to arrays and tables without the returned value being subsequently evaluated. (Normally table and array elements are re-evaluated when the index operator "[ ]" is used; see section 2.4.11.) Sometimes it is desirable to prevent this — under certain conditions very time-consuming — evaluation. For this purpose the function `index_val` can be used.

Similar to the underline function `_index`, `index_val` enables the user access to tables, arrays, lists and expression sequences. (For reasons of consistency lists and expression sequences are treated as with `_index`.) The arguments are also given as with `_index`: the first argument is the data structure, to be accessed (a table, list expression sequence or an array). The other arguments are used as indexes.

If the user wishes to use the function `index_val` for indexed access then this can be achieved with the assignment `_index := index_val;`. After this assignment has been given then the function `index_val` is always used when indexed accessing with the index operator is carried out. `index_val` is a kernel function.

```
>> a := array(1..2, 1..2, (1,2)=x): x := 3:
   a[1,2], index_val(a, 1, 2);

     3, x
```

See also: val

■ **input – interactive input of MuPAD expressions**

Calling sequence:

```
input(expr ...)
input()
```

Parameter:

expr   —   character string or identifier

Summary:

input permits interactive requesting of values of identifiers. The arguments of **input** can be any series of character strings and identifiers. The character strings are displayed in sequence; for each identifier an input prompt in form of a prompt character is given. Here the user can enter a MuPAD expression which is then assigned to the corresponding identifier without the expression being evaluated. The return value of **input** is the last expression entered by the user.

If **input** is called without arguments then only a prompt is given and the user is expected to enter an expression which is then the return value.

Input must occur exactly as on the interactive level and may consist of more than one line. Input is only complete when the expression is syntactically complete. Additionally, the expression must be closed with a semicolon or colon.

Thus, with **input("Input x:", x)** firstly the text "Input **x:**" and then the prompt ">>" are displayed. The user may then enter an expression for **x**, e.g. **12;**, which is then assigned to **x**. The return value is the value given by the user: **12**.

The arguments of **input** are not evaluated. **input** is a kernel function.

```
>> input("Input degree ", deg);
```

```
    Input degree >>2;
```

```
>> input("x ? ", x, "y ? ", y);
```

```
    x ? >>1.2;
```

```
    y ? >>3*x;
```

```
>> a:=input();
```

```
    >>"test":
```

See also: `textinput`

`insert_ordered` – inserts elements in an ordered list

Calling sequence:

`insert_ordered(expr, list ⟨ , func ⟩)`

Parameter:

expr   —   expression
list   —   ordered list
func   —   Boolean function

Summary:

`insert_ordered` inserts the expression `expr` in the ascending ordered list `list` according to the ordering function `func` and returns a new list. The call `func(a,b)` must return TRUE if a < b holds, according to the order of the list.

If no ordering function is given then the system function `_less` is used for the comparison.

The list is not tested to see if it is in ascending order.

```
>> insert_ordered(4, [1, 4, 7]);
```

```
    [1, 4, 4, 7]
```

See also: sort

■ int – definite and indefinite integration

Calling sequence:

```
int(f, x)
int(f, x=a..b)
```

Parameters:

```
f, a, b   —   expressions
x         —   identifier
```

Summary:

int(f,x) returns the antiderivative of the expression **f** with respect to **x**, or the unevaluated function call if no antiderivative in terms of elementary functions exists. int(f,x=a..b) returns the definite integral of **f** from **a** to **b**. If a and b are constants a numerical evaluation of the integral is obtained by `float(int(f,x=a..b))`.

For details of the algorithm see:

M. Bronstein. A Unification of Liouvillian Extension. AAECC Applicable Algebra in Engineering, Communication and Computing. 1:5–24,1990.

M. Bronstein. The Transcendental Risch Differential Equation. Journal of Symbolic Computation. 9:49–60,1990.

H.I. Epstein and B.F. Caviness. A Structure Theorem for the Elementary functions and Its Application to the Identity Problem. International Journal of Computer and Information Science. 8:9-37,1979.

K.O. Geddes, S.R. Czapor and G. Labahn. Algorithms for Computer Algebra. 1992.

```
>> int(1/x/ln(x),x);

    ln(ln(x))


>> int(1/(x^2-8),x);

        1/2                 1/2     1/2                 1/2
       2     ln(x + 2 2   )     2       ln(x - 2 2   )
     - ------------------- + ---------------------
               8                         8
```

```
>> int(int(int(1,z=0..c*(1-x/a-y/b)),y=0..b*(1-x/a)),x=0..a);

   a*b*c*1/6
```

```
>> int(sin(cos(x)),x=0..1);

   int(sin(cos(x)), x = 0..1)
```

```
>> float(%);

   0.7386429979
```

See also: diff

---

isprime – prime number test

Calling sequence:
isprime(integer)

Parameter:
integer   —   integer

Overloading:
all

Summary:

isprime is a stochastical prime number test. The function returns TRUE when the integer integer is either a prime number or a strong pseudo prime number for ten random bases. In all other cases the value FALSE is returned.

isprime returns an error if its argument is a number but not an integer. if the argument cannot be evaluated to a number, isprime returns the unevaluated function call. isprime is a kernel function.

```
>> isprime(989999);

   TRUE
```

```
>> isprime(0);
```

```
    FALSE
```

```
>> isprime(-13);
```

```
    FALSE
```

See also: `nextprime, ithprime`

◼ `iszero` – tests if value is zero

Calling sequence:

```
iszero(poly)
iszero(expr)
```

Parameter:

```
poly  —  polynomial
expr  —  expression
```

Overloading:

1

Summary:

If the argument is a polynomial, `iszero` returns TRUE, if `poly` is the zero polynomial. The result is only valid if the coefficients of the polynomial are in normal form (i.e. if zero has a unique representation in the coefficient ring).

If the argument is not a polynomial then `iszero` returns TRUE if the argument is the integer 0 or the floating-point number 0.0.

`iszero` is a kernel function.

```
>> p := poly(x^2+y, [x]): iszero(p);
```

```
    FALSE
```

```
>> iszero(poly(0,[x,y]));
```

    TRUE

`ithprime` – returns the i-th prime number

Calling sequence:

`ithprime(integer)`

Parameter:

`integer`   —   positive integer

Overloading:

all

Summary:

`ithprime(i)` returns the i-th prime number where i is a positive integer.

Internally, `ithprime` uses a table whose length is dependent on the system used. E.g. on a Sun4 `ithprime` can only find prime numbers smaller than 1,000,000. `ithprime` returns an error when the searched for prime number is not in the table.

`ithprime` also returns an error if the argument is a number but not a positive integer. `ithprime` returns the function call with evaluated arguments if the argument is not a number. `ithprime` is a kernel function.

```
>> ithprime(1);

   2


>> ithprime(2457);

   21911
```

See also: `isprime, nextprime`

`last, %` – accesses the last values calculated

Calling sequence:

```
last(integer)
%integer
```

Parameter:

`integer` — positive integer

Overloading:

all

Summary:

`last(n)` returns the value of the n-th calculation in the history table. Thus, `last(1)` returns the last value, `last(2)` the penultimate one, `last(3)` the last but second one, etc. Instead of `last(1)` or `last(3)` the user can also write more briefly `%1` or `%3`. Instead of `last(1)` or `%1` the user can also write even more briefly `%`.

With `%` the argument must be a positive integer, with `last` it can also be an expression which evaluates to an integer.

The environment variable `HISTORY` gives the number of results that can be accessed.

A value returned by `last` or `%` is not re-evaluated.

Be careful with using `last` in procedures since this is not considered a good style of programming. Furthermore, `last` may not be supported in future releases of MuPAD.

`last` returns an error when the argument is not a positive integer. `last` is a kernel function.

```
>> c:=a+b+a: a:=b: last(2);

   a*2+b

>> c:=a+b+a: a:=b: c: %1;

   b*3
```

See also: `HISTORY, history`

■   **lcm – least common multiple of polynomials**

Calling sequence:

```
lcm(poly ...)
lcm(expr ...)
```

Parameter:

```
poly  —  polynomial
expr  —  expression
```

Overloading:

all

Summary:

lcm calculates the least common multiple of any number of polynomials. The polynomials may either have the rationals, a residue class ring IntMod(p) with a prime number p or a domain as their coefficient ring. If the coefficient ring is a domain then the same conditions as with the function gcd apply.

The arguments of lcm may be either polynomials or expressions. Expressions are converted into polynomials (see function poly). lcm returns FAIL if it is not possible to convert an expression into a polynomial.

```
>> lcm(x^3 - y^3, x^2 - y^2);

   -x^4 - x^3*y + x*y^3 + y^4


>> a:= poly(x^2 - y^2, [x,y], IntMod(17)):
   b:= poly(x^2 - 2*x*y + y^2, [x,y], IntMod(17)):
   lcm(a, b);

   poly(x^3 - x^2*y - x*y^2 + y^3, [x, y], IntMod(17))
```

See also: content, gcd, ilcm, poly

---

lcoeff – the leading coefficient of a polynomial

Calling sequence:

```
lcoeff(poly ⟨ , order ⟩)
lcoeff(expr ⟨ , indets ⟩ ⟨ , order ⟩)
 indets  ⟶  [ expr_1 ...]
```

Parameter:

`poly`	—	polynomial
`expr, expr_1`	—	expressions
order	—	`LexOrder, DegreeOrder` or `DegInvLexOrder`

Overloading:

1

Summary:

`lcoeff` returns the leading coefficient of a polynomial. With the option *order* the desired term ordering can be given, see section 2.3.13. If no order is given then the lexicographical one is used.

The first argument of `lcoeff` is allowed to be a polynomial or an expression and optionally a list of indeterminates. An expression is converted into a polynomial in the specified indeterminates (see function `poly`). `lcoeff` returns `FAIL` if the expression cannot be converted into a polynomial. `lcoeff` is a kernel function.

```
>> lcoeff(3*x^3+x^2*y^2+2, [x,y]);

   3
```

```
>> lcoeff(3*x^3+x^2*y^2+2, [x,y], DegreeOrder);

   1
```

```
>> lcoeff(poly(0, [x,y]));

   0
```

See also: `coeff, degreevec, lterm, nterms, nthcoeff, nthterm, poly, tcoeff`

■ **ldegree – lowest degree of a polynomial**

Calling sequence:

```
ldegree(poly ⟨ , expr_2 ⟩)
ldegree(expr ⟨ , indets ⟩ ⟨ , expr_2 ⟩)
  indets  ⟶  [ expr_1 ...]
```

Parameter:

```
poly                    —  polynomial
expr, expr_1, expr_2    —  expressions
```

Summary:

`ldegree(p)` returns the lowest total degree of the terms of the polynomial p. `ldegree(p,x)` returns the lowest degree of the variable x in p.

The first argument of `ldegree` is allowed to be a polynomial or an expression and optionally a list of indeterminates. An expression is converted into a polynomial in the specified indeterminates (see function `poly`). `ldegree` returns `FAIL` if the expression cannot be converted into a polynomial. `ldegree` is a kernel function.

```
>> ldegree(x^3+x^2*y^2);

    2

>> p := poly(x^2*z+x*z^3, [x,z]): ldegree(p);

    3

>> ldegree(x^2*z+x*z^3, [x,z], z);

    1
```

See also: `degree, degreevec, poly`

---

length — size of an expression

Calling sequence:

```
length(e)
```

Parameter:

```
e  —  expression
```

Summary:

`length(e)` returns an integer that is an indication of the size of the object e. The length of an integer is its number of digits, the length of any variable is one.

```
>> length(435349703);
```

   9

```
>> length(x^3+sin(y)-1/ln(4-t));
```

   16

■ **level – sets the substitution depth for evaluation**

Calling sequence:

```
level(expr)
level(expr, integer)
```

Parameter:

expr	—	expression
integer	—	non-negative integer

Overloading:

2

Summary:

level sets the substitution depth for the evaluation of the expression expr. More exactly, level(expr, n) is evaluated, by expr being evaluated with the substitution depth n. The result of this evaluation is returned. With level(expr), expr is evaluated with the substitution depth $2^{31} - 1$.

In practice, level temporarily overwrites the environment variable LEVEL. The variable MAXLEVEL remains unchanged. When the substitution depth MAXLEVEL is reached an error message is given.

Please remember that in MuPAD on the interactive level an expression is usually completely substituted. In contrast, in a procedure it is only substituted to a depth of one. The most important use of level is the multiple substitution of expressions in procedures. If, for instance, x := y; and y := 1; are entered on the interactive level then x is evaluated to 1, i.e. x is completely evaluated.

If the assignments x := y; and y := 1; are contained in a procedure then only a one-step substitution is carried out, i.e. x is evaluated to y. If the user needs a complete substitution in a procedure then level(x) must be given which is then, as on the interactive level, evaluated to 1.

In contrast to most other functions, arguments which are expression sequences, are not flattened. `level` is a kernel function.

```
>> a:=b: b:=c: c:=13: a, level(a,2);

   13, c


>> p:=proc() local a, b, c; begin
       a:= b; b:= c; c:= 13;
       a, level(a, 2), level(a)
   end_proc:
   p();

   b, c, 13
```

See also: `eval`, `hold`, `LEVEL`, `val`

LEVEL – substitution depth of identifiers

Calling sequence:

```
LEVEL := integer
```

Parameter:

`integer` — positive integer

Summary:

`LEVEL` gives the maximum substitution depth for identifiers in expressions. Before an expression is evaluated the substitution depth is 0. If an identifier is replaced by its value during evaluation then the substitution depth is increased by 1 and the value of the identifier is evaluated with this depth. After the value has been evaluated the substitution depth is decreased by 1.

If the maximum substitution depth of `LEVEL` is reached during evaluation then the identifier is evaluated no further. On the interactive level `LEVEL` has the default value 100. Before the execution of a procedure `LEVEL` is set to 1. Thus, in procedures, variables are only substituted once. This is valid as long as `LEVEL` is not changed. After a procedure is exited `LEVEL` is reset to its original value. The substitution depth can also be influenced by using the function `level`.

If during evaluation the substitution depth MAXLEVEL is reached then an error message is given and the evaluation is terminated. This is an heuristic for recognizing recursive definitions, as in a := NIL; a := a; a;. (Here a is "infinitely" substituted by a.) If MAXLEVEL is greater than LEVEL then recursive definitions are not recognized.

The environment variable LEVEL can be set to any integer between 1 and $2^{31} - 1$. With the assignment LEVEL := NIL;, LEVEL is set to its default value 100 on the interactive level.

```
>> a1:=b1: b1:=c1: c1:=7: a1;

   7


>> a2:=b2: b2:=c2: c2:=7: LEVEL:=2: a2;

   c2


>> a3:=b3: b3:=c3: c3:=7: MAXLEVEL:=2: LEVEL:=2: a3;

   Error: Recursive definition
```

See also: MAXLEVEL, level

■   LIB_PATH, READ_PATH, WRITE_PATH − file search pathes

Calling sequence:

```
LIB_PATH := string ...
READ_PATH := string ...
WRITE_PATH := string
```

Parameter:

string — character string

Summary:

LIB_PATH determines the directories where library files are searched for by the functions loadproc and loadlib. By default, in the UNIX version of MuPAD LIB_PATH is defined as the subdirectory $MuPAD_ROOT_PATH/share/lib, but can be re-defined by using the option -l.

READ_PATH determines the search path for the function **read**. **read** firstly searches for a file in the directories given by READ_PATH, then in the current directory and finally in the directories given by LIB_PATH.

More than one directory can be given by assigning a sequence of character strings to the variables LIB_PATH and READ_PATH.

WRITE_PATH determines the directory, into which the functions **fopen**, **fprint**, **write** and **protocol** write files. If WRITE_PATH is not defined then the files are written into the current directory.

Important: The directories given by LIB_PATH, READ_PATH or WRITE_PATH, when concatenated with a file name, must be valid path names!

Path names are system dependent. Under UNIX a subdirectory is started with a /; on the Macintosh a : is used instead. Use the function pathname to specify system independent pathnames. 

```
>> READ_PATH := "math/lib/","math/local/";
```

See also: finput, fopen, fprint, fread, ftextinput, pathname, protocol, read, write

---

limit − limit of expressions

Calling sequence:

limit(expr, x⟨ =v ⟩ ⟨ ,dir ⟩)

Parameter:

expr   —   expression
x   —   identifier
v   —   expression
dir   —   Left or Right

Summary:

The call limit(expr,x=v) computes the limit of the expression **expr** when **x** goes to **v**. The default point is **v=0**. The call limit(expr,x=v,Right) (resp. Left) gives the limit when **x** goes to **v** from the right (resp. from the left). When both limits (from the left and from the right) are different, it returns **undefined**.

The limit function uses **series** and returns **FAIL** if the order of the expansion computed is not sufficient. In this case, it may be needed to increase the value of the global variable ORDER (see **series**) in order to find a limit.

When the limit can not be computed, it returns unevaluated.

The algorithm based on the dissertation of Dominik Gruntz: "On Computing Limits in a Symbolic Manipulation System", submitted to the Swiss Federal Institute of Technology Zürich in 1995.

```
>> limit((1-cos(x))/x^2,x);

   1/2


>> limit((1+1/n)^n,n=infinity);

   E


>> limit((sin(tan(x))-tan(sin(x)))/x^7,x);

   FAIL


>> ORDER:=8: limit((sin(tan(x))-tan(sin(x)))/x^7,x);

   -1/30


>> limit(1/x,x=0), limit(1/x,x=0,Left), limit(1/x,x=0,Right);

   undefined, -infinity, infinity


>> Ex:= (exp(x*exp(-x)/(exp(-x)+exp(-2*x^2/(x+1))))-exp(x))/x:
   limit(Ex,x=infinity);

   -exp(2)
```

See also: series

■ linsert – inserts elements into a list

Calling sequence:

linsert(list, expr, n)

Parameter:

```
list   —   list
expr   —   expression
n      —   integer
```

Summary:

`linsert` inserts elements into the list `list` before the n-th element of `list` and returns the new list as the result. If `expr` is a list, then all elements of `expr` are inserted, otherwise `expr` itself is inserted. If $n < 1$ then `expr` is inserted in front of the first element of `list`. If $n >$ `nops(list)` then `expr` is appended to `list`.

```
>> linsert([a,b,c], x+y, 2);

   [a, x + y, b, c]
```

```
>> linsert([a,b,c],[x,y,z],0);

   [x, y, z, a, b, c]
```

```
>> linsert([a,b,c],d,1000);

   [a, b, c, d]
```

See also: append

---

linsolve – solves linear systems

Calling sequence:

linsolve(sys, unk)

Parameters:

```
sys   —   set of equations
unk   —   set of unknowns
```

Summary:

`linsolve(sys, unk)` solves the linear system `sys` with respect to the unknowns `unk`. It returns a set of assignments of the unknowns if there is a solution, and `null()` if there is no solution.

This function should be used only for linear systems, as it does not check the linearity of **sys**. For non-linear systems, the function **solve** should be used instead.

```
>> linsolve({x+y=1,2*x+2*y=3},{x,y});


>> linsolve({x+y=1,2*x+y=3},{x,y});

    {y = -1, x = 2}


>> linsolve({cos(x)+sin(x)=1,cos(x)-sin(x)=0},{cos(x),sin(x)});

    {cos(x) = 1/2, sin(x) = 1/2}


>> linsolve({2*a[1]+3*a[2]=5,7*a[2]+11*a[3]=13,
            17*a[3]+19*a[1]=23},{a[1],a[2],a[3]});

    {a[2] = 981/865, a[1] = 691/865, a[3] = 398/865}
```

See also: solve

■ **lmonomial – the leading monomial of a polynomial**

Calling sequence:

```
lmonomial(poly ⟨ , order ⟩)
lmonomial(expr ⟨ , indets ⟩ ⟨ , order ⟩)
 indets  ⟶  [ expr_1 ...]
```

Parameter:

poly	—	polynomial
expr, expr_1	—	expressions
order	—	LexOrder, DegreeOrder or DegInvLexOrder

Overloading:

1

Summary:

lmonomial returns the leading monomial of a polynomial. With the option
order the desired term ordering can be given, see section 2.3.13. If no order is
given then the lexicographical one is used.

The first argument of lmonomial is allowed to be a polynomial or an expression
and optionally a list of indeterminates. An expression is converted into a
polynomial in the specified indeterminates (see function poly). lmonomial
returns FAIL if the expression cannot be converted into a polynomial. The
result is of the same type as the first argument, e.g. if the argument is a
polynomial then the result is a polynomial too. lmonomial is a kernel function.

```
>> lmonomial(3*x^3+6*x^2*y^2+2, [x,y]);

   3*x^3


>> lmonomial(poly(3*x^3+6*x^2*y^2+2, [x,y]), DegreeOrder);

   poly(6*x^2*y^2, [x,y])


>> lmonomial(poly(0, [x,y]));

   poly(0, [x,y])
```

See also: lcoeff, lterm, nterms, nthcoeff, nthmonomial, nthterm, poly,
          tcoeff

■ load – prints identifiers with a value

Calling sequence:

load()

Parameter:

— none

Summary:

load returns all identifiers that have values. The ⌒utput is in the form
of the assignment *identifier*:=*value*;. load returns the element of the type
DOM_NULL as the return value (the "empty" datum). load is a kernel function.

See also: `anames, reset`

■ `loadlib` – loads a library package

Calling sequence:

`loadlib(string)`

Parameter:

`string` — character string

Summary:

`loadlib` loads the procedures of the library package with the name `string`. The MuPAD library packages (e.g. **Type**) are loaded by the system init file at the beginning of a MuPAD session. Therefore the function `loadlib` is only necessary for loading further packages.

The individual procedures of the library are usually loaded with the `loadproc` function. They are only defined at their first call; before this their definition is invalid.

A library is only loaded at the first call of `loadlib`. A subsequent call of `loadlib` does not re-load the same library.

`loadlib` first tries to read the binary file named LIBFILES/BIN/*string*.mb. If such a file does not exist the text file LIBFILES/*string*.mu is searched for. These files are searched for in the directories given by LIB_PATH.

The library file, in its turn, then loads the procedures of the library with `loadproc`. As a model for a library file the file LIBFILES/stdlib.mu under UNIX can be used.

One may cause the system to load user defined library files in addition to those in the system library by defining a local library which contains the same subdirectory LIBFILES as the system library and the new library files. The root directory of the local library must then be inserted into the LIB_PATH.

```
>> loadlib("sharelib"):
```

See also: `fread, loadproc`

■ `loadproc` – **loads an object on demand**

Calling sequence:

```
loadproc(expr, string_1, string_2 ⟨ , expr_1 ... ⟩)
```

Parameter:

```
expr, expr_1        —  expressions
string_1, string_2  —  strings
```

Summary:

`loadproc` returns an element of a special *loadproc-domain*. This domain element is used to read a procedure from file on demand. Thus procedures are not read before they are used in order to save memory space.

Formally an element of the loadproc-domain, when evaluated, does the following: It first tries to read a binary file given by

```
string_1."BIN/".string_2.".mb".
```

If this file doesn't exist the text file

```
string_1.string_2.".mu"
```

is searched for. These files are searched in the directories given by `LIB_PATH`.

The file read must contain the "real" definition of the object `expr`. After the file is read `expr` is evaluated (now to its intended "real" value) and returned as value of the evaluation of the element of the loadproc-domain.

Note that the first argument `expr` is not evaluated; the other arguments are evaluated as usual.

The following statement taken from the `stdlib` library file defines D to be an element of the loadproc-domain:

```
D:= loadproc(D, pathname("STDLIB"), "D"):
```

When D is evaluated for the first time the file `STDLIB/BIN/D.mb` is read (or the text file `STDLIB/D.mu` if the binary file does not exist). This file contains the "real" definition of D:

```
D:= proc(l, e) ...
```

Thus the procedure is assigned to the identifier D when the file is read. The element of the loadproc-domain eventually returns the new value of D, which is the procedure to be loaded.

Any alias definitions existing at the time the file is read are un-defined during the read operation in order to avoid side-effects. The alias definitions are reinstalled again after the file is read.

The optional arguments `expr_1` ... may be used to do some additional computations after the file is read. When the additional arguments `e1,e2,...,en` are given then the function call `e1(e2,...,en)` is evaluated after the file is read. This may be used for example to change the object in a certain way after it is read.

One may cause the system to load user defined files instead of those in the system library by defining a local library which contains the same directories as the system library and the files to redefine. The root directory of the local library must then be inserted into the `LIB_PATH`.

```
>> random:= loadproc(random, pathname("STDLIB"), "random"):
```

See also: `alias`, `fread`, `loadlib`, `pathname`

■ `lterm` – the leading term of a polynomial

Calling sequence:

```
lterm(poly 〈 , order 〉)
lterm(expr 〈 , indets 〉 〈 , order 〉)
 indets  ⟶  [ expr_1 ...]
```

Parameter:

`poly`	—	polynomial
`expr, expr_1`	—	expressions
order	—	`LexOrder`, `DegreeOrder` or `DegInvLexOrder`

Overloading:

1

Summary:

`lterm` returns the leading term of a polynomial. With the option *order* the desired term ordering can be given, see section 2.3.13. If no order is given then the lexicographical one is used.

The first argument of `lterm` is allowed to be a polynomial or an expression and optionally a list of indeterminates. An expression is converted into a polynomial in the specified indeterminates (see function `poly`). `lterm` returns `FAIL` if the expression cannot be converted into a polynomial. The result is of the same type as the first argument, e.g. if the argument is a polynomial then the result is a polynomial too. `lterm` is a kernel function.

```
>> lterm(3*x^3+6*x^2*y^2+2, [x,y]);

   x^3
```

```
>> lterm(poly(3*x^3+6*x^2*y^2+2, [x,y]), DegreeOrder);

   poly(x^2*y^2, [x,y])
```

```
>> lterm(poly(0, [x,y]));

   poly(0, [x,y])
```

See also: `lcoeff`, `lmonomial`, `nterms`, `nthcoeff`, `nthmonomial`, `nthterm`, `poly`, `tcoeff`

■ map – applies a function to operands

Calling sequence:

```
map(expr, func)
map(expr, func, expr_1 ...)
```

Parameter:

```
expr     —   expression
func     —   function
expr_1   —   expression
```

Overloading:

1

Summary:

`map` applies the function `func` to all operands of the expression `expr`. When `func` is called the expressions `expr_1` ... are used as additional arguments.

For each operand `opd` of `expr` the call of `func` has the form

```
    func(opd, expr_1 ...)
```

The results of the function calls are then inserted in the expression `expr` instead of the old operands.

If the first argument `expr` is an expression sequence, then this sequence is not flattend. If the argument is a rational or complex number, then `func` is applied to the number and not to its parts as given by `op`.

If `expr` is a list, set, table or array then the function is applied to the individual elements of the corresponding data structure. `map` is a kernel function.

```
>> map(a+b+3, sin);

   sin(a) + sin(b) + sin(3)


>> map(a+b+3, f, x, y);

   f(a,x,y) + f(b,x,y) + f(3,x,y)


>> map([1,x], _plus, 12);

   [13, x+12]
```

See also: `op, select, subsop, zip`

■ `mapcoeffs` – applies a function to the coefficients of a polynomial

Calling sequence:

```
mapcoeffs(poly, proc ⟨ , expr_2 ... ⟩)
mapcoeffs(expr ⟨ , indets ⟩, proc ⟨ , expr_2 ... ⟩)
  indets  ⟶  [ expr_1 ...]
```

Parameter:

`poly`	—	polynomial
`proc`	—	procedure or function
`expr, expr_1, expr_2`	—	expressions

Overloading:

1

Summary:

`mapcoeffs` applies a function to the coefficients of a polynomial and then returns the polynomial with the new coefficients.

- With a call of the form `mapcoeffs(p,f)`, `f(ci)` is executed for each coefficient `ci` of the polynomial `p`.
- With a call of the form `mapcoeffs(p,f,a1,...,an)`, `f(ci,a1,...,an)` is executed for each `ci` with the additional arguments `a1,...,an`.

The first argument of `mapcoeffs` is allowed to be a polynomial or an expression and optionally a list of indeterminates. An expression is converted into a polynomial in the specified indeterminates (see function `poly`). `mapcoeffs` returns `FAIL` if the expression cannot be converted into a polynomial. The result is of the same type as the first argument, e.g. if the argument is a polynomial then the result is a polynomial too. `mapcoeffs` is a kernel function.

```
>> mapcoeffs(3*x^3+x^2*y^2+2, sin);

    sin(3)*x^3+sin(1)*x^2*y^2+sin(2)


>> mapcoeffs(3*x^3+x^2*y^2+2, _plus, 2, a);

    (a+5)*x^3+(a+3)*x^2*y^2+a+4


>> p := poly(x^3+2):
   f := proc(a,b) begin exp(a+b) end_proc:
   mapcoeffs(p, f, 2);

    poly(exp(3)*x^3 + exp(4), [x])
```

See also: `coeff, lcoeff, map, nterms, nthcoeff, poly, tcoeff`

■ **maprat – executes a function on general expressions**

Calling sequence:

```
maprat(e, f ⟨ ,p ⟩)
rationalize(e ⟨ ,p ⟩)
```

Parameters:

e — expression
f — function
p — procedure

Summary:

maprat(e, f, p) first replaces all subexpressions in e for which p does not return TRUE and which are not constants or variables, by new variables. After that it calls the function f on this new expression. Finally, the result of f is taken and the initial subexpressions are substituted back. If p is not given, a default procedure is used, which replaces all non-rational subexpressions, i.e. those expressions that are neither a sum nor a product nor a power with an integer exponent.

The call `rationalize(e, p)` returns an expression sequence containing the intermediate expression used in `maprat`, and a list of substitutions.

```
>> factor(sin(1)^2-1);

   Error: not a polynomial [factor]


>> maprat(sin(1)^2-1,factor);

   [1, sin(1) - 1, 1, sin(1) + 1, 1]


>> e:=x^3+2*sqrt(3):
   rationalize(e);

   X1*2 + x^3, {X1 = 3^(1/2)}


>> rationalize(e,func(contains({"_plus","_mult"},type(x)),x));

   X2 + X3*2, {X2 = x^3, X3 = 3^(1/2)}
```

See also: genident

■ **match – pattern matching**

Calling sequence:

```
match(expr, mexpr)
```

Parameter:

```
expr, mexpr  —  expression
```

Summary:

match tries to find a match by which mexpr is matched with expr. If match finds such a match, it returns it. Otherwise match returns FAIL.

A match is a table associating names of match-variables with substitution values for these match-variables.

A value x is a match-variable if it is not of basic type and if its domain contains an entry "FREEVARIABLE". The first operand of a match variable is considered the name, which is used to identify the variable. The second operand is considered a function. A match-variable x can be matched with a value y dependent on a match t, if t[extop(x,1)] equals y or if t does not possess an entry with index extop(x,1) and extop(x, 2)(y, t, x) yields TRUE. If the domain has a third operand the match variable is assumed to be able to match a sequence of expressions.

mexpr matches expr by match t, if

1. mexpr equals expr, when every match-variable occurring in mexpr is substituted by the value associated with its name.

2. for every match-variable contained in mexpr the expression sequence
   `y:=t[extop(x,1)]:  t[extop(x,1)]:=NIL: extop(x,2)(y,t,x);`
   yields TRUE.

```
>> d:=domain: d::FREEVARIABLE:=TRUE:
   y:=new(d,"y",TRUE): z:=new(d,"x",TRUE,TRUE):
   match(a+2,z);
   match(a+b+2,z+2);
   match(f(a,b,2),f(a,z));
   match(a+b+2,y+2);
   match(a+2,z*2);
   match(a1+a2+a3+a4+a5,z+y);
   match(a*sin(a),y*sin(y));
   match(a*sin(b),y*sin(y));

     table("x"=a + 2)
     table("x"=(a, b))
     table("x"=(b, 2))
```

```
FAIL
FAIL
table("x"=(a1, a2, a3, a4),"y"=a5)
table("x"=a)
FAIL
```

■ Mathematical Constants and Functions

Summary:

In MuPAD, at the beginning of a session the following mathematical constants and functions are defined:

`PI`	—	pi (3.14159 …)
`E`	—	e (2.71828 …)
`EULER`	—	Euler constant (0.57721 …)
`I`	—	imaginary unit
`infinity`	—	infinity
`abs`	—	absolute value of a numerical argument
`bernoulli`	—	Bernoulli numbers and polynomials
`binomial`	—	binomial coefficients
`eint`	—	exponential integral
`erf`	—	error function
`erfc`	—	complementary error function
`exp`	—	exponential function
`fact`	—	factorial function `fact(n)` = n!
`gamma`	—	gamma function
`igamma`	—	incomplete gamma function
`ln`	—	natural logarithm (base e)
`psi`	—	digamma function
`sign`	—	sign of a numerical argument
`sqrt`	—	square root of a numerical argument
`zeta`	—	zeta function

Furthermore, the trigonometric and hyperbolic functions

```
cos, cot, csc, sec, sin, tan,
cosh, coth, csch, sech, sinh, tanh
```

as well as their inverse functions

```
acos, acot, acsc, asec, asin, atan,
acosh, acoth, acsch, asech, asinh, atanh
```

are defined.

If the argument cannot be simplified the functions return the function call with the evaluated argument.

---

max – maximum of numbers

Calling sequence:

```
max(expr ...)
```

Parameter:

expr   —   expression

Overloading:

all

Summary:

max calculates the maximum of the numerical arguments. If one of the arguments cannot be evaluated to a number then the function call is returned with the maximum of the numerical arguments and the remaining evaluated arguments. max flattens its arguments.

max returns an error when one of its arguments is a complex number. max is a kernel function.

```
>> max(-3/2, 7, 1.4);

    7


>> max(-4, b+2, 1, 3);

   max(b+2, 3)


>> max(a,max(b,c));

   max(a,b,c)
```

See also: min

MAXLEVEL – recognizes recursive definitions

Calling sequence:

```
MAXLEVEL := integer
```

Parameter:

```
integer   —   positive integer
```

Summary:

`MAXLEVEL` is used to recognize recursive definitions, like in `a:=NIL; a:=a; a;`. Here a would be "infinitely" replaced by a if the substitution depth was not limited by the environment variables `LEVEL` and `MAXLEVEL`. (For substitution depth compare `LEVEL`, `level` and section 2.2.4.)

If during evaluation of an expression the substitution depth `MAXLEVEL` has been reached then the error `Recursive Definition` is returned and the evaluation is terminated. If `MAXLEVEL` is higher than `LEVEL` this error is never returned as the maximum substitution depth is set with `LEVEL`.

The highest possible value for `MAXLEVEL` is $2^{31} - 1$. With the assignment `MAXLEVEL := NIL;`, `MAXLEVEL` is set to the default value 100.

`MAXLEVEL` is an environment variable.

```
>> a1:=b1: b1:=c1: c1:=7: MAXLEVEL:=2: LEVEL:=2: a1;

   Error: Recursive definition


>> a2:=b2: b2:=c2: c2:=7: MAXLEVEL:=3: LEVEL:=2: a2;

   c2
```

See also: `LEVEL`, `level`

■ min – minimum of numbers

Calling sequence:

```
min(expr ...)
```

Parameter:

```
expr   —   expression
```

Overloading:

all

Summary:

min calculates the minimum of the numerical arguments. If one of the arguments cannot be evaluated to a number then the function call with the minimum of the numerical arguments and the remaining evaluated arguments is returned. min flattens its arguments.

min returns an error if one of its arguments is a complex number. min is a kernel function.

```
>> min(-3/2, 7, 1.4);

    -3/2


>> min(-4, b+2, 1, 3);

    min(b+2, -4)


>> min(a,min(b,c));

    min(a,b,c)
```

See also: max

---

modp, mods — modulo functions

Calling sequence:
```
modp(rational, integer)
mods(rational, integer)
```

Parameter:
```
rational  —  rational number
integer   —  integer
```

Overloading:
all

Summary:

modp and **mods** calculate the positive or respectively the symmetrical modulo functions. The second argument represents the modulus and must be a integer not equal to zero. If the first argument is a rational number then the modular inverse of the denominator is calculated and then multiplied with the numerator.

The functions **modp** and **mods** can be used to re-define the modulo operator **mod**. After the assignment **_mod := mods;** the **mod** operator calculates the symmetrical modulo function. After the assignment **_mod := modp;** the operator again calculates the default positive modulo function.

The functions return an error when the arguments are numbers of a wrong type. If an argument cannot be evaluated to a number then the function call with evaluated arguments is returned. **modp** and **mods** are kernel functions.

```
>> modp(4/3, 7);

    6


>> mods(4/3, 7);

    -1


>> 11 mod 7; _mod := mods: 11 mod 7;

    4, -3
```

■ **multcoeffs** – **multiplies the coefficients of a polynomial with a factor**

Calling sequence:

```
multcoeffs(poly, expr_2)
multcoeffs(expr 〈 , indets 〉, expr_2)
  indets  ⟶  [ expr_1 ...]
```

Parameter:

```
poly                  —  polynomial
expr, expr_1, expr_2  —  expressions
```

Overloading:

1

Summary:

`multcoeffs(p,e)` multiplies all coefficients of the polynomial p with the expression e. The new polynomial is returned.

The first argument of `multcoeffs` is allowed to be a polynomial or an expression and optionally a list of indeterminates. An expression is converted into a polynomial in the specified indeterminates (see function `poly`). `multcoeffs` returns `FAIL` if the expression cannot be converted into a polynomial. The result is of the same type as the first argument, e.g. if the argument is a polynomial then the result is a polynomial too. `multcoeffs` is a kernel function.

```
>> multcoeffs(3*x^3+x^2*y^2+2, a);

    a*3*x^3 + a*x^2*y^2 + a*2


>> p := poly(x^3+2): multcoeffs(p, sin(y));

    poly(sin(y)*x^3 + 2*sin(y), [x])
```

See also: `coeff, lcoeff, nterms, nthcoeff, mapcoeffs, poly, tcoeff`

---

new – creates a domain element

Calling sequence:

`new(dom ⟨ ,expr ... ⟩)`

Parameter:

dom   —   domain
expr   —   expression

Summary:

A domain element consists of a reference to the corresponding domain and a list of values which represent the contents of the domain element. `new` creates a domain element of the given domain `dom` whose content is a sequence of the remaining arguments of the function. `new` is a kernel function.

```
>> d:=domain(): a:=new(d,1,2,3,4): extop(a);
```

```
    1, 2, 3, 4
```

See also: `extnops`, `extop`, `extsubsop`

■ **nextprime** – next prime number

Calling sequence:
`nextprime(integer)`

Parameter:
`integer` — integer

Overloading:
all

Summary:
`nextprime` returns the smallest prime number that is larger than or equal to the absolute value of `integer`.

`nextprime` returns an error when `integer` is a number but not an integer. If the argument cannot be evaluated to a number, `nextprime` returns the unevaluated function call. `nextprime` is a kernel function.

```
>> nextprime(11);

   11

>> nextprime(-13);

   2

>> nextprime(56475767478567);

   56475767478601
```

See also: `isprime`, `ithprime`

■ **nops** – number of operands in an expression

Calling sequence:

```
nops(expr)
```

Parameter:

`expr` — expression

Overloading:

all

Summary:

nops returns the number of operands in the expression **expr**. nops can be
overloaded but the function used for overloading must return an integer. In
contrast to most other functions, arguments which are expression sequences,
are not flattened. nops is a kernel function.

```
>> nops(a*b+3*c+d^2);

    3


>> nops({a,b,c,d});

    4


>> a:=[3,4,5,8]: nops(a);

    4


>> nops(f(3*x,4,y+2));

    3


>> nops(g());

    0


>> a:=x,y: b:= s,t: f(a,b), nops(a); nops(a,b);
```

```
f(x,y,s,t), 2
Error: Wrong number of parameters in function call [nops]
```

See also: op, subsop

■ **norm – norm of a polynomial**

Calling sequence:

```
norm(poly ⟨ , number ⟩)
norm(expr ⟨ , indets ⟩ ⟨ , number ⟩)
 indets  ⟶  [ expr_1 ...]
```

Parameter:

poly	—	polynomial
expr, expr_1	—	expressions
number	—	number not equal to 0

Overloading:

1

Summary:

norm calculates numerically the norm of a polynomial whose coefficients are numbers or expressions which may be converted to numbers via **float** or domain elements.

- norm(p) returns the maximum norm of the polynomial p. This is defined as $\max_{i\geq0} |c_i|$, where the c_i are the coefficients of p.

- norm(p,n) returns the n-norm of the polynomial p. This is defined as

$$\left(\sum_{i\geq0} |c_i|^n\right)^{1/n},$$

where the c_i are the coefficients of p. n may be any complex number not equal to 0.

If the coefficient ring of the polynomial is a domain, then this must have the method **norm**. The method must return the norm of a domain element as a number or an expression which may be converted to a number via **float**.

The first argument of **norm** is allowed to be a polynomial or an expression and optionally a list of indeterminates. An expression is converted into a polynomial in the specified indeterminates (see function `poly`). `norm` returns `FAIL` if the expression cannot be converted into a polynomial. `norm` returns an error if the argument `number` is not a number or is zero. `norm` is a kernel function.

```
>> norm(6*x*y+9*y^2+2, [x,y]);

    9
```

```
>> p := poly(2*x^3+2*y, [x,y]): norm(p, 2);

    2.828427124
```

See also: `float`, `poly`

`normal` – **normal form of a rational expression**

Calling sequence:
`normal(expr)`

Parameter:
`expr` — expression

Overloading:
all

Summary:
`normal` returns the normal form of the rational expression `expr`. This has the form numerator/denominator, whereby the numerator and denominator are expanded polynomial expressions whose greatest common divisor is 1.

If the expression `expr` contains non-rational sub-expressions, e.g. `sin(x)` or `x^(-1/3)`, then these are replaced by auxiliary variables before being normalized. After normalization, these variables are replaced by the original sub-expressions. Algebraic dependencies of the sub-expressions are not taken into account.

```
>> normal(x^2 - (x+1)*(x-1));
```

```
   1

>> normal((x^2-1) / (x+1));

   x - 1

>> normal(1/sin(x)^2 + y/sin(x));

   sin(x)^(-2) * (y*sin(x) + 1)
```

See also: denom, gcd, numer

■ **nterms – the number of terms of a polynomial**

Calling sequence:

```
nterms(poly)
nterms(expr ⟨ , indets ⟩)
 indets   ⟶   [ expr_1 ...]
```

Parameter:

poly	—	polynomial
expr, expr_1	—	expressions

Overloading:

1

Summary:

nterms returns the number of non-zero terms of a polynomial. A zero polynomial has no terms and, therefore, returns 0.

The first argument of nterms is allowed to be a polynomial or an expression
and optionally a list of indeterminates. An expression is converted into a
polynomial in the specified indeterminates (see function poly). nterms returns
FAIL if the expression cannot be converted into a polynomial. nterms is a
kernel function.

```
>> nterms(x^2*y^2 + x^2 + y + 2, [x,y]);

   4
```

```
>> nterms(x^2*y^2 + x^2 + y + 2, [x]);
```

$$2$$

```
>> nterms(poly(0, [x]));
```

$$0$$

See also: `coeff, nthcoeff, nthmonomial, nthterm, poly`

---

nthcoeff − the n-th coefficient of a polynomial

Calling sequence:

```
nthcoeff(poly, integer)
nthcoeff(expr ⟨ , indets ⟩, integer)
 indets  ⟶  [ expr_1 ...]
```

Parameter:

poly	—	polynomial
expr, expr_1	—	expressions
integer	—	positive integer

Overloading:

1

Summary:

`nthcoeff(p, n)` returns the n-th non-zero coefficient of the polynomial p. The coefficients are ordered according to the lexicographical order of the terms. If n is larger than the number of terms of the polynomial then the function returns `FAIL`. A zero polynomial has no terms and, therefore, also returns `FAIL`.

The first argument of `nthcoeff` is allowed to be a polynomial or an expression and optionally a list of indeterminates. An expression is converted into a polynomial in the specified indeterminates (see function `poly`). `nthcoeff` returns `FAIL` if the expression cannot be converted into a polynomial. `nthcoeff` is a kernel function.

```
>> nthcoeff(3*x^3+x^2*y^2+2, [x,y], 3);
```

$$2$$

```
>> nthcoeff(3*x^3+x^2*y^2+2, 4);

    FAIL
```

See also: `coeff`, `degreevec`, `lcoeff`, `lterm`, `nterms`, `nthterm`, `poly`, `tcoeff`

■ **nthmonomial – the n-th monomial of a polynomial**

Calling sequence:

```
nthmonomial(poly, integer)
nthmonomial(expr ⟨ , indets ⟩, integer)
 indets  ⟶  [ expr_1 ...]
```

Parameter:

```
poly            —   polynomial
expr, expr_1    —   expressions
integer         —   positive integer
```

Overloading:

1

Summary:

`nthmonomial(p, n)` returns the n-th monomial of the polynomial p. The monomials are ordered lexicographically. If n is larger than the number of monomials in the polynomial then the function returns **FAIL**. A zero polynomial has no monomials and, therefore, always returns **FAIL**.

The first argument of `nthmonomial` is allowed to be a polynomial or an expression and optionally a list of indeterminates. An expression is converted into a polynomial in the specified indeterminates (see function `poly`). `nthmonomial` returns **FAIL** if the expression cannot be converted into a polynomial. The result is of the same type as the first argument, e.g. if the argument is a polynomial then the result is a polynomial too. `nthmonomial` is a kernel function.

```
>> nthmonomial(3*x^3+6*x^2*y^2+2, [x,y], 2);

    6*x^2*y^2
```

```
>> nthmonomial(poly(3*x^3+6*x^2*y^2+2, [x,y]), 3);
```

```
    poly(2, [x,y])
```

```
>> nthmonomial(poly(3*x^3+6*x^2*y^2+2, [x,y]), 4);
```

```
    FAIL
```

See also: lcoeff, lmonomial, lterm, nterms, nthcoeff, nthterm, poly, tcoeff

---

nthterm − the n-th term of a polynomial

Calling sequence:

```
nthterm(poly, integer)
nthterm(expr ⟨ , indets ⟩, integer)
 indets  ⟶  [ expr_1 ...]
```

Parameter:

poly	—	polynomial
expr, expr_1	—	expressions
integer	—	positive integer

Overloading:

1

Summary:

nthterm(p, n) returns the n-th term of the polynomial p. The terms are ordered lexicographically. If n is larger than the number of terms in the polynomial then the function returns FAIL. A zero polynomial has no terms and, therefore, always returns FAIL.

The first argument of nthterm is allowed to be a polynomial or an expression and optionally a list of indeterminates. An expression is converted into a polynomial in the specified indeterminates (see function poly). nthterm returns FAIL if the expression cannot be converted into a polynomial. The result is of the same type as the first argument, e.g. if the argument is a polynomial then the result is a polynomial too. nthterm is a kernel function.

```
>> nthterm(3*x^3+6*x^2*y^2+2, [x,y], 2);
```

```
    x^2*y^2
```

```
>> nthterm(poly(3*x^3+6*x^2*y^2+2, [x,y]), 3);

     poly(1, [x,y])

>> nthterm(poly(3*x^3+6*x^2*y^2+2, [x,y]), 4);

     FAIL
```

See also: `lcoeff`, `lmonomial`, `lterm`, `nterms`, `nthcoeff`, `nthmonomial`, `poly`, `tcoeff`

■ **null – returns the element of the type *DOM_NULL***

Calling sequence:

`null()`

Parameter:

—   none

Summary:

`null` returns the element of the type *DOM_NULL* (the "empty" value). `null` is a kernel function.

```
>> domtype(null());

     DOM_NULL
```

■ **numer – numerator of a rational expression**

Calling sequence:

`numer(expr)`

Parameter:

`expr`   —   expression

Summary:

numer returns the numerator of the rational expression **expr**. For this, **expr** is converted into its normal form with **normal**. The numerator is an expanded polynomial expression. (See function **normal**.)

```
>> numer(-3/4);

   -3
```

```
>> numer(x + 1/(2/3*x - 2/x));

   -3*x + 2*x^3
```

```
>> numer((cos(x)^2 - 1) / (cos(x) - 1));

   cos(x) + 1
```

See also: denom, gcd, normal

O – order term

Calling sequence:

O(expr)

Parameter:

expr   —   expression

Summary:

O gives the order term of a series expansion. The argument of the order term has to be a polynomial. It is defined by the expression **expr**.

The arithmetical operations +, −, * and ^ are defined for order terms. When calculating with order terms it is assumed that the limit value is finite, i.e. an expansion "at infinity" cannot be represented by O.

```
>> O(x^4 + 2*x^2) - O(x)^2;

   O(x^2)
```

See also: `taylor`

■ ode – ordinary differential equations

Calling sequence:

```
ode(eq, y(x))
ode(sys, unks)
```

Parameters:

eq	—	equation or expression
y, x	—	unknowns
sys	—	set of equations
unks	—	set of unknown functions

Summary:

`ode(eq,y(x))` creates the ordinary differential equation defined by the equation `eq`, or `eq=0` if `eq` is an expression, with respect to the function `y` of the variable `x`. Similarly, `ode(sys, unks)` creates a system of equations.

An ordinary differential equation (or a system) created by `ode` is an element of the domain `ode`. It can then be solved with the command `solve`. In the case of one single equation, `solve` returns a list of solutions, where each solution is an expression. In the case of a system of equations, `solve` returns a list of assignments for the unknown functions. Initial conditions can be added in the set `sys` using the differential operator `D`, for example `f(1)=2, D(f)(0)=0, D(D(f))(0)=1`.

```
>> loadlib("ode");
   ode(x^2*diff(y(x),x)+3*x*y(x)=sin(x)/x,y(x));

       /                    2                 sin(x)        \
   ode | 3 x y(x) + x  diff(y(x), x) = ------, y(x) |
       \                                      x             /

>> solve(%);

   [C1*x^(-3)-x^(-3)*cos(x)]
```

```
>> ode({diff(f(t),t,t)+4*f(t)=sin(2*t),f(0)=0,D(f)(0)=0},f(t)):
   solve(%);

   [sin(t*2)*1/8 + t*cos(t*2)*(-1/4)]

>> sys := {diff(x(t),t)-x(t)+y(t)=0,diff(y(t),t)-x(t)-y(t)=0}:
   solve(ode(sys, {x(t),y(t)}));

   [[x(t) = (I) C4 exp((1 + I) t) + (- I) C5 exp((1 - I) t), y

   (t) = C4 exp((1 + I) t) + C5 exp((1 - I) t)]]
```

See also: D, solve

■ **op** – returns the operands of an expression

Calling sequence:

```
op(expr)
op(expr, position)
op(expr, [ position ...])
```

Parameter:

```
expr        —   expression
position    —   non-negative integer or range
```

Overloading:

1

Summary:

op returns the operands of the expression **expr** individually or as a sequence.

op(**expr**, i) returns the i-th operand of **expr**. For this i must be a non-negative integer. The 0th operand op(**expr**,0) is a special case: op(**expr**,0) is, for instance, the operator or function in arithmetical expressions or function calls. For lists or sets op(**expr**,0) is not defined.

op(**expr**, i..j) returns the i-th to j-th operands of **expr** as an expression sequence. i and j must be non-negative integers, with i smaller or equal to j. This expression is equivalent to op(**expr**, k) \$ k=i..j.

Equivalent to the cumbersome nested call op(...op(op(expr,p1),p2)...pn) is op(expr, [p1,p2,...,pn]), if all pi's are numbers. This expression returns the pn-th operand of the ... of the p2-th operand of the p1-th operand of expr. With this the user can descend on any path along the operands of an expression. Due to this the expression [p1,p2,...,pn] is also called *path expression*. Instead of numbers, ranges can also be used in a path expression. In this case the corresponding operands are returned as a sequence.

Finally, op(expr) returns all operands of expr except the zero one. op(expr) is equivalent to op(expr, 1..nops(expr)).

op has two further important special cases for rational and complex numbers. If num is a rational number then op(num) returns the denominator and numerator. If num is a complex number then op(num) returns the real and imaginary parts of the number.

If access to a specific operand is not possible, e.g. in op(expr, nops(expr) + 1), then FAIL is the result. In contrast to most other functions, arguments which are expression sequences, are not flattened. op is a kernel function.

```
>> op(a*b+b^d);

   a*b, b^d


>> expr:= a+e*f+c*F(d+12):
   op(expr, [3,2]), op(expr, [3,2,1,2]);

   F(d+12), 12


>> op(a*b+b^d, 0..1);

   _plus, a*b


>> op(f(12,x,y), 0..2);

   f, 12, x


>> op(a+e*f+c*(d+12), [2..3,2]);

   f, d+12
```

```
>> op(a+b+c, 4);

    FAIL
```

See also: nops, subsop

`partfrac` – **partial fraction decomposition**

Calling sequence:

`partfrac(f ⟨ , x ⟩)`

Parameters:

f — fraction
x — unknown

Summary:

partfrac(f,x) returns the partial fraction decomposition of the expression f with respect to the variable x. The second argument can be omitted when f has only one indeterminate.

```
>> partfrac(x^4/(x^3-3*x+2));

    x + 1/(x-1)*11/9 + 1/(x + 2)*16/9 + (x-1)^(-2)*1/3
```

```
>> f := x^2/(x^2-a^2):
   partfrac(f,x);

    a/(a + x)*(-1/2) + a/(a-x)*(-1/2) + 1
```

```
>> partfrac(f,a);

    x/(a + x)*1/2 + x/(a-x)*(-1/2)
```

See also: normal

`pathname` – **returns hardware dependent path names**

Calling sequence:

```
pathname(⟨ Root, ⟩ string ...)
```

Parameter:

```
Root     —  option
string   —  string
```

Summary:

`pathname` is used to create hardware dependent path names. These path names are necessary to be able to specify file names that are independent of the operating system, for examples for loading libraries.

The names of the subdirectories are given as strings. If the option `Root` is given then an absolute path name is created. Otherwise a relative path name is created.

In order to create valid path names for every operating system supported by MuPAD the following conventions must be complied with:

- The name must not contain the characters "/", "\", ":" or blanks " ".
- The name must not be longer than 8 characters.

Compliance with these conventions is tested by `pathname`.

In MS-DOS it is not possible with `pathname` to give a volume name for a path name. The created names are always relative to the current volume. Examples for names created with `pathname` are:

`pathname("lib","linalg")`	`"lib/linalg/"`	UNIX
	`"lib\linalg\"`	MS-DOS
	`":lib:linalg:"`	MacOS
`pathname(Root,"lib","linalg")`	`"/lib/linalg/"`	UNIX
	`"\lib\linalg\"`	MS-DOS
	`"lib:linalg:"`	MacOS

```
>> pathname("lib", "linalg");

   "lib/linalg/"


>> pathname(Root, "lib", "linalg") . "solve.mu";

   "/lib/linalg/solve.mu"
```

See also: `sysname`

■ pdivide – pseudo-division of polynomials

Calling sequence:

```
pdivide(poly_1, poly_2 〈 , opt 〉)
pdivide(expr_1, expr_2 〈 , indet 〉 〈 , opt 〉)
 indet ⟶ [ expr ]
```

Parameter:

poly_1, poly_2	—	polynomials
expr, expr_1, expr_2	—	expressions
opt	—	one of the options Quo or Rem

Overloading:

1,2

Summary:

pdivide(p,q) calculates the pseudo-division of the univariate polynomial p by q and returns b, s, r which satisfy the equation $bp = sq + r$, whereby $b = $ lcoeff$(q)\hat{\ }$(degree(p) – degree(q) + 1). If no option is given then the result is the sequence consisting of b, s and r.

- If the identifier Quo is given as the option *opt* then only the pseudo-quotient s is calculated.

- If the identifier Rem is given as the option then only the pseudo-remainder r is returned.

Coefficient rings that are domains do not need special methods for this function because the coefficients are not divided by pdivide.

The first two arguments may be polynomials or expressions with an optional list of unknowns. Expressions are converted into polynomials (see function poly). pdivide returns FAIL if the expressions cannot be converted. Both polynomials must have the same type, i.e. their variables and coefficient rings must be identical. The result has the same type as the first two arguments. pdivide is a kernel function.

```
>> p := poly(x^3+x+1): q := poly(3*x^2+x+1):
   pdivide(p, q);

      9, poly(3*x-1, [x]), poly(7*x+10, [x])
```

```
>> pdivide(x^3+x+1, a*x^2+x+1, [x], Quo);

    a*x - 1
```

See also: content, divide, multcoeffs, poly

■ **pdioe – solves polynomial diophantine equations**

Calling sequence:

```
pdioe(a, b, c ⟨ , x ⟩)
```

Parameters:

a, b, c	—	polynomials or expressions
x	—	variable

Summary:

pdioe(a,b,c) solves the equation $au + bv = c$ over the polynomials and returns the expression sequence u,v. This is only possible if gcd(a,b) divides c; otherwise FAIL is returned. The arguments can be either expressions, in which case the variable x must be given, or polynomials defined with the poly function.

```
>> pdioe(x,13*x+22*x^2+18*x^3+7*x^4+x^5+3,x^2+1,x);

      19 x         2   7 x     x
    - ---- - 6 x  - ---- - -- - 13/3 , 1/3
       3             3     3
```

```
>> pdioe(x+1,x^2-1,x,x);

    FAIL
```

```
>> pdioe(poly(a+1,[a]),poly(a^2+1,[a]),poly(a-1,[a]));

    poly(a, [a]), poly(- 1, [a])
```

See also: gcdex, poly

■ **phi – the Euler phi function**

Calling sequence:

phi(integer)

Parameter:

integer   —   integer not equal to 0

Overloading:

all

Summary:

phi calculates the Euler phi function of the argument integer, i.e. the number of numbers smaller than abs(integer) which are relatively prime to integer.

phi returns an error if the argument is a number but not an integer unequal to 0. phi returns the function call with evaluated arguments if the argument is not a number. phi is a kernel function.

>> phi(-7);

   6

---

plot2d – plots two-dimensional graphical scenes

Calling sequence:

```
plot2d(⟨ scene-option ... , ⟩ [ object ⟨ , object-option ... ⟩ ] ...)
  scene-option    ⟶    identifier = expr
  object          ⟶    Mode = Curve, values, range
  object          ⟶    Mode = List, primitive_list
  values          ⟶    [expr_x, expr_y]
  range           ⟶    var = [expr_1, expr_2]
  primitive_list  ⟶    [primitive ⟨ , ... ⟩]
  primitive       ⟶    point2D
  primitive       ⟶    polygon2D
  object-option   ⟶    identifier = expr
```

Parameter:

```
identifier       —   identifier
expr             —   expression
expr_x, expr_y   —   expressions
var              —   identifier
expr_1, expr_2   —   expressions
point2D          —   two-dimensional point generated by point
polygon2D        —   polygon consisting of two-dimensional points
```

Summary:

`plot2d` is used to plot two-dimensional curves and lists of points and polygons. Any number of two-dimensional objects can be grouped into one scene and represented. `plot2d` describes such a scene.

In a `plot2d` command a variety of attributes can be used to specify the graphical representation of the whole scene and the individual objects. In order not to compel the user to specify each attribute in a `plot2d` command, there exists a default value (see also section 4.4.6) for each option, which is used if the corresponding attribute is not specified in a `plot2d` command.

Before describing the objects in a `plot2d` command, the attributes of the scene have to be given. These attributes are optional, i.e. they do not need to be given. An individual option is to be given in the form:

```
identifier = expr
```

Here, `identifier` is an identifier for the attribute and `expr` is an expression that gives the value of the attribute. The following attributes are available for two-dimensional scenes: `PlotDevice`, `Title`, `TitlePosition`, `Axes`, `Ticks`, `Arrows`, `Labeling`, `ViewingBox`, `Scaling`, `FontFamily`, `FontStyle`, `FontSize`, `LineStyle`, `LineWidth`, `ForeGround`, `BackGround`, `PointWidth`, `PointStyle`, `Labels` and `AxesOrigin`. For a more detailed list together with the valid parameters and default values please refer to table 4.1 and table 4.3 in section 4.2.

After the attributes of the scene have been given, the different objects belonging to the scene are to be determined. Each object is to be specified in a separate list. As explained above a two-dimensional object can be either a two-dimensional curve or a list consisting of two-dimensional points and polygons.

In a `plot2d` call a two-dimensional curve is described by the following expression sequence:

```
Mode=Curve, [expr_x,expr_y], var=[expr_1,expr_2]
```

Here, `expr_x` and `expr_y` are any MuPAD expressions, depending on the variable `var`. These two expressions are used to describe the points of the curve, i.e. the x and y co-ordinates of the curve. The variable `var` has to be an identifier. The expressions `expr_1` and `expr_2` describe the lower and upper bound of the range, in which the co-ordinate expressions are to be evaluated. Please note that also explicitly defined curves have to be given in parametrized form. In the expression sequence above the curve is parametrized through the variable `var`.

Please note that the expressions `expr_x` and `expr_y` are evaluated before `plot2d` is called. In many cases this is not desirable, because an argument of the form `MyCurveX(var)` is evaluated with the current value of `var` when `plot2d` is called. This "premature" evaluation can be prevented by embedding the argument in a `hold` function. In the example above `hold(MyCurveX(var))` has to be given as the argument. 

A two-dimensional list of points and polygons can be given in a `plot2d` call by the following expression sequence:

```
Mode = List, primitive_list
```

Here, `primitive_list` describes a list of two-dimensional points and polygons, which can be created by use of the functions `point` and `polygon` (see also section 4.3).

In contrast to the above mentioned attributes for a scene, the mode of an object as well as the parametrization or the list of points and polygons are not optional, i.e. these attributes have to be given by the user. Having specified these necessary attributes of an object further options can be given in order to influence the graphical representation of the object. Like the options of the scene these additional attributes are optional and can be given in the form:

```
identifier = expr
```

The available attributes are: `Title`, `TitlePosition`, `Color`, `Style`, `Grid` and `Smoothness` whereby the specifications of `Title`, `TitlePosition` and `Color` are identical for all objects. Therefore, they are summarized in table 4.4 in section 4.2. The valid values for the other options depend on the mode of the object. For an object with `Mode = Curve` the options are listed in table 4.5 in section 4.2. The available options for an object with `Mode = List` are listed in the help pages of `point` and `polygon`.

```
>> plot2d([Mode = Curve,
              [u, cos(u)], u = [-PI, PI]
```

```
          ]);
```

```
>> alias(fsin = funcattr(sin, "float")):
   alias(fcos = funcattr(cos, "float")):

   point_list := [point(fsin(i*PI/80),
                        fcos(i*PI/80)) $ i = -80..80]:

   plot2d([Mode = List, point_list]);
```

```
>> plot2d(Axes = Box, Ticks = 0,
          [Mode = Curve,
              [u*cos(u), u*sin(u)],
              u = [0, 2*PI], Grid = [50]]);
```

```
>> plot2d(Axes = Box, Ticks = 0,
          [Mode = Curve,
              [sin(u), cos(u)], u = [0, 2*PI],
              Grid = [50], Style = [Points]],
          [Mode = Curve,
              [2*sin(u), cos(u)], u = [0, 2*PI],
              grid = [50], Style = [Lines]],
          [Mode = Curve,
              [sin(u), 1.5*cos(u)], u = [0, 2*PI],
              Grid = [50], Style = [LinesPoints]]);
```

See also: plot3d, point, polygon

■ plot3d − plots three-dimensional graphical scenes

Calling sequence:

plot3d(⟨ *scene-option* ... , ⟩ [*object* ⟨ , *object-option* ... ⟩] ...)

scene-option	$\longrightarrow$	`identifier = expr`
object	$\longrightarrow$	`Mode = Curve,` *values, range*
object	$\longrightarrow$	`Mode = Surface,` *values, range, range*
object	$\longrightarrow$	`Mode = List,` *primitive_list*
values	$\longrightarrow$	`[expr_x, expr_y, expr_z]`
range	$\longrightarrow$	`identifier_1 = [expr_1, expr_2]`
primitive_list	$\longrightarrow$	`[`*primitive* $\langle$ `,` ... $\rangle$`]`
primitive	$\longrightarrow$	`point3D`
primitive	$\longrightarrow$	`polygon3D`
object-option	$\longrightarrow$	`identifier = expr`

Parameter:

`identifier`	—	identifier
`expr`	—	expression
`expr_x, expr_y, expr_z`	—	expressions
`identifier_1`	—	identifier
`expr_1, expr_2`	—	expressions
`point3D`	—	3D point generated by `point`
`polygon3D`	—	polygon consisting of 3D points

Summary:

`plot3d` is used to plot three-dimensional curves, surfaces and lists of points and polygons. Any number of three-dimensional graphical objects can be grouped into one scene and plotted. `plot3d` describes such a scene.

In a `plot3d` command a variety of attributes can be used to specify the graphical representation of the whole scene and the individual objects. In order not to compel the user to specify each attribute in a `plot3d` command, there exists a default value (see also section 4.4.6) for each option, which is used if the corresponding attribute is not specified in a `plot3d` command.

Before describing the objects in a `plot3d` command, the attributes of the scene have to be given. These attributes are optional, i.e. they do not need to be given. An individual option is to be given in the form:

```
identifier = expr
```

Here, `identifier` is an identifier for the attribute and `expr` is an expression that gives the value of the attribute. The following attributes are available for two-dimensional scenes: `PlotDevice`, `Title`, `TitlePosition`, `Axes`, `Ticks`, `Arrows`, `Labeling`, `ViewingBox`, `Scaling`, `FontFamily`, `FontStyle`, `FontSize`, `LineStyle`, `LineWidth`, `ForeGround`, `BackGround`, `PointWidth`, `PointStyle`, `Labels`, `AxesOrigin`, `CameraPoint` and `FocalPoint`. For a more

detailed list together with the valid parameters and default values please refer
to table 4.1 and table 4.3 in section 4.2.

After the attributes of the scene have been given, the different objects belonging to the scene are to be determined. Each object is to be specified in
a separate list. As explained above a three-dimensional object can be either
a three-dimensional curve, a surface or a list consisting of three-dimensional
points and polygons.

In a `plot3d` call a three-dimensional curve is described by the following expression sequence:

```
Mode=Curve, [expr_x,expr_y,expr_z], var=[expr_1,expr_2]
```

Here, `expr_x`,`expr_y` and `expr_z` are any MuPAD expressions, depending on
the variable `var`. These three expressions are used to describe the points of
the curve, i.e. the x, y and z co-ordinates of the curve. The variable `var` has
to be an identifier. The expressions `expr_1` and `expr_2` describe the lower
and upper bound of the range, in which the co-ordinate expressions are to be
evaluated. Please note that also explicitly defined curves have to be given in
parametrized form. In the expression sequence above the curve is parametrized
through the variable `var`.

In contrast, a three-dimensional surface is described by the following expression sequence:

```
Mode=Curve, [expr_x,expr_y,expr_z],
var_u=[expr_1,expr_2], var_v=[expr_3,expr_4]
```

Similar to the specification of a curve, `expr_x`,`expr_y` and `expr_z` are expression that now depend on the variables `var_u` and `var_v`, i.e. the x, y and z
co-ordinates of the surface are parameterized through `var_u` and `var_v`.

 Please note that the expressions `expr_x`, `expr_y` and `expr_z` are evaluated
before `plot3d` is called. In many cases this is not desirable, because an ar-
gument of the form `MyCurveX(var_u)` is evaluated with the current value of
`var_u` when `plot3d` is called. This "premature" evaluation can be prevented by embedding the argument in a `hold` function. In the example above
`hold(MyCurveX(var_u))` has to be given as the argument.

A three-dimensional list of points and polygons can be specified in a `plot3d`
call by the following expression sequence:

```
Mode = List, primitive_list
```

Here, `primitive_list` describes a list of three-dimensional points and polygons, which can be created by use of the functions `point` and `polygon` (see also section 4.3).

In contrast to the above mentioned attributes for a scene, the mode of an object as well as the parametrization or the list of points and polygons are not optional, i.e. these attributes have to be given by the user. Having specified these necessary attributes of an object further options can be given in order to influence the graphical representation of the object. Like the options of the scene these additional attributes are optional and can be given in the form:

```
identifier = expr
```

The available attributes are: `Title`, `TitlePosition`, `Color`, `Style`, `Grid` and `Smoothness` whereby the specifications of `Title`, `TitlePosition` and `Color` are identical for all objects. Therefore, they are summarized in table 4.4 in section 4.2. The valid values for the other options depend on the mode of the object. For an object with `Mode = Curve` the options are listed in table 4.5 in section 4.2. For an object with `Mode = Surface` the options are listed in table 4.6 in section 4.2.

The available options for an object with `Mode = List` are listed in the help pages of `point` and `polygon`.

```
>> plot3d(Axes = Box, Ticks = 0,
          Title = "Plot of Spacecurve sin(u)",
          TitlePosition = Below,
          [Mode = Curve,
               [u, u, sin(u*PI)],
               u = [-3.0, 3.0],
               Grid = [50],
               Style = [Impulses]]);


>> plot3d(Axes = Box, Ticks = 0,
          Title = "Spiral in form of a Sphere",
          TitlePosition = Below,
          [Mode = Curve,
               [(-3+abs(u))*cos(3*u*PI),
                (-3+abs(u))*sin(3*u*PI),
                3*cos((u+3)*1/6*PI)],
               u = [-3.0, 3.0],
               Grid = [50],
```

```
                        Smoothness = [5]]);


>> plot3d(Title = "Transparent sphere",
          TitlePosition = Below,
          Axes = None, Ticks = 0,
          [Mode = Surface,
                [sin(u)*cos(v),
                 sin(u)*sin(v),
                 cos(u)],
                u = [0, PI], v = [-PI, PI],
                Grid = [20, 20], Smoothness = [1, 1],
                Style = [Transparent, AndMesh]]);


>> plot3d(Axes = None, Ticks = 0,
          Title = "Plot of sin(u^2+v^2)",
          TitlePosition = Below,
          [Mode = Surface,
                [u, v, 1/2*sin(u*u+v*v)],
                u = [0, PI], v = [0, PI],
                Grid = [30, 30],
                Style = [HiddenLine, Mesh]]);


>> plot3d(Axes = None, CameraPoint = [14.4, 14.4, 12.0],
          Title = "Spiral surrounding a Sphere",
          TitlePosition = Below,
          [Mode = Curve,
                [(-3+abs(u))*cos(3*u*PI),
                 (-3+abs(u))*sin(3*u*PI),
                 3*cos((u+3)*1/6*PI)],
                u = [-3, 3],
                Grid = [50], Smoothness = [5],
                Title = "surrounding spiral"],
          [Mode = Surface,
                [2*sin(u)*cos(v),
                 2*sin(u)*sin(v),
                 2*cos(u)],
                u = [0, PI], v = [-PI, PI],
                Grid = [20, 20], Title = "sphere",
                Style = [ColorPatches, AndMesh]]);
```

```
>> plot3d(Axes = None,
          Title = "Three different Surfaces",
          TitlePosition = Below,
          CameraPoint = [13.0, -24.0, 20.0],
          [Mode = Surface,
               [(4+cos(v))*cos(u),
                (4+cos(v))*sin(u),
                sin(v)],
               u = [0, 2*PI], v = [0, 2*PI],
               Grid = [30, 30],
               Style = [HiddenLine, Mesh]],
          [Mode = Surface,
               [2*sin(u)*cos(v),
                2*sin(u)*sin(v),
                2*cos(u)],
               u = [0, PI], v = [-PI, PI],
               Grid = [20, 20],
               Style = [ColorPatches, AndMesh]],
          [Mode = Surface,
               [u, v, -3.0],
               u = [-5.0, 5.0], v = [-5.0, 5.0],
               Grid = [20, 20],
               Style = [ColorPatches, AndMesh]]);
```

See also: plot2d, point, polygon

---

point – creates graphical primitve of the type DOM_POINT

Calling sequence:

point(*coords* ⟨ , *option* ⟩)
```
  coords       ⟶    coords_2d
  coords       ⟶    coords_3d
  option       ⟶    Color = [r, g, b]
  coords_2d    ⟶    x, y
  coords_3d    ⟶    x, y, z
```

Parameter:

x, y, z — real evaluable expressions
r, g, b — real evaluable expressions between 0.0 and 1.0

Summary:

`point` defines a 2D- or 3D-point, which can be displayed graphically. By use of this function the simplest structure of a graphical primitives is created. `point` can be used to create both 2D- and 3D-points. Hereby **x** and **y** respectively **x**, **y** and **z** correspond to the 2D- or 3D-coordinates of the point. In addition this point can be colored by use of the option `Color = [r, g, b]`. The three parameters **r**, **g** and **b** have to be real values between `0.0` and `1.0`.They are used to describe the amount of red, green and blue of the desired color.

```
>> a := point(0.0, 0.0, 0.0);

   point(0.0, 0.0, 0.0)
```

```
>> b := point(0.0, 1.0, 1.0, Color = [0.0, 1.0, 1.0]);

   point(0.0, 1.0, 1.0, Color = [0.0, 1.0, 1.0])
```

See also: `plot2d`, `plot3d`, `polygon`

■ poly – creates polynomials

Calling sequence:

```
poly(expr ⟨ , [ expr_1 ...] ⟩ ⟨ , ring ⟩)
poly(pol ⟨ , [ expr_1 ...] ⟩ ⟨ , ring ⟩)
  ring   ⟶   Expr
  ring   ⟶   IntMod( integer )
  ring   ⟶   domain
```

Parameter:

`expr, expr_1`	—	expressions
`pol`	—	polynomial
`integer`	—	integer larger than 1
`domain`	—	domain

Summary:

`poly` converts an expression or a polynomial into a polynomial (basic type *DOM_POLY*).

- `poly(e, [x1,...,xn], r)` converts the expression `e` into a polynomial with the indeterminates `x1,...,xn`. The coefficient ring of the polynomial is given by `r`. The expression need not have been previously expanded; it is internally expanded by `poly`.

  If no indeterminate is given then the indeterminates are searched for in the first argument, similar to the function `indets`. If the coefficient ring is not given then the ring `Expr` is used. In this case any MuPAD expression is permitted as the coefficient ring.

  `poly` returns `FAIL` if the expression cannot be converted into a polynomial.

- `poly(p, [x1,...,xn], r)` converts the polynomial `p` into a polynomial with the indeterminates `x1,...,xn`. The coefficient ring is once more given by `r`.

  If no indeterminates are given then only the coefficient ring of the polynomial is changed. If the coefficient ring is not given then the ring of the polynomial `p` is used.

The order of the indeterminates is given by their position in the list. This is `x1 > x2 > ... xn`. If the indeterminates are searched for in an expression then the order of the indeterminates is random.

The indeterminates need not be identifiers. Any expression can be used as an indeterminate as long as it is not rational. Thus, `sin(x)`, `f(x)` or `y^1/3` are all possible valid indeterminates.

Apart from any MuPAD expression, remainder classes of integers and domains are also permitted as coefficient rings.

- The identifier `Expr` means that any expressions can be used as coefficients.

- An expression of the form `IntMod(n)` stands for the remainder class ring $\mathbb{Z}/n\mathbb{Z}$, whereby a symmetrical representation is used. In this expression `IntMod` is only a dummy and n must be an integer greater than 1.

- A domain (Typ *DOM_DOMAIN*) means that only elements of this domain are permitted as coefficients.

A polynomial is only then in normal form when its coefficients are in normal form, i.e. when the 0 is unambiguously represented.

If the coefficient ring is a domain then this domain must contain certain entries:

- The entry `zero` must contain the element of the domain that represents the 0 (the neutral element in respect to addition).

- The entry **one** must contain the element that represents the 1 (the neutral element in respect to multiplication).

- The method **_plus** must add two domain elements.

- The method **negate** must negate a domain element (i.e. calculate the inverse in respect to addition).

- The method **_mult** must multiply two domain elements.

- The method **_power** must calculate the integer power of a domain element. The first argument is the domain element, the second argument is an integer.

Furthermore, the following methods can be useful. They are necessary for the converting of polynomials with **poly** and by the functions **gcd**, **diff**, **divide** and **norm**:

- The method **diff** must differentiate a domain element with respect to a variable.

- The method **gcd** must return the greatest common divisor of any number of domain elements.

- The method **intmult** must calculate the integer multiple of a domain element.

- The method **divex** must divide two domain elements. If division cannot be carried out then the method must return **FAIL**

- The method **norm** must calculate the norm of a domain element and return it as a number.

- The method **convert** must convert an expression into a domain element. If this is not possible then the method must return **FAIL**.

- The method **expr** must convert a domain element into an expression.

If the coefficient ring of a polynomial is a domain then **convert** is used by **poly** to convert an expression into a coefficient. If the method **convert** does not exist then only domain elements can be used as coefficients.

If a polynomial over a domain is to be converted into a polynomial over the coefficient ring **Expr** with **poly** then the method **expr** is used to convert the coefficients into expressions. If the method **expr** does not exist then the domain elements are simply inserted in the expression.

poly is a kernel function.

```
>> poly(2*x*(x+1)^2);
```

```
   poly(2*x^3 + 4*x^2 + 2*x, [x])

>> poly((x*(y+1))^2, [x,y]);

   poly(x^2*y^2 + 2*x^2*y + x^2, [x,y])

>> poly((x*(y+1))^2, [x]);

   poly((y+1)^2 * x^2, [x])

>> poly(f(x)*(f(x)+x^2));

   poly(f(x)^2 + f(x)*x^2, [f(x),x])

>> poly(9*x^3+4*x-7, [x], IntMod(7));

   poly(2*x^3 - 3*x, [x], IntMod(7))

>> p := poly(((a+b)*x-a^2)*x, [x]);
   poly(p, [a,b]);

   poly((a+b)*x^2 - a^2*x, [x]),
   poly(-x*a^2 + x^2*a + x^2*b, [a,b])

>> p := poly(-4*x+5*y-5, [x,y], IntMod(7));
   poly(p, IntMod(3));

   poly(3*x - 2*y + 2, [x,y], IntMod(7)),
   poly(y - 1, [x,y], IntMod(3))
```

See also: `coeff`, `expr`, `indets`, `nterms`, `nthcoeff`, `nthmonomial`, `nthterm`

`Poly` – returns a domain of polynomials

Calling sequence:

```
Poly( [ expr ...] ⟨ , ring ⟩)
  ring   ⟶   Expr
  ring   ⟶   IntMod( integer )
  ring   ⟶   domain
```

Parameter:

`expr`	—	expression
`integer`	—	integer greater than 1
`domain`	—	domain

Summary:

`Poly([x1,...,xn], r)` creates the domain of the polynomials in the variables `x1, ..., xn` over the coefficient ring `r`. The domain elements are polynomials of the type *DOM_POLY*. The variables and the coefficient ring must be given as for the function `poly`. If no ring is given then `Expr` is used.

A domain created with `Poly` can be used as a coefficient ring for the function `poly`. Thus, with `Poly` polynomials can be created that have polynomials as their coefficients. With this the user can create a recursive representation of polynomials.

```
>> e:=x * (y^2*2 + y) + 3*y:
   poly(e, [x,y]);
   poly(e, [x], Poly([y]));

    poly(x*y^2*2 + x*y + y*3, [x, y])
    poly(x*(y + y^2*2) + y*3, [x], Poly([y], Expr))
```

See also: `poly`

■ `polygon` – **creates the graphical primitive of the type** `DOM_POLYGON`

Calling sequence:

`polygon(point_1 ⟨ , point_2, ... ⟩ ⟨ , option ... ⟩)`
  option ⟶ `identifier = value`

Parameter:

`point_i` — graphical primitive created by `point`

Summary:

`polygon` is used to create a 2D- or 3D polygon, which can be displayed graphically.

These polygons are built of points, created by use of the function `point`. In case of a 2D polygon only 2D points are permitted and in case of a 3D

polygon only 3D points are permitted. The following options are available for
the definition of a polygon:

Option/Value	Meaning	Default value
`Color`	Color of the polygon	color of the object
`[r, g, b]`	The real values between 0.0 and 1.0 are used to describe the amount of red, green and blue of the desired color.	
`Closed`		FALSE
`TRUE`	The polygon will be closed. This option is currently only available for polygons consisting of three points, i.e. triangles.	
`FALSE`	The polygon is not closed.	
`Filled`		FALSE
`TRUE`	The polygon will be filled. Again it is to be noticed, that currently only polygons consisting of three points can be filled.	
`FALSE`	The polygon will not be filled.	

Triangles, which are both closed and filled, will be displayed filled. In addition
the edges of that triangles will be rendered in the actual foreground color.

```
>> a := point(0.0, 0.0):
   b := point(1.0, 1.0):
   c := polygon(a,b);

   polygon(point(0.0, 0.0), point(1.0, 1.0))

>> a := point(0.0, 0.0, 0.0):
   b := point(1.0, 1.0, 1.0):
   c := point(1.0, 1.0, 0.0):
```

```
d := polygon(a, b, c, Closed = TRUE):
e := polygon(a, b, c, Filled = TRUE):
f := polygon(a, b, c, Closed = TRUE, Filled = TRUE):
```

See also: `plot2d, plot3d, point`

■ `powermod` – power modulo of a number or polynomial

Calling sequence:

`powermod(expr_1, integer, expr_2)`

Parameter:

`expr_1, expr_2` — expressions
`integer`   — non-negative integer

Summary:

`powermod(b,e,m)` calculates the power `b^e mod m` for a non-negative integer exponent `e`. It is much more efficient as the direct calculation of such an expression.

- If the modulus `m` is a positive integer then the base `b` must be a rational number or an expression which may be converted to an `IntMod`-polynomial.

  If `b` is a rational number then the modular inverse of the denominator is calculated and then multiplied with the numerator.

- In any other case the modulus `m` must be a univariate polynomial or polynomial expression. The base `b` must be an expression which may be converted to a polynomial of the same type as `m`. The polynomials are then divided by `divide`.

If the modulus `m` is an integer and the base `b` a rational then dependent on the current definition of the system function `_mod`, either the positive or symmetrical integer modulus is calculated. However, when using `powermod`, `_mod` may not be a user-defined procedure or function environment. Only the system functions `modp` and `mods` are permitted as values of `_mod`. If the the base `b` is a polynomial then always the symmetrical modulus is calculated.

`powermod` returns `FAIL` if one of the expressions can not be converted to polynomials.

```
>> powermod(3,3,7);

   6

>> powermod(x^2 + 7*x - 3, 2, 7);

   x^2 + x^4 + 2

>> powermod(x^2 + 7*x - 3, 2, x^2 + 1);

   -56*x - 33
```

See also: `divide`, `modp`, `mods`, `poly`

PRETTY_PRINT – controls the formatting of output

Calling sequence:

```
PRETTY_PRINT := Boolean
```

Parameter:

`boolean` — Boolean values

Summary:

`PRETTY_PRINT` controls the Pretty-Printer which is responsible for a formatted output. If `PRETTY_PRINT` has the value `TRUE` then the Pretty-Printer is used for output.

The default value of `PRETTY_PRINT` is `TRUE`. `PRETTY_PRINT` is an environment variable.

```
>> PRETTY_PRINT := FALSE;

   FALSE
```

See also: `TEXTWIDTH`

---

primpart – primitive part of a polynomial

Calling sequence:

```
primpart(poly)
primpart(expr ⟨ , indets ⟩)
 indets  ⟶  [ expr_1 ...]
```

Parameter:

```
poly            —  polynomial
expr, expr_1    —  expressions
```

Summary:

primpart calculates the primitive part of a polynomial, i.e. the greatest common divisor of the coefficients of the polynomial is removed. The content must be able to be calculated using the function **gcd**.

The first argument of **primpart** is allowed to be a polynomial or an expression and optionally a list of indeterminates. An expression is converted into a polynomial in the specified indeterminates (see function **poly**). **primpart** returns **FAIL** if the expression cannot be converted into a polynomial.

```
>> primpart(poly(6*x^3*y + 3*x*y + 9*y, [x,y]));

   poly(3*y + x*y + 2*x^3*y, [x, y])


>> primpart(poly(6*x^3*y + 3*x*y + 9*y, [x]));

   poly(x + 2*x^3 + 3, [x])


>> normal(primpart(4*x*y + 6*x^3 + 6*x*y^2 + 9*x^3*y, [x]));

   2*x*y + 3*x^3
```

See also: content, gcd, poly

■ **print – prints expressions on the screen**

Calling sequence:

```
print(⟨ Unquoted, ⟩ expr ...)
```

Parameter:

```
Unquoted  —  option
expr      —  expression
```

Overloading:

all

Summary:

print is used for output on screen. With the statement **print(expr)**, the argument **expr** is evaluated and the result is displayed on the monitor. More than one argument can be given. These are then evaluated in sequence and the results are displayed. The individual outputs are separated by a comma and at the end of the output a new line is started.

If the option **Unquoted** is given as the first argument then character strings are printed without their enclosing quotation marks.

print returns the element of the type *DOM_NULL* as its return value (the "empty" datum). **print** is a kernel function.

```
>> print("Hello", "World" . " !");

    "Hello", "World !"


>> print(Unquoted, "Hello", "World" . " !");

    Hello, World !


>> d := 5: print("3 times d: ", d*3);

    "3 times d: ", 15


>> print((a := 3), (b := a));

    3, 3
```

See also: fprint

PRINTLEVEL – controls automatic output of assignments and expressions

Calling sequence:

```
PRINTLEVEL := integer
```

Parameter:

`integer`   —   non-negative integer

Summary:

`PRINTLEVEL` is primarily used to search for errors.

In order to understand the `PRINTLEVEL` mechanism the term *output depth* must be explained. For interactively executed statements the output depth is 1. If a further statement block is executed within a statement then for this statement block the output depth is increased by 1. Thus, e.g. the statements in the **then** part of an interactively entered **if** statement have the output depth 2. When a procedure is called the output depth is additionally increased to the next multiple of 10.

If, during evaluation, the value of `PRINTLEVEL` is greater or equal to the actual output depth then evaluated statements and expressions are output (in so far as they are contained in a statement sequence).

With the assignment `PRINTLEVEL := NIL;`, the variable `PRINTLEVEL` is set to the default value 0. `PRINTLEVEL` is an environment variable.

```
>> a:= 3: if a > 2 then a:= 2; a end_if;

   2


>> PRINTLEVEL:=2: a:=3: if a > 2 then a:=2; a end_if;

   PRINTLEVEL := 2
   a := 3
   a := 2
   2
   2
```

■   `profile` – generates a runtime protocol

Calling sequence:

`profile(expr)`

Parameter:

`expr`   —   expression

Summary:

In order to be able to analyse the runtime behavior of a program in detail it is necessary to protocol the execution of the procedures that a program calls. The number of procedure calls gives further information about the overall behavior of a program. By using this information time-critical or runtime-consuming program parts can be found. Thus, the user can easily recognize those program parts that have to be optimized in order to reduce the runtime. Checking the efficiency of the remember option of a procedure is a further aspect of the runtime analysis of a program.

The profiles of the following programs that each calculate the n-th Fibonacci number show the structure of a runtime protocol.

```
>> fib1:=proc(n)
   begin
      if n<2 then n else fib1(n-1)+fib1(n-2) end_if:
   end_proc:
   profile(fib1(10)):

   Total time: 200 ms
   ------------------
   fib1: 100 % 200 ms total 177 call(s) 0 lookup(s) 1.1 ms/call

   <fib1> calls
      fib1 : 176 time(s)
```

Calculation of the 10th Fibonacci number takes 200ms. Because, in this example, the program only consists of the procedure `fib1`, the overall runtime is equal to the runtime of the procedure `fib1`. Therefore, the runtime of `fib1` takes 100% of the overall runtime. The procedure `fib1` is called 177 times. The mean runtime per call of fib1 is 1.1ms. The second part of the protocol shows the procedures directly called from *<proc>* for every called procedure *<proc>*. Additionally the number of calls is noted. The output above shows that the procedure `fib1` calls itself recursively.

```
>> fib2:=proc(n)
       option remember;
   begin
      if n<2 then n else fib2(n-1)+fib2(n-2) end_if:
   end_proc:
   profile(fib2(10)):

   Total time: 20 ms
```

```
------------------
fib2: 100 % 20 ms total 19 call(s) 8 lookup(s) 1.0 ms/call

<fib2> calls
   fib2 : 18 time(s)
```

In contrast to the procedure **fib1**, **fib2** uses the option **remember**. The calculation of the 10th Fibonacci number now only requires 19 calls of **fib2**, whereby 8 calls of the result are calculated by being looked-up in the remember table and not by executing the procedure body.

To illustrate this here is a more complicated example (the first run of this example will produce a different statistic because the function has to be loaded from a file).

```
>> profile(gcd(6*x^3+9*x^2*y^2, 2*x+2*x*y+3*y^2+3*y^3)):

Total time: 30 ms
------------------
gcdlib::expr2polys   : 33.3 % 10 ms total 1 call(s)
                         0 lookup(s) 10.0 ms/call
gcdlib::special_cases: 33.3 % 10 ms total 1 call(s)
                         0 lookup(s) 10.0 ms/call
gcd                  : 33.3 % 10 ms total 1 call(s)
                         0 lookup(s) 10.0 ms/call
gcdlib::coeff_type   :  0.0 % 0 ms total 1 call(s)
                         0 lookup(s) 0.0 ms/call
gcdlib::norm_intp    :  0.0 % 0 ms total 1 call(s)
                         0 lookup(s) 0.0 ms/call


<gcd> calls
   gcdlib::coeff_type    : 1 time(s)
   gcdlib::special_cases : 1 time(s)
   gcdlib::expr2polys    : 1 time(s)

<gcdlib::special_cases> calls
   gcdlib::norm_intp : 1 time(s)
```

Only user-defined procedures are protocolled, not system functions. In order to protocol system functions the user must re-define them as a procedure. See the second example below for details.

This processing can be illustrated using the heuristic algorithm for calculating the greatest common divisor of two primitive polynomials with integer coefficients (again the first run produces a different statistic).

```
>> a := poly(
     9*x^5 + 2*x^4*y*z - 189*x^3*y^3*z + 117*x^3*y*z^2 + 3*x^3 -
     42*x^2*y^4*z^2 + 26*x^2*y^2*z^3 + 18*x^2 - 63*x*y^3*z +
     39*x*y*z^2 + 4*x*y*z + 6,
     [x, y, z]
   ):

   b := poly(
     6*x^6 - 126*x^4*y^3*z + 78*x^4*y*z^2 + x^4*y + x^4*z +
     13*x^3 - 21*x^2*y^4*z - 21*x^2*y^3*z^2 + 13*x^2*y^2*z^2 +
     13*x^2*y*z^3 - 21*x*y^3*z + 13*x*y*z^2 + 2*x*y + 2*x*z + 2,
     [x,y, z]
   ):

   profile(gcdlib::heu_gcd(a, b)):

    Total time: 100 ms
    -------------------
    gcdlib::heu_gcd: 99.9 % 100 ms total 3 call(s)
                       0 lookup(s) 33.3 ms/call

    <gcdlib::heu_gcd> calls
       gcdlib::heu_gcd : 2 time(s)
```

To examine the runtime more exactly now the system functions `igcd`, `genpoly` and `evalp` are protocolled, too:

```
>> igcd_sys := igcd:
   evalp_sys := evalp:
   genpoly_sys := genpoly:
   igcd := proc() begin igcd_sys(args()) end_proc:
   evalp := proc() begin evalp_sys (args()) end_proc:
   genpoly := proc() begin genpoly_sys(args()) end_proc:

   profile(gcdlib::heu_gcd(a, b));

   igcd := igcd_sys:
```

```
evalp := evalp_sys:
genpoly := genpoly_sys:

Total time: 70 ms
-----------------
gcdlib::heu_gcd: 71.4 % 50 ms total 3 call(s)
                        0 lookup(s) 16.6 ms/call
evalp          : 28.5 % 20 ms total 6 call(s)
                        0 lookup(s) 3.3 ms/call
igcd           :  0.0 %  0 ms total 1 call(s)
                        0 lookup(s) 0.0 ms/call
genpoly        :  0.0 %  0 ms total 3 call(s)
                        0 lookup(s) 0.0 ms/call

<igcd> calls
   evalp : 2 time(s)

<gcdlib::heu_gcd> calls
   evalp             : 4 time(s)
   genpoly           : 3 time(s)
   igcd              : 1 time(s)
   gcdlib::heu_gcd : 2 time(s)
```

See also: debug, trace

■ **protocol** – protocols a session

Calling sequence:

```
protocol()
protocol(integer)
protocol(string)
```

Parameter:

```
integer   —   positive integer
string    —   character string
```

Summary:

protocol writes a protocol of all input and output of a session to a text file.
The file can be specified by its name or through a file descriptor:

- If a positive integer **integer** is given as the first argument then this is considered the file descriptor. The corresponding file must already have been opened with **fopen** as a text file for writing.

- If a character string is given with the argument **string** then the file with this name is opened. If this file does not already exist then a new one with this name is created. If there is any data in this file then it is overwritten in this modus.

  If **protocol** is called in MacMuPAD on the interactive level then the user is warned before the file is overwritten. The operation can then be cancelled.

  If the variable **WRITE_PATH** has a value then the file is created in the given directory, in all other cases it is created in the current directory. (This is the directory in which MuPAD was started.)

- If, in MacMuPAD, an empty character string is given with the argument **string** then a dialog is opened in which the user can choose a file.

If a new protocol is started while a protocol is running then the old one is terminated and the corresponding file is closed. A call of **protocol** without arguments terminates a running protocol and closes the corresponding file. If the protocol file is closed with **fclose** then the protocol is also terminated. **protocol** is a kernel function.

```
>> protocol("test-protocol");
```

See also: **fclose, fopen, fprint, write**

---

psi – digamma and polygamma functions

Calling sequence:

psi(x ⟨ , n ⟩)

Parameter:

x   —   expression
n   —   non-negative integer

Overloading:

1

Summary:

The digamma function is defined as

$$\mathtt{psi(x)} = \psi(x) = \frac{d}{dx}\ln(\,\Gamma(x)\,) = \frac{\frac{d}{dx}\Gamma(x)}{\Gamma(x)}$$

The polygamma function `psi(x,n)` is the n-th derivative of the digamma function, that is `diff(psi(x),x$n)`. `psi` returns the function call with evaluated arguments, if the argument cannot be evaluated to a number. `psi` is a kernel function.

```
>> psi(2.4);

   0.6529011697
```

```
>> psi(2+I);
   float(%);

   psi(2+I)
   0.5946503103 + 0.5766740444*I
```

```
>> expand(psi(x+1,2));

   psi(x, 2) + x^(-3)*2
```

See also: gamma

■ `radsimp` – simplifies expressions with radicals

Calling sequence:

`radsimp(e)`

Parameter:

`e` — expression

Summary:

`radsimp(e)` returns a simpler expression which is equivalent to `e`, or `e` otherwise. It applies specially to constant expression with square roots. This function implements the algorithms described in *Decreasing the Nesting Depth of Expressions Involving Square Roots*, by Borodin, Fagin, Hopcroft and Tompa, JSC 1, 1985, pages 169-188.

```
>> radsimp(sqrt(4+2*sqrt(3)));

    3^(1/2) + 1

>> radsimp(sqrt(14+3*sqrt(3+2*sqrt(5-12*sqrt(3-2*sqrt(2)))))));

    2^(1/2) + 3

>> radsimp(3/(sqrt(7)-2));

    7^(1/2) + 2

>> TT := sqrt(1+sqrt(3))+sqrt(3+3*sqrt(3))-sqrt(10+6*sqrt(3)):
   radsimp(TT);

    0
```

See also: `combine, simplify`

---

random – creates random numbers

Calling sequence:

```
random()
random(range)
random(integer)
```

Parameter:

```
range    —  integer range
integer  —  non-negative integer
```

Summary:

If **random** is called without arguments then a 12 digit non-negative number is returned.

If the argulment is a range then **random** returns a procedure whose call returns a random number from this range.

The call **random(n)** is a short form of **random(0..n-1)**.

Because random returns procedures it is possible to use several random number generators. However, all generators created with the same range return the same series of random numbers.

The global variable SEED is used for changing the sequence of random numbers. Any integer not equal to 0 can be assigned to SEED.

```
>> SEED := 1: random();

   427419669081


>> r1 := random(-5..5): r1();

   -1


>> r2 := random(10): r2();

   7
```

■ **randpoly – creates a random polynomial**

Calling sequence:

```
randpoly(⟨ [ expr ...] ⟩ ⟨ , ring ⟩ ⟨ , option ⟩...)
  ring      ⟶   Expr
  ring      ⟶   IntMod( integer )
  ring      ⟶   domain
  option    ⟶   Terms = integer_1
  option    ⟶   Degree = integer_1
  option    ⟶   Coeffs = func
```

Parameter:

```
expr        —   expression
integer     —   integer larger than 1
domain      —   domain
integer_1   —   positive integer
func        —   function or procedure
```

Summary:

randpoly returns a random polynomial with variables specified by the list
[expr...] and the coefficient domain **ring**. See **poly** for a detailed description of possible variables and coefficient domains.

If the argument **ring** is missing **Expr** is used as default. If the list of variables
is missing [x] is used as default.

The polynomial is created by adding a given number of terms, where the
degrees of the variables and the coefficients are in a given range:

- With the option **Terms** = n the number of terms added may be defined.
  The default value is 6.

- With the option **Degree** = n the maximal degree of the variables may be
  defined. The default value is 5.

- With the option **Coeffs** = f a function may be defined which is used
  to create random coefficients. The function will be called without any
  arguments, the result must be a valid coefficient.

  If this option is missing and **ring** is **Expr**, the coefficients will be random
  integers in the range -999 ... 999. If **ring** is a user-defined domain it
  must have a method "**random**" to create the coefficients if no function is
  given.

```
>> randpoly([z], Expr);

   poly(z^2*902 + z^3*470 + z^5*(-494) + 26, [z])
```

See also: poly

Re – real part of an expression

Calling sequence:

Re(expr)

Parameter:

expr   —   expression

Overloading:

All

Summary:

Re returns the real part of any expression. For this the method "Re" from the domain of **expr** is used. If this domain does not contain such a method then this function returns the expression **Re(expr)**.

Exceptions are expressions of the basic types **DOM_INT**, **DOM_RAT**, **DOM_FLOAT**, **DOM_COMPLEX** and **DOM_EXPR** which are directly processed by the function **Re**.

For expressions of the type **DOM_EXPR** the determination of the real part of an expression can be defined under the function attribute "Re" in the form of a procedure or function. If the operator of **expr** has such an attribute then the procedure or function under the entry of the attribute "Re" is called with the operands of the argument **expr** and the result is returned.

If such a function attribute is not defined and the operator of **expr** is one of the function environments **_plus**, **_mult** or **_power**, then this is directly processed by the function **Re** and the result is returned. In other cases the expression **Re(expr)** is returned.

Use the function **rectform** if you have indeterminates which represent real or complex quantifiers (see examples below).

```
>> Re( -5*sin(1+2*I) );

   sin(1)*cosh(2)*(-5)


>> Re( u+v*I );

   -Im(v) + Re(u)


>> rectform( tan(x+I*y),{x,y} );
         sin(2 x)              /        sinh(2 y)          \
   -------------------- + | -------------------- | I
   cos(2 x) + cosh(2 y)      \ cos(2 x) + cosh(2 y) /


>> Re( % );

   sin(x*2)/(cos(x*2) + cosh(y*2))
```

See also: `conjugate, funcattr, Im, rectform`

■ **read – searches, reads and executes a file**

Calling sequence:

```
read(integer)
read(string)
```

Parameter:
`integer` — positive integer
`string` — character string

Summary:
`read` reads-in the contents of a file and executes the statements in the file. (Naturally the file must contain valid MuPAD statements.) `read` returns the value of the final executed statement as its result.

The file can be specified by its name or through a file descriptor:

- If a positive integer `integer` is given as the first argument then this is considered a file descriptor. The corresponding file must already have been opened with `fopen`.

- If a character string is given with the argument `string` then the file with this name is opened.
  At first the file is searched for in the directories given by the system variable `READ_PATH`. Then it is searched for in the current directory. (This is the directory in which MuPAD was started.) Finally the file is searched for in the directories given by the variable `LIB_PATH`.
  After the data have been read-in the file is closed.

- If, in MacMuPAD, an empty character string is given with the argument `string` then a dialogue is opened in which the user can choose a file.

`read` returns the error "Can't read from file" when the data cannot be read-in.

```
>> a:=3: b:=5: write("testfile.mb",a,b): reset():
   a, b; read("testfile.mb"): a, b;

   a, b
   3, 5
```

See also: `fclose`, `finput`, `fopen`, `fread`, `ftextinput`, `LIB_PATH`, `READ_PATH`, `write`

■ **readpipe** – reads data from a pipe

Calling sequence:

```
readpipe(expr, integer)
readpipe(expr, integer, Block)
```

Parameter:

```
expr     —   expression
integer  —   number of a processor node
```

Summary:

`readpipe(expr, integer)` reads a datum from the pipe with the name `expr` that comes from the cluster with the number `integer`. `expr` can be any datum. If no value is in the pipe at this time then the function returns the element of the type *DOM_NULL*.

`readpipe(expr, integer, Block)` reads a datum from the pipe with the name `expr` that comes from the cluster with the number `integer`. If the pipe contains no value at the present time then the function waits until a datum is written into the pipe and then returns this. This function is described in more detail in section 2.8.2.4. `readpipe` is a kernel function.

See also: `writepipe, readqueue, writequeue`

■ **readqueue – reads data from a queue**

Calling sequence:

```
readqueue(expr)
readqueue(expr, Block)
```

Parameter:

```
expr   —   expression
```

Summary:

`readqueue(expr)` reads a datum from the queue with the name `expr`. `expr` can be any datum. If at this time there is no value in the queue then the function returns the element of the type *DOM_NULL*.

`readqueue(expr, Block)` reads a datum from the queue with the name `expr`. If at the present time there is no datum in the queue then the function waits until a datum is written into the queue and then returns this. This function is described in more detail in section 2.8.2.2. `readqueue` is a kernel function.

See also: `writequeue, readpipe, writepipe`

`rectform` – rectangular form of complex expressions

Calling sequence:

`rectform(expr ⟨ ,rl ⟩)`

Parameter:

expr   — expression
rl   — list of identifiers

Overloading:

1

Summary:

`rectform(x)` tries to split **x** into its real and imaginary parts and returns x – if possible – in the form $\Re(x) + I\Im(x)$. If this is not possible then `rectform(x)` tries to find for each subexpression in **x** such a complex canonical form.

The result of `rectform` is an element of the domain `rectform`. Such an element consists of four operands. The first two operands give the real and imaginary parts of the expression. The third operand represents all subexpressions of **x** for which such a rectangular form can not be computed (possible 0 if there are not such subexpressions). The fourth operand will be described later.

The functions `Re` and `Im` are overloaded. If an element of the type `rectform` is passed to `Re` or `Im` it is checked if the third operand is 0. In this case, `Re` returns the first operand and `Im` the second operand of an element of the type `rectform`. Otherwise the new expressions which results by `Re` and `Im` will be computed to their rectangular form.

`rectform` maps onto sets, lists and arrays, and also onto polynomials and series, where rectform maps onto the coefficients.

Arithmetical operations between ordinary expressions and elements of the type `rectform` are possible, whereby the results will be of the type `rectform`.

The function `expr` is overloaded for arguments of the type `rectform`. If x has the type `rectform` then `expr(x)` will return the expression `extop(x,1) + I*extop(x,2) + extop(x,3)`, i.e. an expression of the type `DOM_EXPR`.

The general assumption of `rectform` is that unknown variables represent complex-valued quantities. But there are two possibilities to change this assumption:

- First – let us call this the "global" method – all unknown variables defined in the set `rectform::globalReal` will represent real variables. For all subsequent operations, e.g. arithmetical operations, `rectform` will test if a corresponding unknown variable can be found in the set `rectform::globalReal` and treat this variable as real if true.

- The second method – let us call this the "local" method – is to give a set `rl` of unknowns that represent real-valued quantities. Then for this expression returned by the call of `rectform` the variables in `rl` will ever be treated as real. This set is part of any element of the type `rectform` and it is stored in the fourth operand of such an element. The fourth operand of an element `x` of `rectform` can be changed by calling `rectform` with `x` by stating `rl`. Then the expression will be recalculated in respect to the new variables in `rl` defined to be real. If `rectform` is called with `x` without the second argument `rl`, then `rectform` will return `x` without changes. The method `real` returns the fourth operand of `x`.

The main difference of the global method and the local method is that the global method will not change the internal representation of an element of the type `rectform`, but will influence all further operations, where this element is be a part of.

For arithmetical operations with different sets of unknown variables set to be real, the result will be computed for the intersection of both sets (see examples). Thus the local method may not influence further operations.

For expressions, for which the operator is of the type `DOM_FUNC_ENV` and for which the function attribute `"rectform"` is defined, `rectform` calls the entry of this attribute with the operands of `x` and the argument `rl`, if this was given to `rectform`. See examples below how to define his own function attribute `"rectform"`.

`rectform` treats expressions of the type `"_plus"`, `"_mult"` and `"_power"` and system functions explicitly.

```
>> rectform( sin(x) );

   sin(Re(x))*cosh(Im(x)) + cos(Re(x))*sinh(Im(x))*I


>> Im( % );

   cos(Re(x))*sinh(Im(x))
```

```
>> rectform( tan(x) );

        sin(2 Re(x))             /        sinh(2 Im(x))          \
   --------------------------+| --------------------------- | I
   cos(2 Re(x))+cosh(2 Im(x))  \cos(2 Re(x))+cosh(2 Im(x)) /


>> a:= rectform( x );

   Re(x) + Im(x)*I


>> b:= rectform( y,{y} );

   y


>> b::real( b );

   {y}


>> c:= a+b;

   (Re(x) + Re(y)) + (Im(x) + Im(y))*I


>> rectform::globalReal := {x,y}:
   a, b;

   Re(x) + Im(x)*I, y


>> c;

   (Re(x) + Re(y)) + (Im(x) + Im(y))*I


>> rectform( c );

   x + y
```

```
>> sin:= funcattr( sin, "rectform", proc(x)
       local a, b, rl;
   begin
       # compute the rectangular form of x #
       if args(0) = 2 then
       # the argument rl was given #
           x:= rectform::new( x,args(2) )
       else
           x:= rectform::new( x )
       end_if;
       rl:= extop(x,4);

       if extop(x,3) <> 0 then
       # unable to compute the rectangular form of sin(x) #
           new( rectform,0,0,hold(sin)(rectform::expr(x)),rl )
       else
           a:= extop(x,1);
           b:= extop(x,2);
           if domtype(a+I*b) = DOM_COMPLEX
               and domtype(a) = DOM_FLOAT
               or domtype(b) = DOM_FLOAT then
           # float evaluation #
               rectform::new( sin(a+I*b) )
           else
           # split into their real and imaginary part #
               new(rectform,sin(a)*cosh(b),cos(a)*sinh(b),0,rl)
           end_if
       end_if
   end_proc):

   rectform( sin(z) );

    sin(Re(z))*cosh(Im(z)) + (cos(Re(z))*sinh(Im(z)))*I

>> rectform( sin(y) );

    sin(y)
```

See also: Re, funcattr, Im, sign

◼ reset – initializes a MuPAD session

Calling sequence:

```
reset()
```

Parameter:

— none

Summary:

reset re-initializes a MuPAD session. **reset** deletes the values of all identifiers and then resets the environment variables to their default values. Finally the initializing files .**mupadsysinit** and .**mupadinit** are read again.

reset is only permitted on the interactive level. In a procedure, a call of **reset** returns the error `'reset' forbidden in procedures`. **reset** is a kernel function.

```
>> a:=1: LEVEL:=5: reset(): a, LEVEL;

    a, 100
```

---

resultant – resultant of two polynomials

Calling sequence:

```
resultant(poly_1, poly_2 ⟨ , expr ⟩)
resultant(expr_1, expr_2 ⟨ , indets ⟩ ⟨ , expr ⟩)
 indets  ⟶  [ expr_3 ...]
```

Parameter:

```
poly_1, poly_2            —   polynomials
expr, expr_1, expr_2, expr_3  —   expressions
```

Overloading:

1,2

Summary:

resultant returns the resultant of the poynomials **poly_1** and **poly_2** with respect to the variable **expr**.

If no variable is given then the resultant is determined with respect to the main variable of the polynomials (this is the first element in the list of variables in the polynomials).

The first two arguments may be either polynomials or expressions with an optional list of variables. If the arguments are expressions then these are converted into polynomials (see function `poly`). `resultant` returns `FAIL` if the expressions cannot be converted. Both polynomials must be of the same type, i.e. their variables and coefficient rings must be identical.

If the arguments are expressions, then the result is an expression. If the arguments are univariate polynomials, then the result is an element of the coefficient ring. If the arguments are multivariate polynomials, then the result is a polynomial not containing the variable `expr`.

```
>> resultant(a*x+b, c*x+d, x);

   a*d - b*c
```

```
>> resultant(poly(x^2 - 1), poly(x + 1));

   0
```

See also: gcd, poly

■ return – exits from a procedure

Calling sequence:

return(⟨ expr ... ⟩)

Parameter:

expr   —   expression

Summary:

return terminates a procedure. The arguments of the call are the return values of the procedure.

Remember: Normally MuPAD ends a procedure after all statements of the procedure body have been processed. In this case the return value of the procedure is the result of the last statement.

In contrast, **return** effects an immediate exit from the procedure. The arguments are evaluated and returned as the return value of the procedure in the form of an expression sequence. If **return** is called without arguments then the element of the type *DOM_NULL* is returned.

Note that the parentheses after **return** may *not* be omitted like with the statements **break** or **next** or in the programming language C.

If **return** is called outside a procedure then the arguments are evaluated and these are returned as the result of the call. **return** is a kernel function.

```
>> maximum:=proc(x, y) begin
      if x > y then return(x) end_if; y
   end_proc:
   maximum(3, 2), maximum(4, 5);

   3, 5
```

See also: error

◼ **round – rounds a number**

Calling sequence:

round(expr)

Parameter:

expr — expression

Overloading:

all

Summary:

round rounds a numerical argument that is not a complex number, to the next integer. Complex arguments are rounded component for component. If the argument can not be evaluated to a number, **round** returns the unevaluated function call.

If the exponents of real numbers are very large then a call of **round** can lead to a loss of accuracy. This is not recognized as an error. **round** is a converting function of the kernel.

```
>> round(8/3);

   3
```

```
>> round(-2.6);

   -3
```

```
>> round(4/7 - 2.2*I);

   1 - 2*I
```

See also: `frac, ceil, trunc, floor`

■ `rtime` – measures real time

Calling sequence:

```
rtime(statement)
rtime()
```

Parameter:

`statement` — statement

Summary:

`rtime(statement)` executes the statement **statement** and returns the consumed real time. Any statement can be given as argument. `rtime()` returns the entire real time used since starting the MuPAD session. `rtime` returns the time in milliseconds. Note that at present the last three digits are always 0, i.e. the function only measures the time in seconds.

This function has been included because the results of the CPU time measurement are sometimes not very informative. This is especially true of computers that use the parallelism of MuPAD. `rtime` is a kernel function.

```
>> rtime((a := isprime(fact(90)-1))); a;

   1000, FALSE
```

See also: `time`

. `select` – selects operands

Calling sequence:

```
select(expr, func)
select(expr, func, expr_1 ...)
```

Parameter:

`expr`	—	list, set or expression
`func`	—	function
`expr_1`	—	expression

Summary:

`select` applies the function `func` to all operands of `expr`. Only those operands of `expr`, for which the call returns `TRUE`, remain in the returned object. `expr` has to be a list, set or expression. Note that if the first argument `expr` is an expression sequence, then this sequence is not flattend.

With the call of `func`, the expressions `expr_1` ... are used as additional parameters. The call of `func` has the form

```
func(opd, expr_1 ...)
```

for each operand `opd` of the expression `expr`. The function call has to return either `TRUE` or `FALSE`. If `FALSE` is returned then the corresponding operand is removed. `select` is a kernel function.

```
>> select({ {a,x,b}, {a}, {x,1} }, contains, x);

    {{x, 1}, {a, x, b}}


>> select([sin(x), x^2, y, 11], has, x);

    [sin(x), x^2]


>> select(sin(x) + x^2 + y + 11, has, x);
```

```
    sin(x) + x^2
```

See also: `map`

■ **seq – generates expression sequences**

Calling sequence:

```
seq(expr, equation)
 equation  ⟶  index = expr_1
```

Parameter:

`expr`	—	algebraic expression
`index`	—	expression
`expr_1`	—	expression

Summary:

`seq` generates an expression sequence as follows:

For each element of `expr_1`, `index` is subsituted in `expr` by this element. The results of these substitutions are then concatenated to the final expression sequence.

Typically, `expr_1` is a set or a list, however any expressions are permitted.

```
>> R := {a,b,c,d}: seq(f(i), i=R);

    f(a),f(b),f(c),f(d)


>> seq(f(u,v), u=[t1,t2,t3]);

    f(t1,v),f(t2,v),f(t3,v)
```

See also: sequence operator

■ **series – series expansions**

Calling sequence:

```
series(f, x ⟨ ,n ⟩)
series(f, x=v ⟨ ,n ⟩)
```

Parameters:

```
f, v   —   expressions
x      —   identifier
n      —   nonnegative integer
```

Summary:

`series(f,x=v,n)` computes the series expansion of the expression `f` with respect to `x` at the point `v`, to order `n`. The default point is 0, and the default order is the value of the global variable `ORDER`, which is 6 initially.

The function `series` returns a domain object that can be manipulated with the standard arithmetic operations. The following methods are available too: `ldegree` gives the exponent of the leading term, `order` gives the exponent of the error term, `expr` converts to an expression, removing the error term, `coeff(s,n)` gives the coefficient of exponent n, `lcoeff` gives the leading coefficient, `revert` reverts the expansion (i.e. computes the inverse for the composition), `diff` differentiates a series expansion, and `map` maps a function onto the coefficients.

```
>> s := series(sin(x),x);

           3     5
          x     x            6
    x  -  --  + ---  +  O(x )
          6     120
```

```
>> coeff(s,5);

    1/120
```

```
>> s * s;

            4       6
     2     x     2 x            7
    x   -  --  + ----  +  O(x )
           3      45
```

Below we compute the composition of `s` by itself, i.e. the series expansion of $\sin\sin(x)$.

```
>> s @ s;
```

```
        3    5
       x    x          6
   x - -- + -- + O(x )
       3    10
```

```
>> series(sin(x),x)/series(cos(x),x)=series(tan(x),x);
```

```
       3     5                  3     5
      x   2 x         6        x   2 x         6
   x + -- + ---- + O(x ) = x + -- + ---- + O(x )
      3    15                  3    15
```

```
>> bool(%);
```

```
   TRUE
```

See also: `limit`

■ **setuserinfo** – sets the information level of a procedure

Calling sequence:

`setuserinfo(expr ⟨ , posint ⟩)`

Parameter:

`expr`	—	expression
`posint`	—	positive integer

Summary:

setuserinfo is used to set either the global "information level" or the information level of a certain procedure given by **expr** to the value given by **posint**.

The information level determines whether informations are printed via the function **userinfo** or not. **userinfo** normally is used to print progress informations during the execution of procedures. **userinfo** prints informations if either the information level of the procedure calling **userinfo** or the global information level is greater than or equal to the level given in the call of **userinfo**.

The global information level is changed if **expr** is the identifier **Any**. Otherwise **expr** is assumed to be the name of the procedure whose level is to be changed. The name of the procedure must be the value returned by **procname** inside of the procedure.

The new information level is given by the second argument **posint**. If the second argument is missing the level is set to 0, i.e. no further informations are printed.

Note that by changing the global information level with **Any** the information level of explicitly defined procedures is not changed.

```
>> f:= proc(x) begin userinfo(2,"Argument:",x); end_proc:
   setuserinfo(f,2):
   f(12);

   f, Argument:, 12
```

See also: help, info, procname, userinfo

---

sign – **sign of a real or complex expression**

Calling sequence:

sign(x)

Parameter:

x   —   expression

Overloading:

All

Summary:

sign determines the sign of a real or complex expression. It is defined by $\frac{x}{|x|}$ for $x \neq 0$.

The function call **sign(0)** returns 0 by default, but the value of **sign(0)** can be redefined by direct assignment (e.g. **sign(0) := 42;**).

If the type of **x** is one of **DOM_INT**, **DOM_RAT** or **DOM_FLOAT** then the kernel function **sign** will be used to determine the sign of **x**. The return value of this is −1 for negative signs, 1 for positive signs and 0 if the argument is zero.

If the argument is the floating-point number 0.0 then 0 is returned if the number is zero at all internal positions. In some cases a floating-point number that returns 0.0 can have internal digits unequal to zero, because internally more positions are calculated as set with the environment variable DIGITS. In this case the result is 1 or -1.

For expressions of the type DOM_EXPR the determination of the sign of an expression can be defined under the function attribute "sign" in the form of a procedure or function. If the operator of x has such an attribute then the procedure or function under the entry of the attribute "sign" is called with the operand of x and the result is returned.

If such a function attribute is not defined and the operator of x is one of the function environments _plus, _mult or _power, then this is directly processed by the function sign and the result is returned. In other cases the expression sign(x) is returned.

```
>> sign( -8/3 );

   -1
```

```
>> sign( 2+3*I );

   13^(-1/2)*(2 + 3*I)
```

```
>> sign( PI*x );

   sign(x)
```

```
>> sign( exp(sin(2+3*I)) );

   exp(cos(2)*sinh(3)*I)
```

See also: abs, funcattr

■ simplify – simplifies expressions

Calling sequence:

simplify(e ⟨ , fn ⟩)

Parameters:

e   —   expression
fn —   function name (for example `exp` or `sqrt`)

Overloading:

1

Summary:

`simplify(e)` returns a simplified expression which is mathematically equivalent to `e`. This function can be overloaded for user-defined domains: if the user has defined a domain D and if `e` is an object of that domain, then the function `D::simplify` will be called if it exists. To invoke special simplification routines, a second argument may be given to specify the routine to be used. By now special routines exist for the exponential function `exp` and the square-root `sqrt`.

```
>> simplify(sqrt(4+2*sqrt(3)),sqrt);

   3^(1/2) + 1


>> simplify((exp(x)-1)/(exp(x/2)+1),exp);

   exp(x*1/2)-1


>> D:=domain():
   e:=new(D,x^2-1):
   D::simplify:=proc(x)
   begin
      extsubsop(x,1=Factor(extop(x,1)))
   end_proc:
   simplify(e);

   new(domain(), (x-1)*(x + 1))
```

See also: `combine, expand, radsimp`

`solve` – **general solver**

Calling sequence:

```
solve(eq, x)
solve(sys, unks)
solve(object)
```

Parameter:

```
eq      —   equation or expression
x       —   indeterminate
sys     —   set of equations
unks    —   set of indeterminates
object  —   domain element
```

Overloading:

1

Summary:

`solve(eq, x)` solves the equation `eq` with respect to the variable `x` and returns a list of solutions. The output `[]` means that there is no solution; the output `[1,1,2]` means that 1 is solution with multiplicity 2 and 2 is a single solution; the output `[x]` means that any value of `x` is solution.

The call `solve(sys, unks)` solves the system `sys` with respect to the variables in `unks`. It returns a list of solutions. Each solution is a list of assignments of the indeterminates that solves the system. If the user wants to assign the solutions to the indeterminates, these assignments have to be carried out in the same order as they appear in the list.

The notation `RootOf(...)` is used in the output for irreducible polynomial equations of degree 5 or higher and in systems for equations of degree 2 or higher.

```
>> solve(x^4-5*x^2+6*x=2,x);

    [3^(1/2)-1, -3^(1/2)-1, 1, 1]


>> solve(a*x^2+b*x+c,x);

    --                      2 1/2                        2 1/2 --
    |  - b + (- 4 a c + b )         - b - (- 4 a c + b )       |
    |  ---------------------,    ---------------------------   |
    --          2 a                          2 a              --
```

```
>> op(solve({u+v+w=1, 3*u+v=3, u^2-2*v-w=0},{u,v,w}));

   [v = u*(-3) + 3, w = u*2-2, u = RootOf(u*4 + u^2-4)]
```

See also: decompose, factor, linsolve

---

sort – sorts the elements of a list

Calling sequence:

sort(list ⟨ , proc ⟩)

Parameter:

list — list
proc — procedure

Summary:

sort sorts the elements of a list in ascending order. Optionally a procedure
can be given, which is used to specify a function which is used to compare the
elements of the list. If no such procedure **proc** is given, then the list is sorted
as follows:

- A list with real numbers is sorted numerically.
- A list with strings is sorted lexicographically.
- In other cases the list is sorted according to the system's internal order.
  This order is unambiguous but not documented. (Such sorting is useful
  when e.g. the user wants to compare two lists in which the sequence of
  the elements need not be identical.)

When strings are being compared it is important to note that in general capital
letters are sorted in front of small letters, thus "ZZ" is smaller than "a".

In order to sort arbitrary lists, a procedure **proc** which is used to compare
the elements of the lists may be passed to **sort** as the second argument. This
procedure must have two parameters and return a Boolean value. More exactly
the expression **proc(a,b)** must return TRUE, if **a** is to be sorted in front of **b**.

Sorting is not stable and elements with the same order may be swapped. The
mean run time for sorting n elements is $\mathcal{O}(n \log n)$. **sort** is a kernel function.

```
>> sort([4, -1, 2/3, 0.5]);
```

```
   [-1, 0.5000000000, 2/3, 4]

>> sort(["chip", "alpha", "Zip"]);

   ["Zip", "alpha", "chip"]

>> cmp := proc(a,b) begin bool(abs(a) < abs(b)) end_proc:
   sort([-2, 1, -3, 4], cmp);

   [1, -2, -3, 4]
```

See also: `insert_ordered`

■ **sqrfree** – square-free factorization of polynomials

Calling sequence:

```
sqrfree(poly)
sqrfree(expr)
```

Parameter:

poly   —   polynomial
expr   —   expression

Overloading:

1

Summary:

`sqrfree` calculates the square-free factorization of a multivariate polynomial. The coefficient ring of the polynomial may be either the rationals, a residue class ring `IntMod(p)` with prime number p or a domain representing a unique factorization domain of characteristic 0.

The argument of `sqrfree` may be either a polynomial or an expression. An expression is converted into a polynomial (see function `poly`). `sqrfree` returns `FAIL` if the expression cannot be converted.

`sqrfree(a)` returns the list [u, f1, e1, ..., fn, en] with a = u * f1^e1 * ...* fn^en, where the fi are the square-free factors of a and u is the content of a. The fi are primitive and pairwise different.

```
>> sqrfree(2-2*x-6*x^4+6*x^5+6*x^8-6*x^9-2*x^12+2*x^13);

    [2, x + x^2 + x^3 + 1, 3, x-1, 4]

>> sqrfree(x^6+x^4*y*6+x^2*y^2*9);

    [1, x*(y*3 + x^2), 2]

>> sqrfree(poly(2+5*x+4*x^2+x^3));

    [1, poly(x + 2, [x]), 1, poly(x + 1, [x]), 2]
```

See also: poly, factor

---

strlen – length of a string

Calling sequence:

strlen(string)

Parameter:

string   —   string

Overloading:

all

Summary:

strlen returns the number of characters in the character string string. strlen returns an error if the argument is not a character string. strlen is a kernel function.

```
>> strlen("Hello World");

    11
```

See also: substring, strmatch

---

strmatch – compares two character strings

Calling sequence:

`strmatch(string1, string2)`

Parameter:

`string1, string2`   —   character strings

Overloading:

all

Summary:

`strmatch` returns TRUE if the character string `string2` is identical to the character string `string1`, otherwise FALSE is returned. The arguments may contain the "wildcards" \? and *: \? stands for any or a missing character, * stands for any (also an empty) series of characters.

Due to its use as a wildcard, the sign \ must be masked with \, i.e. written as \\. `strmatch` returns an error if the arguments are not character strings. `strmatch` is a kernel function.

```
>> strmatch("Mississippi", "Mis\?i\*pi\*");

    TRUE
```

```
>> strmatch("Missi\\ssippi", "Miss\?i\\ssippi\?");

    TRUE
```

```
>> strmatch("a\?b", "a\?cb"), strmatch("a\?cb", "a\?b");

    FALSE, TRUE
```

See also: `strlen, substring`

■   **subs – substitutes complete subexpressions**

Calling sequence:

```
subs(expr, subsexpr... ⟨ , Unsimplified ⟩)
   subsexpr        ⟶    [ subsequation ...]
   subsexpr        ⟶    subsequation
   subsequation    ⟶    expr_1 = expr_2
```

Parameter:

```
expr              —   expression
expr_1, expr_2    —   expressions
Unsimplified      —   option
```

Overloading:

1

Summary:

subs substitutes subexpressions of the expression **expr**. The simplest call of **subs** is in the form **subs(expr, expr_1=expr_2)** with the substitutions specified in form of equations **expr_1=expr_2**. With this every subexpression of the form **expr_1** in expression **expr** is substituted by the expression **expr_2**.

The arguments of **subs** are substitution equations or lists of substitution equations. The arguments are processed in sequence from left to right:

- If the argument is a single substitution equation then the substitution is carried out and the result of this substitution is then processed further.

- However, if the argument is a list of substitution equations then all substitutions in the list (seen logically) are carried out simultaneously. The result of these substitutions is then processed further with the next argument. (See examples 2 and 3.) This is also called *parallel substitution*.

With **subs** only *complete* subexpressions are substituted. These are those subexpressions which are accessible by the function **op**. If the user wishes to substitute an incomplete subexpression, like **a+3** in **a+b+3**, then the function **subsex** has to be used. 

In general, a simplification of the expression is carried out after every substitution. Sometimes it is desirable to avoid simplification. This can be achieved with the option **Unsimplified**. **subs** uses a copy of **expr**, thus the argument **expr** remains unchanged. **subs** is a kernel function.

```
>> subs(a*(b+c), b+c=a);


   a^2


>> subs(a+x, a=x+y, x=z);
```

```
    y + z*2

>> subs(a+x, [a=x+y, x=z]);

    x + y + z

>> subs(a+x, [a=x+y, x=z], x=y);

    y*2 + z

>> subs(a+b+c, a+b=x);

    a + b + c

>> subs(a+b+2, a=1, b=0, Unsimplified);

    1 + 0 + 2
```

See also: op, subsop, subsex

■   **subsex – expanded substitution of subexpressions**

Calling sequence:

subsex(expr, *subsexpr*... ⟨ , Unsimplified ⟩)
  subsexpr   ⟶   [*subsequation* ...]
  subsexpr   ⟶   *subsequation*
  subsequation ⟶   expr_1 = expr_2

Parameter:

expr   —   expression
expr_1, expr_2   —   expressions
Unsimplified   —   option

Overloading:

1

Summary:

subsex substitutes subexpressions of **expr**. In contrast to the function **subs**, **subsex** also replaces *incomplete* subexpressions, e.g. a+3 in a+b+3. However, in contrast to **subs**, **subsex** is much slower, because the expressions have to be looked through entirely.

The simplest call of **subsex** is in the form **subsex(expr, expr_1 = expr_2)** of the substitution equation **expr_1 = expr_2**. Here all subexpressions with the form **expr_1** in the expression **expr** are substituted by the expression **expr_2**.

As in the function **subs**, the arguments of **subsex** are substitution equations or lists of them. The arguments are processed in sequence from left to right:

- If the argument is a single substitution equation then the substitution is carried out and the result of this substitution is then processed further.
- If the argument is a list of substitution equations then all the substitutions in the list (seen logically) are carried out simultaneously (parallel substitution).

In general, after every substitution the expression is simplified. Sometimes it is desirable to avoid this simplification. This can be achieved by using the option **Unsimplified**. **subsex** uses a copy of **expr**, thus the argument **expr** remains unchanged. **subsex** is a kernel function.

```
>> subsex(a+b*c*d+b*d, b*d=c);

    a + c + c^2


>> subsex(a+b*c*x, b*c=x, x=a);

    a + a^2


>> subsex(a+b*c*x, [b*c=x, x=a]);

    a + a*x


>> subsex(a+b+c, [a+c=x, b+c=y]);

    b + x


>> subsex(a+b+c, a+c=0, b=1, Unsimplified);
```

```
1 + 0
```

See also: op, subs, subsop

■ **subsop – substitutes operands**

Calling sequence:

```
subsop(expr, subsequation... 〈 , Unsimplified 〉)
  subsequation   ⟶   integer = expr_1
  subsequation   ⟶   [ integer ...] = expr_1
```

Parameter:

expr	—	expression
integer	—	non-negative integer
expr_1	—	expression
Unsimplified	—	option

Overloading:

1

Summary:

subsop substitutes subexpressions of the expression expr whereby the subexpressions are specified by their position as in the function op. Due to the position being specified directly subsop is faster as subs or subsex.

The simplest call of subsop is subsop(expr, i = expr_1) with the substitution equation i = expr_1. With this the i-th subexpression in the expression expr is substituted by the expression expr_1.

The operand to be substituted can also be specified through a path, as with the function op. E.g. with subsop(a+b*c, [2, 1]=x*y) the first operand b of the second operand b*c in the expression a+b*c is substituted by x*y. Thus the result is a+c*x*y. In contrast to the function op no ranges are permitted as operands.

The arguments of subsop are processed in sequence from left to right: The substitution is carried out and the result is processed further with the next argument. Parallel substitution as in subs or subsex is not possible.

In general, after every substitution the expression is simplified. Sometimes it is desirable to avoid this simplification. This can be achieved by using the option `Unsimplified`. `subsop` uses a copy of `expr`, thus the argument `expr` remains unchanged.

If access to the specified operand is not possible, e.g. with the expression `subsop(expr, nops(expr) + 1 = x)`, an error occurs if `ERRORLEVEL` equals 3. If the `ERRORLEVEL` is lesser than 3 then `FAIL` is the result. `subsop` is a kernel function.

```
>> subsop(2*c+a^2, 2=d^5);

    c*2 + d^5


>> subsop(a+b, 0=_mult);

    a*b


>> subsop(b+a^2, [2,2]=4, 1=x*y);

    x*y + a^4


>> subsop(a*b+c^2, 1=x*y, [1,2]=z);

    x*z + c^2


>> subsop(a+b+c, 1=0, 3=2, Unsimplified);

    0 + b + 2
```

See also: op, subs, subsex

---

substring – extracts a substring

Calling sequence:

`substring(string, integer_1, integer_2)`

Parameter:

`string`	—	string
`integer_1, integer_2`	—	non-negative integers

Overloading:

all

Summary:

`substring(string, pos, len)` returns `len` characters of the character string `string` starting at position `pos`. The first character of `string` has the position 0. The sum `pos + len` has to be smaller or equal to the length of `string`, i.e. smaller or equal to `strlen(string)`.

`substring` returns an error if the arguments `integer_1` and `integer_2` are not non-negative integers. If these arguments do not specify a proper substring then the error `Invalid index` is returned. `substring` is a kernel function.

```
>> substring("Hello World", 0, 5);

    "Hello"
```

```
>> substring("This is a string", 5, 4);

    "is a"
```

See also: `strlen, strmatch`

■ **sum – definite and indefinite summation**

Calling sequence:

```
sum(f, k)
sum(f, k=a..b)
sum(f, k=RootOf(p,x))
```

Parameters:

f, a, b	—	expressions
k	—	indeterminate
p	—	polynomial

Summary:

`sum(f, k)` computes the indefinite sum of `f` with respect to `k`, that is an expression `g(k)` such that $f = g(k + 1) - g(k)$. The call `sum(f, k=a..b)` computes the definite sum of `f` from `a` to `b`. The function `sum` implements

Abramov's algorithm for rational functions, Gosper's algorithm for hypergeometric function, and Zeilberger's algorithm for the definite summation of holonomic functions.

The special form `sum(f, k=RootOf(p,x))` normalizes the sum of the rational function `f` in `k` over the roots of the polynomial `p` in `x`.

```
>> sum(1/(x^2-1),x);

      1         1
  - --- - ----------
    2 x    2 (x - 1)
```

```
>> sum(1/k/(k+2)^2,k);

      1     psi(k + 2, 1)         1
  - --- + ------------- + ----------
    4 k          2          - 4 k - 4
```

```
>> sum(binomial(n+k,k),k);

    k*binomial(k + n, k)/(n + 1)
```

```
>> sum(1/(n^2+21*n),n=1..infinity);

    18858053/108636528
```

```
>> sum(binomial(n,k)/2^n-binomial(n+1,k)/2^(n+1),k);

    - 2 k binomial(n, k) + k binomial(n + 1, k)
    -------------------------------------------------
                  n         n         n
            - 2 2   + 4 k 2   - 2 n 2
```

```
>> sum(k^2,k=RootOf(x^3+a*x^2+b*x+c,x));
```

```
    b*(-2) + a^2
```

See also: `psi`

■ **sysname** – returns the name of the operating system

Calling sequence:

`sysname()`

Parameter:

—   none

Summary:

`sysname` returns a string containing the name of the operating system in which
MuPAD is currently executed. At present the following return values are
possible:

- `"UNIX"` for UNIX
- `"MSDOS"` for MS-DOS
- `"MACOS"` for the Macintosh operating system

`sysname` is a kernel function.

```
>> sysname();

   "UNIX"
```

See also: `system`

■ **System Variables, Constants, Operators and Functions**

Environment-Variable:

```
DIGITS         ERRORLEVEL     EVAL_STMT      HISTORY
LEVEL          LIB_PATH       MAXLEVEL       PRINTLEVEL
PRETTY_PRINT   READ_PATH      TEXTWIDTH      WRITE_PATH
```

System Constants:

```
E   I   FALSE  NIL   PI    TRUE
```

System Operators:

```
+   -   *   /   ^   @   $   ..   .     ::   <   >   >=   <=
=   <>  div mod union   intersect    minus   not  and  or
```

System Functions:

Type Converting	—	bool, ceil, float, floor, frac, round, trunc
Transcendental Functions	—	exp, ln
— Trigonometric Functions	—	sin, cos, tan
— Arc Functions	—	asin, acos, atan
— Hyperbolic Functions	—	sinh, cosh, tanh
— Area Functions	—	asinh, acosh, atanh
Arithmetic Functions	—	abs, fact, frac, igcd, igcdex, ifactor, isprime, ithprime, ilcm, max, min, modp, mods, nextprime, phi, powermod, random, sign, sqrt
Symbolic Manipulation	—	conjugate, D, denom, diff, expand, Im, normal, numer, O, Re, taylor
Polynomial Operations	—	coeff, content, degree, degreevec, divide, evalp, expr, gcd, genpoly, icontent, iszero, lcm, lcoeff, lmonomial, lterm, mapcoeffs, multcoeffs, norm, nterms, nthcoeff, nthmonomial, nthterm, pdivide, poly, Poly, tcoeff
Type Identification	—	anames, domtype, testtype, type
Manipulation of Expressions	—	append, contains, domattr, funcattr, has, id, indets, index_val, map, extnops, extop, extsubsop, nops, op, sort, subs, subsop, subsex
Evaluation of Expressions	—	eval, evalassign, hold, level, val
Creating Data Structures	—	array, built_in, domain, func_env, genident, new, null, seq, table
Input and Output	—	fclose, finput, fopen, fprint,

	—	`ftextinput, input, print, protocol, read, textinput, write`
Manipulation of Strings	—	`expr2text, strmatch, strlen, substring, tbl2text, text2expr, text2list, text2tbl`
Parallelism	—	`global, readpipe, readqueue, topology, writepipe, writequeue`
Program Statistic	—	`history, load, bytes, time, rtime`
Graphics	—	`plot2d, plot3d`
Procedure Specific Functions	—	`args, error, func, return, testargs, traperror`
Searching for Errors	—	`debug, trace`
Miscellaneous	—	`contains, help, last, loadlib, loadproc, pathname, reset, sysname, system`

■ **`system` – executes an operating system command**

Calling sequence:

`system(string)`

Parameter:

`string` — character string

Summary:

`system` executes the operating system command given by `string` is executed. `system` returns an integer with the status value of the command. This value is system dependent. Under UNIX the status value 0 means that the command has been successfully executed.

This function is not available in all MuPAD implementations, e.g. not in MacMuPAD. The form of output is still dependent on the user's interface, e.g. in XMuPAD output is directed to a terminal window. `system` is a kernel function.

```
>> system("date");
```

```
    Wed Aug 16 11:42:00 MET DST 1995
    0
```

See also: `sysname`

`table` – defines a table

Calling sequence:

```
table()
table(equation ...)
  equation   ⟶   index = value
```

Parameter:

```
index   —   expression
value   —   expression
```

Summary:

`table` is used to define tables. The parameters of the function `table` consist of equations in which the left side is interpreted as the index of an entry and the right side is interpreted as the corresponding value. The table is initialized with the entry of these equations. The index and value of an entry are subject to no conditions.

In general an explicit definition of tables is not necessary, because an assignment to an indexed identifier that cannot be evaluated to a list or array assigns this identifier automatically a table. A detailed description of tables can be found in section 2.3.11. `table` is a kernel function.

```
>> T := table(a = b, c = d): T[a];

    b


>> T := table( (a,b) = 100,  a+b = 200): T[a,b]+T[a+b];

    300


>> S[4] := 42:
   S;
```

```
    table(4=42)
```

See also: `array`

■ `taylor` – Taylor series

Calling sequence:

```
taylor(expr, equ ⟨ , int ⟩)
taylor(expr, ident ⟨ , int ⟩)
```

Parameter:

`expr`	—	expression
`ident`	—	identifier
`equ`	—	equation
`int`	—	positive integer

Summary:

`taylor` returns the Taylor series of the expression given by `expr`.

- As second argument an equation in the form `x=p` can be given. `x` has to be an unknown and `p` has to be an expression. The Taylor expansion takes place relative to `x` around the point `p`.

- Alternatively a simple identifier can be given as the second argument. In this case the series relative to the unknown is expanded around the point 0.

Optionally as the third argument the desired order of the Taylor series can be given. If no order is given then the value of the global variable `DEGREE` is used as the order. The default value of `DEGREE` is 6.

The remainder term of the Taylor series is printed as an order term in the form `O(...)`.

`taylor` returns an error if `expr` cannot be expanded into a Taylor series.

```
>> taylor(exp(x^2),x);

    O(x^6) + x^2 + x^4*1/2 + 1


>> taylor(ln(x), x=1, 4);
```

```
O(x^4) + x + (x-1)^2*(-1/2) + (x-1)^3*1/3 - 1
```

See also: `diff, O`

---

tbl2text — converts a table into a character string

Calling sequence:

`tbl2text(table)`

Parameter:

`table` — table

Overloading:

all

Summary:

`tbl2text` converts the table `table` into a character string. The table has to have exactly n entries with the indices `1..n`. These entries have to be strings. The result is one character string created through the concatenation of all table entries from `table[1]` to `table[n]`.

`tbl2text` returns an error if the indices of the entries do not form a range `1..n` with n greater or equal to 1. `tbl2text` is a kernel function.

```
>> tbl2text(table(1="Text ",2="in ",3="a ",4="table."));

    "Text in a table."
```

See also: `text2expr, text2list, text2tbl`

---

tcoeff — the lowest coefficient of a polynomial

Calling sequence:

```
tcoeff(poly)
tcoeff(expr ⟨ , indets ⟩)
  indets  ⟶  [ expr_1 ...]
```

Parameter:

```
poly           —  polynomial
expr, expr_1   —  expressions
```

Overloading:

1

Summary:

`tcoeff` returns the lowest coefficient of a polynomial with respect to the lexicographical order. (The lowest coefficient with respect to an other order cannot be determined.)

The first argument of `tcoeff` is allowed to be a polynomial or an expression and optionally a list of indeterminates. An expression is converted into a polynomial in the specified indeterminates (see function `poly`). `tcoeff` returns `FAIL` if the expression cannot be converted into a polynomial. `tcoeff` is a kernel function.

```
>> tcoeff(3*x^3+x^2*y^2+2, [x,y]);

     2


>> tcoeff(poly(3*x^3+x^2*y^2, [x,y]));

     1


>> tcoeff(0, [x,y]);

     0
```

See also: `coeff`, `lcoeff`, `nterms`, `nthcoeff`, `poly`

■ testargs – allows control of parameter testing

Calling sequence:

```
testargs(⟨ bool ⟩)
```

Parameter:

```
bool   —   Boolean value
```

Summary:

testargs is used to control parameter testing in procedures. Very often when a procedure is called by the user its parameters are to be tested, whereas when procedures are called by other procedures this is not necessary (assuming that the programming is error-free).

With **testargs** the user can determine if a program was directly called by the user or by another procedure and then carry out the relevant test. The expression **testargs()**, in general, returns TRUE when it was executed in a procedure called on the interactive level, i.e. by the user. However, if the procedure was called by another procedure then **testargs()** returns FALSE. The user is able to control the testing of parameters as follows:

```
p := proc(x)
   local ...;
begin
   if testargs() then
       test parameter x...
   end_if;
   The actual algorithm follows here...
end_proc;
```

Above we mentioned the condition "in·general": **testargs** has two states, the *normal state* as described above and a *test state*. In the test state the expression **testargs()** always returns TRUE, i.e. the parameters are always tested. This is useful during trouble shooting when programming.

- The test state is set with **testargs(TRUE)**. The call returns FALSE if **testargs** was already in normal state.

- The normal state is reset with **testargs(FALSE)**. The call returns TRUE if **testargs** was already in test state.

testargs is a kernel function.

```
>> p1 := proc() begin
        if testargs() then "test" else "no test" end_if
   end_proc:
   p2 := proc() begin p1() end_proc:
   p1(), p2();

   "test", "no test"
```

```
>> testargs(TRUE), p1(), p2();

    FALSE, "test", "test"
```

See also: args

■ **testtype – compares types**

Calling sequence:

testtype(expr_1, expr_2)

Parameter:

expr_1, expr_2   —   expressions

Summary:

testtype is used to determine if expression expr_1 is of the type expr_2, or if expr_1 can be converted into this type. If this is the case then the function returns TRUE, in other cases it returns FALSE.

For this, firstly, the function tests if the domain that contains the element expr_1 has a method with the index "testtype". If this is the case then this method is called with the arguments of testtype as its arguments and the result is returned. If, however, the method returns the value FAIL or such a method does not exist, then domattr(expr_2, "testtype")(expr_1, expr_2) is called — if expr_2 is a domain then the method is searched for here otherwise it is searched for in the domain that contains expr_2 — and this result is returned.

In the pre-defined basic domains this method is implemented so that the user can determine with testtype if expr_1 is an element of the domain expr_2. Furthermore there is the domain NUMERIC which enables the user to check if expr_1 is a number.

The basic types defined in MuPAD are listed in the table A.1 in the appendix. testtype is a kernel function.

```
>> testtype(x + y, DOM_EXPR);

    TRUE
```

```
>> testtype(2^3, DOM_INT);
```

```
    TRUE
```

```
>> testtype(2.3 + 4*I, NUMERIC);
```

```
    TRUE
```

See also: type, domtype

TeX — TeX-formatted output

Calling sequence:

```
TeX(expr)
```

Parameter:

expr — expression

Overloading:

1

Summary:

TeX returns a TeX-formatted string representing expr. This string may be printed to a file using fprint. Use the printing option Unquoted to remove quotes and to expand special characters like newlines and tabs.

The output string may be used in the math-mode of TeX. Note that TeX currently doesn't break large formulas into smaller ones.

```
>> TeX(int(exp(x^2)/x, x));
```

```
    "\\int \\frac{1}{x} \\mbox{exp}\\left(x^2\\right) d x"
```

```
>> print(Unquoted,%);
```

```
    \int \frac{1}{x} \mbox{exp}\left(x^2\right) d x
```

See also: fprint

■ **text2expr** – **converts character strings into MuPAD statements**

Calling sequence:

`text2expr(string)`

Parameter:

`string` — character string

Overloading:

all

Summary:

`text2expr` converts the character string `string` into a MuPAD statement if the character string contains a syntactically correct statement sequence. In contrast to the function `input`, the statement sequence need not be closed with a semicolon or a colon.

The statement is returned without being evaluated further. It can be evaluated by use of the function `eval`. `text2expr` is a kernel function.

```
>> text2expr("x:= 3; x+2+1");

    (x:= 3; x+2+1)


>> text2expr("a:= 3");
   a;
   eval(%2), a;

    a:= 3
    a
    3, 3
```

See also: `input`, `eval`

■ **text2list** – **converts a string into a list**

Calling sequence:

`text2list(string, separators, ⟨ Cyclic ⟩ )`

Parameter:

string	—	character string
separators	$\longrightarrow$	Mout [string_1, ...]
string_1	—	character string
Cyclic	—	option

Overloading:

all

Summary:

text2list decomposes the character string **string** into substrings and stores them in a new list. This list is then returned.

For the decomposition of **string** the strings in the list [string_1 ...] are used as separators.

If the option **Cyclic** is not given then **string** is decomposed as follows: The first occurrence of one of the *separators* is located in **string**. If a separator is found, the substring up to this separator is entered as the first element of the list and the separator is entered as the second element. The remaining character string is processed as above until there are no more characters in the string. If no separator can be located the function terminates.

If the option **Cyclic** is given then **string** is decomposed as follows: The first occurrence of the first separator **string_1** is located in string. If **string_1** is found, the substring up to this separator is entered as the first element of the list and the separator **string_1** is entered as the second element. The remaining character string is processed as above, except that now the second separator in the list *separators* is searched. If a separator can not be found in the character string, the function terminates. After the last separator has been found, search continues with the first separator again.

text2list is a kernel function.

```
>> text2list("Oh,a text,how nice!", ["an","how","a"]);

    [ "Oh,", "a", " text,", "how", " nice!" ]


>> text2list("Oh,a text,how nice!", [" ",","],Cyclic);

    [ "Oh,a", " ", "text", ",", "how", " ", "nice!"  ]
```

See also: `tbl2text, text2expr, text2tbl`

■ `text2tbl` – converts a string into a table

Calling sequence:

`text2tbl(string, [ string_1, ...], ⟨ Cyclic ⟩ )`

Parameter:

`string`   — character string
`string_1` — character string
`Cyclic`   — option

Overloading:

all

Summary:

`text2tbl` decomposes the character string `string` in substrings and stores these as character strings in a new table. This table is then returned. The table contains n entries with the indices 1..n.

For the decomposition of `string` the strings in the list `[string_1 ...]` are used as separators.

If the option `Cyclic` is not given then `string` is decomposed as follows: The first occurrence of one of the *separators* is located in `string`. If a separator is found, the substring up to this separator is stored under the index 1 in the table and the separator is stored under the index 2. The remaining character string is processed as above until there are no more characters in the string. If no separator can be located the function terminates.

If the option `Cyclic` is given then `string` is decomposed as follows: The first occurrence of the first separator `string_1` is located in string. If `string_1` is found, the substring up to this separator is entered is stored under the index 1 in the table and the separator is stored under the index 2. The remaining character string is processed as above, except that now the second separator in the list *separators* is searched. If a separator can not be found in the character string, the function terminates. After the last separator has been found, search continues with the first separator again.

In both cases a concatenation of the table entries give the original character string. `text2tbl` is a kernel function.

```
>> text2tbl("Oh,a text,how nice!",["an","how","a"]);

   table(1="Oh,",2="a",3=" text,",4="how",5=" nice!")

>> text2tbl("Oh,a text,how nice!", [" ",","],Cyclic);

   table(1="Oh,a",2=" ",3="text",4=",",5="how",6=" ",7="nice!")
```

See also: `tbl2text, text2expr, text2list`

`textinput` – **interactive input of text**

Calling sequence:
```
textinput(expr ...)
textinput()
```

Parameter:

`expr`   —   character string or identifier

Summary:

`textinput` allows the user to enter strings interactively. The arguments of `textinput` can be any series of character strings or identifiers. The character strings are printed in sequence, with identifiers a prompt is printed and the user is expected to enter a character string. This character string is then assigned to the identifier. The return value of `textinput` is the last character string entered by the user.

If `textinput` is called without arguments then only a prompt is printed and a character string is expected as input. This is then the return value of the function.

When a character string is entered the enclosing " should not be given. The character string may extend over several lines. The individual lines are then separated in the character string by a \n. By use of `textinput` it is not possible to mask the end of a line with a backslash. In contrast to the usual input mode a backslash can be followed by any characters. In the terminal version of MuPAD input is closed with <Ctrl-D>.

More information about input can be found in the description of the user interfaces. `textinput` is a kernel function.

```
>> textinput("Name", n);

   Name >>MuPAD
   "MuPAD"


>> inp := textinput();

   >>test
   "test"
```

See also: input

■ **TEXTWIDTH – number of characters per line**

Calling sequence:

TEXTWIDTH := integer

Parameter:

integer — integer greater than 9

Summary:

TEXTWIDTH defines the length of a text line used when printing output. Any integer between 9 and $2^{31} - 1$ can be assigned to TEXTWIDTH. Through the assignment TEXTWIDTH := NIL;, TEXTWIDTH is set to its default value of 75.

TEXTWIDTH is an environment variable.

■ **time – measures execution time**

Calling sequence:

time(expr)
time()

Parameter:

expr — expression

Summary:

`time(expr)` evaluates the expression `expr` and returns the CPU time needed. `time()` returns the total CPU time passed since starting the MuPAD session. The time is given in milliseconds.

On some types of computer (those that do not have "time-sharing"), e.g. the Macintosh, the CPU time is roughly equal to the real time. The times for input and output are also included in the complete time. `time` is a kernel function.

```
>> time((a := isprime(1234567891))), a;

   10, TRUE
```

See also: `bytes, rtime`

---

topology – information about parallel structures

Calling sequence:

```
topology()
topology(integer)
topology(Cluster)
```

Parameter:

```
integer   —   positive integer
Cluster   —   identifier
```

Overloading:

all

Summary:

`topology()` returns the number of clusters that MuPAD currently consists of. `topology(integer)` returns the number of processes that the cluster with the number `integer` makes available for the micro-parallelism. Finally, the call `topology(Cluster)` returns the cluster number of the cluster. This function is described in more detail in section 2.8.2.5. `topology` is a kernel function.

---

trace – protocols the execution of procedures

Calling sequence:

```
trace()
trace(identifier ...)
```

Parameter:

`identifier` — identifier

Summary:

`trace` enables the user to name procedures whose execution is to be protocolled. `trace` can only be used in the terminal version of MuPAD. A detailed description can be found in chapter 3. Please note:

- MuPAD must be in trace mode.
- Only user-defined procedures can be protocolled.
- `trace` only affects procedures that are already defined.
- interactively entered procedures cannot be protocolled.

The call `trace()` without arguments, has the effect that no procedure is protocolled. The call `trace(op(anames(1)))` has the effect that all user-defined procedures are protocolled. `trace` is a kernel function.

See also: `debug, PRINTLEVEL`

■ `traperror` – intercepts errors

Calling sequence:

`traperror(expr)`

Parameter:

`expr` — expression

Summary:

`traperror` intercepts errors caused by system functions or the function `error`. Normally an evaluation is completely aborted if an error occurs and the interactive level is entered. However, when an error occurs during the evaluation of the argument `expr` of `traperror`, then only the evaluation of `expr` is aborted and program execution continues after the statement of `traperror`. In this case `traperror` returns an error code which is a positive integer. If no error occurs, traperror returns 0. `traperror` is a kernel function.

```
>> a := 1/x: traperror(subs(a, x=0));

      1025

>> traperror((error("My error");1/0));

      1028

>> x:= 0: traperror((x:= 1/x)), x;

      1025, 0
```

See also: error

---

trunc – truncates a number

Calling sequence:

trunc(expr)

Parameter:

expr — expression

Overloading:

all

Summary:

trunc returns the integral part of expr if the argument is a real number.

The function returns an error if the argument is a complex number. if the argument cannot be evaluated to a number, trunc returns the unevaluated function call.

If the exponents of real numbers are very large then a call of trunc can lead to a loss of accuracy that is not identified as an error. trunc is a converting function of the kernel.

```
>> trunc(8/3);

      2
```

```
>> trunc(-2.6);

   -2
```

See also: frac, ceil, round, floor

■ **type – type of an expression**

Calling sequence:

type(expr)

Parameter:

expr — expression

Overloading:

1

Summary:

type returns the expression type of expr. If expr is an element of the basic domain *DOM_EXPR*, type usually returns a string identifying the operator of the expression. If expr is no element of *DOM_EXPR* then type normally returns the domain which expr belongs to.

One may define the type of expressions formed by user defined operators by adding the attribute "type" to the function environment of the operator. One may also define the type of own domain elements by adding a method "type" to the domain.

The pre-defined expression types are listed in table A.2 in the appendix. Note that expressions of the form a-b and a/b are represented internally as expressions a+(-1)*b and a*b^(-1). Thus they have the expression types "_plus" and "_mult". To determine the domain an object belongs to one should use the function domtype.

In contrast to most other functions, arguments which are expression sequences, are not flattened. type is a kernel function.

```
>> type(x + y*z);

   "_plus"
```

```
>> type(x - y);
```

```
"_plus"
```

See also: `testtype`, `domtype`

MuPAD options under UNIX

Calling sequence:

mupad [−c] [−g] [−G filedescr] [−h helppath] [−i] [−l libpath] [−m mampath]
       [−p stacksize] [−r] [−s syspath] [−S] [−t] [−u userpath] [−v] [file]
xmupad [−L language] *mupad-options*

Summary:

All MuPAD options under UNIX at a glance:

−c	Connect modus	(only for the debugger mdx)
−g	Debug modus	
−G	File descriptor for graphics data	(only for the graphics tool)
−h	Path name for the help file index	(Default: `$R/share/help`)
−l	Path name for the library	(Default: `$R/share/lib`)
−m	Path name for the MAMMUT init file	(Default: library path)
−p	Size of the PARI stack	(Default: 250000)
−r	Raw modus	(only for XMuPAD)
−s	Path name for the system init file	(Default: library path or .)
−S	Start silently without printing the banner	
−t	Trace modus	
−u	Path name for the user init file	(Default: ~)
−v	Verbose modus	

With `$R` the value of the (UNIX) environment variable `MuPAD_ROOT_PATH` is
meant. The names of the initialization files are:

.mupadinit	user init file
.mupadsysinit	system init file
.mupadhelpindex	help file index
.mupadhelpindextty	help file index
.MMMinit	MAMMUT init file

More detailed information can be found on the manual pages **mupad** and **xmupad**. These can be called with e.g. **man mupad**.

◼ **unalias – delete alias definitions**

Calling sequence:

```
unalias(expr...)
unalias()
```

Parameter:

expr — expression

Summary:

unalias deletes the abbreviations given by **expr...**. If **unalias** is called without any arguments all existing alias definitions are deleted.

```
>> alias(sqr(x) = RootOf(x^2+1)): sqr(y);
   unalias(sqr): sqr(y);

   RootOf(y^2 + 1)
   sqr(y)
```

See also: `alias`

◼ **userinfo – prints progress information in procedures**

Calling sequence:

```
userinfo(posint, expr ...)
```

Parameter:

posint	—	positive integer
expr	—	expression

Summary:

userinfo may be used to print progress informations during the execution of procedures. These informations may be algorithms used, intermediate data computed and the like. They should give the user some insight into the computation executed.

An information is printed if the "information level" of either the procedure containing the call of **userinfo** or the global information level is greater or equal to the level given by the first argument **posint** (see function **setuserinfo**).

The output consists of the name of the procedure containing the call of **userinfo** followed by the arguments **expr**. The output is printed without quotes.

```
>> f:= proc(x) begin userinfo(2,"Argument:",x); end_proc:
   setuserinfo(f,2):
   f(12);

     f, Argument:, 12
```

See also: help, info, setuserinfo

---

val – value of an expression (obsolete)

Calling sequence:

val(expr)

Parameter:

expr   —   expression

Summary:

The function **val** has become unnecessary with MuPAD version 1.2 and is only included here in for backwards compatibility. It will not be included in future versions. 

val replaces all identifiers in the expression **expr** by their values and returns the corresponding expression, which is not further evaluated. Especially the depth of substitution is not taken into consideration. Thus, the resulting expression is also not simplified. The only exceptions are sets: after execution of **val**, double set elements are removed. **val** is a kernel function.

```
>> a:=0: val(a*b+4+0);

     0*b + 4 + 0

>> a:=b: val({a, b, a*0});

     {b, b*0}
```

See also: `eval`, `hold`

■ `version` – returns the actual **MuPAD** version number

Calling sequence:

`version()`

Parameter:

— none

Summary:

`version` returns a list with three non-negative integers. The first number is the actual version number of MuPAD, the second is the actual major release number and the third is the actual minor release number.

```
>> version();

     [1, 2, 2]
```

■ `write` – writes the values of variables into a file

Calling sequence:

```
write(integer ⟨ , identifier ... ⟩ )
write( ⟨ type, ⟩ string ⟨ , identifier ... ⟩ )
```

Parameter:

`integer`	—	non-negative integer
type	—	one of the options **Bin** or **Text**
`string`	—	character string
`identifier`	—	identifier

Overloading:

1

Summary:

write saves the values of the variables given by **identifier** into a file. If no variable is given then all variables are saved, including the library routines and environment variables. With this a complete session can be saved. The stored values can be read with **read**.

The file can be specified either by its name or by a file descriptor:

- If the first argument is a positive integer **integer** then this is considered as a file descriptor. The corresponding file must already have been opened with **fopen**.

- If the first argument is 0 then the output is printed on the screen.

- Is the first argument is a string, it is considered a file name and the corresponding file will be opened. Already stored data in this file are overwritten. If the file does not exist it will be created.

  If **write** is called on the interactive level in MacMuPAD then the user is warned before the file is overwritten. The operation can then be cancelled.

  If the variable **WRITE_PATH** has a value then the file is created in this directory; if not it is created in the current working directory. (This is the directory in which MuPAD was started.)

  If, additionally, one of the options **Bin** or **Text** is given then the file is opened as a binary or text file. If this option is not given the file is opened as a binary file.

  After the output is written the file is closed.

- In MacMuPAD, if an empty character string is given by the argument **string** then a dialog is opened in which the user can choose a file

When reading in a complete session with **read**, please remember that the new session must be started in the same environment as the stored session was started in. (The stored data sometimes contain path names of files still to be read.)

With textual storage the variable values are stored as assignments of the form

```
identifier := hold(expr):
```

For instance after a:= b+1; b:= 3; the variables a and b are stored as a:= hold(b+1): b:= hold(3):. **write** is a kernel function.

```
>> a:= b+1: write(Text,"test",a):
   ftextinput("test");

     "a:= hold(b+1):"


>> reset(): read("test"): a;

     b + 1
```

See also: `fclose`, `fopen`, `fprint`, `protocol`, `read`, `WRITE_PATH`

■ **writepipe** – writes into a pipe

Calling sequence:

`writepipe(expr_1, integer, expr_2)`

Parameter:

`expr_1`	—	expression
`integer`	—	number of a processor node
`expr_2`	—	expression

Overloading:

2

Summary:

`writepipe(expr_1, integer, expr_2)` writes the expression `expr_2` into the pipe with the name `expr_1` that leads to the cluster with the number `integer`. This function is described in more detail in section 2.8.2.4. `writepipe` is a kernel function.

See also: `readpipe`, `readqueue`, `writequeue`

■ **writequeue** – writes into a queue

Calling sequence:

`writequeue(expr_1, integer, expr_2)`

Parameter:

```
expr_1   —   expression
integer  —   number of a processor node
expr_2   —   expression
```

Overloading:

2

Summary:

`writequeue(expr_1, integer, expr_2)` writes the expression `expr_2` into the queue with the name `expr_1` of the cluster with the number `integer`. This function is described in more detail in the section 2.8.2.2. `writequeue` is a kernel function.

See also: `readpipe, readqueue, writepipe`

---

zeta – Riemann zeta function

Calling sequence:

`zeta(x)`

Parameter:

x   —   expression

Overloading:

1

Summary:

The zeta function is defined as

$$\texttt{zeta(x)} = \zeta(x) = \sum_{i=0}^{\infty} i^{-x}$$

If x is a negative, even integer, then

$$\zeta(x) = 0,$$

if x is a negative, odd integer, then

$$\zeta(x) = -\frac{\text{bernoulli}(x)}{1-x},$$

if x is zero, then

$$\zeta(x) = -\frac{1}{2},$$

if x is a positive, even integer, then

$$\zeta(x) = \frac{(2\pi)^x}{2} \left| \frac{\text{bernoulli}(x)}{x!} \right| .$$

zeta returns the function call with evaluated arguments, if the argument cannot be evaluated to a number. zeta is a kernel function.

```
>> zeta(-23);

    236364091/65520
```

```
>> zeta(3-I);
   float(%);

    zeta(3 - I)
    1.107214408 + 0.1482908671*I
```

See also: bernoulli

■ **zip – combines lists element-by-element**

Calling sequence:

```
zip(list_1, list_2, func)
zip(list_1, list_2, func, expr)
```

Parameter:

```
list_1, list_2  —   lists
func            —   function
expr            —   expression
```

Overloading:

1,2

Summary:

zip combines the lists list_1 and list_2 element-by-element. The function func is called with all pairs list_1[i], list_2[i] as arguments one after the other and the results of the function calls are combined into a new list.

The call of func has the form

```
func(list_1[i], list_2[i])
```

for each pair of list elements. If the two lists are of different lengths then the functionality is dependent on the optional fourth parameter **expr**:

- If the fourth parameter is missing then **func** is applied only to the first n list elements, whereby n is the length of the shorter list.
- If a fourth parameter is given then this is used as the default value instead of the "missing" list elements in the shorter list.

zip is a kernel function.

```
>> zip([a, b, c, d], [x, y], _plus);

   [a + x, b + y]

>> zip([a, b, c, d], [x, y], _plus, 0);

   [a + x, b + y, c, d]
```

See also: map

Appendix A

Tables

Data Type	Operand Number	Operands
DOM_ARRAY	Number of subarrays	subarrays
DOM_BOOL	1	object
DOM_COMPLEX	2	real and imaginary part
DOM_DOMAIN	number of entries and methods	entries and methods
DOM_EXEC	4	evaluation functions, name, remember table
DOM_EXPR	dependent on the expression type, see table A.2 and table A.3	
DOM_EXT	at least 1	corresponding domain, arbitary data type
DOM_FAIL	1	object
DOM_FLOAT	1	object
DOM_FUNC_ENV	3	evaluation function, output function, attribute table
DOM_IDENT	1	object
DOM_INT	1	object
DOM_NIL	1	object
DOM_NULL	0	
DOM_POINT	3 or 4	coordinates (2 or 3) and color

Table A.1: Data Types and their Operands

Data Type	Operand Number	Operands
`DOM_POLY`	3	polynomial expression, list of indeterminates, coefficient ring
`DOM_POLYGON`	number of points and options	points and options
`DOM_PROC`	6	parameter list, list of the local variables, options, body, remember table, name
`DOM_RAT`	2	numerator and denominator
`DOM_SET`	number of set elements	element of the set
`DOM_LIST`	number of elements in the list	entries of the list
`DOM_STRING`	1	object
`DOM_TABLE`	number of table entries	equations of the form `index = entry`

Table A.1: Data Types and their Operands

Expression type	Operand number	Operands
`"_and"`	number of objects linked by `"_and"`	objects linked by `"_and"`
`"_break"`	0	
`"_concat"`	number of objects linked by `"_concat"`	objects linked by `"_concat"`
`"_div"`	2	divisor and dividend
`"_equal"`	2	left and right hand side of the equation
`"_exprseq"`	number of expressions in the sequence	expressions in the sequence
`"_fconcat"`	number of objects linked by `"_fconcat"`	objects linked by `"_fconcat"`
`"function"`	number of parameters	parameter in the function call
`"_index"`	size of the index	operands of the index

Table A.2: Expression Types and their Operands

Expression type	Operand number	Operands
`"_intersect"`	2	objects intersected by `"_intersect"`
`"_leequal"`	2	left and right hand side of the inequality
`"_less"`	2	left and right hand side of the inequality
`"_minus"`	number of objects linked by `"_minus"`	Objects linked by `"_minus"`
`"_mod"`	2	divisor and dividend
`"_mult"`	number of factors linked by `"_mult"`	factors linked by `"_mult"`
`"_next"`	0	
`"_not"`	1	negated object
`"_or"`	number of objects linked by `"_or"`	objects linked by `"_or"`
`"_plus"`	number of summands	summands
`"_power"`	2	base and exponent
`"_quit"`	0	
`"_range"`	2	lower and upper limit
`"_seqgen"`	number of operands of the sequence generator (1 or 2)	operands of the sequence generator
`"_unequal"`	2	left and right hand side of the inequality
`"_union"`	number of objects linked by `"_union"`	objects linked by `"_union"`

Table A.2: Expression Types and their Operands

Statement type	Number of operands	Operands
`"_assign"`	2	left and right hand side of the assignment
`"_case"`	dependent on the number of of branches	comparison expressions, Statements of the of branches and the otherwise branch
`"_for"`	5	loop variable, lower limit, upper limit, step width, body
`"_for_down"`	5	loop variable, lower limit, upper limit, step width, body
`"_for_in"`	3	loop variable, expression, body
`"_for_in_par"`	4	loop variable, expression, private variables, body
`"_for_par"`	6	loop variable, lower limit, upper limit, step width, private variables, body
`"_if"`	3	condition, then branch, else branch
`"_parbegin"`	2	private variables, body
`"_repeat"`	2	body, condition
`"_seqbegin"`	1	body
`"_stmtseq"`	number of statements in the sequence	statements in the sequence
`"_while"`	2	condition, body

Table A.3: Expression Types for Statements and their Operands

Operator	Priority
::	18
.	17
@	16
not	15
&<ident>	14
^	13
* / mod div	12
+ -	11
intersect	10
minus	9
union	8
..	7
< <= > >= = <>	6
$	5
and	4
or	3
,	2
; :	1

Table A.4: Binding Priority of the Operators (in descending order)

Environment variable	Value
DIGITS	10
LEVEL	100
ERRORLEVEL	0
PRINTLEVEL	0
MAXLEVEL	100
HISTORY	[20, 3]
TEXTWIDTH	75
LIB_PATH	
READ_PATH	
WRITE_PATH	
EVAL_STMT	TRUE
PRETTY_PRINT	TRUE

Table A.5: Environment Variables and their Default Setting

Entry	Function	see p.
`"convert"`	converts expressions into domain elements	46
`"divex"`	divides domain elements	46
`"domattr"`	returns the entries of a domain	64
`"elemattr"`	returns the entries of a domain element	64
`"evaluate"`	evaluation of a domain element	64
`"func_call"`	function call with a domain element as the operator	65
`"_index"`	indexed access of a domain element	65
`"intmult"`	multiplication of a domain element with an integer	46
`"invert"`	invert a domain element	66
`"negate"`	negate a domain element	66
`"name"`	name of the domain	63
`"new"`	creates a new domain element	69
`"norm"`	norm of a domain element	46
`"not"`	boolean complement of a domain element	66
`"one"`	the 1 of the domain	46
`"posteval"`	evaluation of a domain element using the substitution depth 1	65
`"print"`	prints the domain element	67
`"set_func_call"`	function call as the left side of an assignment	65
`"set_index"`	indexed access as the left side of an assignment	66
`"testtype"`	tests, if a datum can be converted into a domain element	140
`"zero"`	the 0 of the domain	46

Table A.6: Special Domain Entries

Button	Command
Next	n
Step	s
Cont	c
Quit	q
Where	w
Up	u
Down	d
Clear all	a
Clear	C *<filename>* *<line>*
Goto proc	g *<name>*
Print	p *<expr_1>* ... *<expr_n>*
Display	D *<expr_1>* ... *<expr_n>*
Stop at	S *<filename>* *<line>* [*<cond>*]
Execute	e *<expr>*

Table A.7: Command Syntax of the Debuggers

<expr> marks an expression, *<filename>* a file name, *<name>* a MuPAD procedure name and *<line>* a non-negative integer. *<cond>* marks a MuPAD expression, whose Boolean evaluation has to return TRUE or FALSE.

Menus on the Macintosh

File Menu

New	opens a new edit window
Open	opens an existing document
Open as Text...	reads-in a session document as a text document
Close	closes the current window. The session window cannot be closed
Save	saves the current window as a document under the current name
Save as...	saves the current window as a document under a new name
Export Session...	saves the session window as a text document
Page Setup...	configurates the printer settings
Print	prints the current window
Preferences	configurates the pre-settings (see *Preferences Menu*)
Quit...	ends MacMuPAD

Edit Menu

Undo	undoes the last change. After a MuPAD calculation Undo is not possible
Cut	cuts out selected a text and saves it in the temporary memeory
Copy	copies a selected text in the temporary memory
Paste	replaces the selected text with the contents of the temporary memeory
Clear	deletes the selected text
Select All	selects the whole document
Set Tabs & Font...	sets the tabulators, font, character size as well as the character style for the current window
Show Clipboard	shows the contents of the temporary memory

Search Menu

Find	calls the dialogue for search and replace
Enter Selection	choses a selected text as a search pattern
Find Again	finds the next occurence of the search pattern
Replace	replaces the selected text with the replace text
Replace & Find again	replaces the selected text with the replace text and searches for the next occurence of the search pattern
Replace all	replaces all occurences of the search text in the text with the replace text
Goto Line	jumps to a given line

Session Menu

Evaluate	evaluates the current input field
Evaluate following	evaluates the current and all the following input fields
Evaluate all	evaluates all input fields
New Input	opens a new input field
New Text	opens a new text field
Change to Input	changes the type of the current field into *new input*
Change to Text	changes the type of the current field into *text*
Interrupt	interrupts the current MuPAD calculation
Separate by Lines	turns the input of speration lines between results on or off
Replace old Results	turns the replacement of old output by new on or off
Pretty Print	turns the Pretty-Printer on or off
Set Textwidth	sets the text width of the MuPAD output
Status	shows the existing memory

Help Menu

Open Manual	opens the selected manual
Next Page	jumps to the next page
Previous Page	jumps to the previous page
First Page	jumps to page 1
Last Page	jumps to the last page
Go to Page	jumps to a given page
Go Back	jumps to the previously chosen page
Other Input	jumps to the corresponding cross reference

Debug Menu

Cont	starts or continues execution
Next	Executes the next statement
Step	Like "Next" without displaying function calls
Up	Goes up one level in the call-chain
Down	Goes down one level in the call-chain
Go to Proc	Lets the user specify a MuPAD function to be displayed
Print	Prints MuPAD expressions
Display	Displays an expression every step
Undisplay	Stops displayin an expression
Execute	Lets the user execute any MuPAD command
Clear all breakpoints	Clears all breakpoints in all source files
Arrange windows	Arranges the debugger windows

Windows Menu

Session	selects the session window
Other Entries	selects the corresponding edit window

Preferences Menu

MuPAD	sets the global pre-settings for the kernel (see pre-settings)
Colors	sets the color pre-setting for the session window
Session	sets other pre-settings for the session window
Debugger	sets the pre-settings for the MuPAD debugger

Appendix B

PARI

The basic arithmetic of the MuPAD system is based on the *PARI* software package. The integration of the package was motivated by the following charateristics of PARI which are identical to the design aims of MuPAD:

- High processing speed

- Extensive software library

- Portable software package

- Data types are transparent for the user (internal typification)

MuPAD contains the whole "basic kernel" and part of the "generic kernel" of PARI. This enables arithmetic of integers, rational numbers, floating point numbers and complex numbers as well as the basic mathematical operations addition, subtraction, division and exponentation. Furthermore, MuPAD uses the algorithms for the calculation of the transcendental functions `abs`, `exp`, `ln`, `sqrt`, the trigonometrical functions and their inverse functions `sin`, `cos`, `tan`, `asin`, `acos`, `atan`, the area functions and their inverse functions `sinh`, `cosh`, `tanh`, `asinh`, `acosh`, `atanh`, the arithmetical functions `eint`, `erfc`, `fact`, `gamma`, `ifactor`, `igamma`, `igcd`, `igcdex`, `ilcm`, `isprime`, `nextprime`, `psi`, `zeta` as well as the functions `ceil`, `floor`, `frac`, `round`, `trunc` for type conversion.

At this point we would like to draw your attention to the fact that PARI has far more basic types than those used in MuPAD. PARI offers, among others, data types and standard operations over finite fields, algebraic number fields, polynomial rings and formal power series. MuPAD only uses a minimal set of the functions that the PARI library makes available. A large part of the

PARI library is concerned with the area of number theory and linear algebra.

Literature about PARI:

1. N.-P. Skoruppa, The PARI-GP package, mathPAD Journal, Vol. 1, No. 3, September 1991.

2. C. Batut, D. Bernardi, H. Cohen, M. Olivier, User's Guide to PARI-GP.

Address:
Prof. Henri Cohen
UFR de Mathematiques et Informatique
Universite Bordeaux I
351 Cours de la Liberation
33405 TALANCE CEDEX, France

E-mail:
pari@alioth.greco-prog.fr
pari@ceremab.u-bordeaux.fr

Appendix C

Netpbm

Netpbm is a package for converting graphics into different formats. Netpbm has been ported to many architectures and was tested for UNIX, VMS and Amiga OS. Netpbm is based on the package Pbmplus, which was improved and to which new converting routines has been added.

Extended Portable Bitmap Toolkit

Distribution of 10dec91

Previous distribution 30oct91

Copyright ©1989, 1991 by Jef Poskanzer.

E-mail addresses:

Jef Poskanzer	:	jef@well.sf.ca.us
Netpbm	:	netpbm@fysik4.kth.se
		oliver@fysik4.kth.se

Appendix D

MuPAD-Syntax in BNF

```
<start>            :    <interact_seq> .

<interact_seq>     :    <interact_expr>
                   |    <interact_seq> <interact_expr> .

<interact_expr>    :    <stmt> <sep>
                   |    help
                   |    system
                   |    quit.

<sep>              :    :
                   |    ; .

<stmt>             :    <assignment>
                   |    <if_stmt>
                   |    <case_stmt>
                   |    <for_stmt>
                   |    <while_stmt>
                   |    <repeat_stmt>
                   |    <seqbegin_stmt>
                   |    <parbegin_stmt>
                   |    next
                   |    break
                   |    quit
                   |    <expr>
                   |    .

<stmt_seq>         :    <stmt>
```

```
                        |   <stmt_seq> <sep> <stmt> .

<if_stmt>          :    if <expr> then <stmt_seq> <else_part>
                        end_if
                   |    if <expr> then <stmt_seq> end_if .

<else_part>        :    else <stmt_seq>
                   |    elif <expr> then <stmt_seq> <else_part>
                   |    elif <expr> then <stmt_seq> .

<case_stmt>        :    case <expr> <of_part> <other_part>
                        end_case .

<of_part>          :    of <expr> do <stmt_seq> <of_part>
                   |    of <expr> do <stmt_seq> .

<other_part>       :    otherwise <stmt_seq>
                   |    .

<for_stmt>         :    for <ident> from <expr> to <expr>
                        <step_part> do <stmt_seq> end_for
                   |    for <ident> from <expr> downto <expr>
                        <step_part> do <stmt_seq> end_for
                   |    for <ident> from <expr> to <expr>
                        <step_part>    parallel    <private>
                        <stmt_seq> end_for
                   |    for <ident> in <expr> do <stmt_seq>
                        end_for
                   |    for <ident> in <expr> parallel
                        <private>
                        <stmt_seq> end_for .

<step_part>        :    step <expr>
                   |    .

<private>          :    private <ident_seq> ;
                   |    .

<while_stmt>       :    while <expr> do <stmt_seq> end_while .

<repeat_stmt>      :    repeat <stmt_seq> until <expr>
                        end_repeat .

<seqbegin_stmt>    :    seqbegin <stmt_seq> end_seq .

<parbegin_stmt>    :    parbegin <private> <stmt_seq> end_par .

<assignment>       :    <name_lhs> := <expr> .
```

```
<expr>              :      <single_expr>
                    |      <expr> , <single_expr> .

<single_expr>       :      <and_expr>
                    |      <single_expr> or <and_expr> .

<and_expr>          :      <seq_expr>
                    |      <and_expr> and <seq_expr> .

<seq_expr>          :      <relation>
                    |      $ <relation>
                    |      <relation> $ <relation> .

<relation>          :      <range>
                    |      <relation> <rel_op> <range> .

<rel_op>            :      < | > | <= | >= | = | <> .

<range>             :      <union_expr>
                    |      <union_expr> .. <union_expr> .

<union_expr>        :      <minus_expr>
                    |      <union_expr> union <minus_expr> .

<minus_expr>        :      <intersect_expr>
                    |      <minus_expr> minus <intersect_expr> .

<intersect_expr>    :      <md_expr>
                    |      <intersect_expr> intersect <md_expr> .

<md_expr>           :      <math_expr>
                    |      <md_expr> <md_op> <math_expr> .

<md_op>             :      mod | div .

<math_expr>         :      <sign_term>
                    |      <math_expr> <add_op> <term> .

<add_op>            :      + | - .

<sign_term>         :      <add_op> <term>
                    |      <term> .

<term>              :      <power>
                    |      <term> <mult_op> <power> .

<mult_op>           :      * | / .

<power>             :      <new_op>
                    |      <power> ^ <new_op> .
```

```
<new_op>        :    <neg_factor>
                |    <new_op> & <ident> <neg_factor> .

<neg_factor>    :    <at_sign_expr>
                |    not <at_sign_expr> .

<at_sign_expr>  :    <point_expr>
                |    <at_sign_expr> @ <point_expr> .

<point_expr>    :    <factor>
                |    <point_expr> . <factor>
                |    <point_expr> <dot_float> .

<factor>        :    NIL
                |    TRUE
                |    FALSE
                |    <set>
                |    <string>
                |    <list>
                |    <number>
                |    ( <stmt_seq> ) <name>
                |    <proc_def> <name>
                |    <ident> <name>
                |    <last> <name> .

<name>          :    :: <ident> <index_call>
                |    <index_call> .

<index_call>    :    <index_call> ( <expr_seq> )
                |    <index_call> [ <expr_seq> ]
                |    .

<name_lhs>      :    <ident> :: <ident> <index_call>
                |    <ident> <concat_ident> <index_call>
                |    ( <expr> ) <concat_ident> <index_call>
                |    <last> <index_call> .

<concat_ident>  :    . <factor> <concat_ident>
                |    .

<number>        :    <int>
                |    <float>
                |    <dot_float>
                |    I .

<list>          :    [ <expr_seq> ] .
```

```
<set>              :     { <expr_seq> } .

<expr_seq>         :     <expr_seq> , <single_expr>
                   |    <single_expr> .

<proc_def>         :     proc ( <ident_seq> ) <decl> <proc_body>
                         .

<decl>             :     <decl> <local_part>
                   |    <decl> <option_part>
                   |    <decl> <name_part>
                   |    .

<local_part>       :     local <ident_seq> ; .

<option_part>      :     option <ident_seq> ; .

<name_part>        :     name <expr> ;
                   |    name ; .

<proc_body>        :     begin <stmt_seq> end_proc .

<ident_seq>        :     <id_seq>
                   |    .

<id_seq>           :     <ident> , <id_seq>
                   |    <ident> .
```

Scanner Symbols

The following notation is according to the Lex-notation for regular expressions:

last	:	(%[1–9][0–9]*)|%
int	:	[0–9]+
float	:	[0–9]+(\\.[0–9]+)?(e(\\+|\\-)?[0–9]+)?
dot_float	:	\\.[0–9]+(e(\\+|\\-)?[0–9]+)?
ident	:	[a–zA–Z_][a–zA–Z_0–9]* (max. 512 characters)
spec_ident	:	These identifiers include all keywords, that is the keywords are recognized and treated as usual identifiers.
string	:	Any sequence of characters enclosed in quotation marks (max. $2^{31} - 1$ characters)
system	:	! followed by any sequence of characters, which are read until the end of line.
help	:	? followed by an optional identifier and an optional colon or semicolon. The input is read unitil the end of line. Keywords are interpreted as identifiers.

Appendix E

Generating Commands of the Color Plots

In this section we have listed the `plot2d` and `plot3d` commands belonging to the graphics shown in the color section.

Picture 1 *The command*

```
>> plot2d(Axes = None,
          [Mode = Curve,
               [sin(u)*sin(2*u)*sin(3*u),
                1/2*sin(4*u)*sin(5*u)*sin(6*u)],
               u = [0, 2*PI], Grid = [300],
               Style = [LinesPoints],
               Color = [Height]]);
```

creates the first color image, which shows a curve parametrized by use of sin(u)*sin(2*u)*sin(3*u) *and* 1/2*sin(4*u)*sin(5*u)*sin(6*u). *In order to draw both lines and points the style* LinesPoints *has been chosen.*

Picture 2 *The second graphics example on the color pages is drawn with the command*

```
>> fexp := funcattr(exp, "float"):

   kdv_two := proc(x, t, c1, c2)
   local    e1, e2, e12, f, fx, fxx;
   begin
```

```
   e1     := fexp(c1*x - c1^3*t):
   e2     := fexp(c2*x - c2^3*t):
   e12    := ((c1-c2)/(c1+c2)^2)*e1*e2:

   f      := 1 +       e1 +       e2 +               e12:
   fx     :=          c1*e1 +   c2*e2 +   (c1+c2)*e12:
   fxx    :=        c1^2*e1 + c2^2*e2 + (c1+c2)^2*e12:

 2*(fxx/f - (fx/f)^2):
end_proc:


plot2d(Axes = None,
       [Mode = Curve,
            [u+i/2, 20*hold(kdv_two)(u,i,1.0,0.6)+i/2],
            u = [-20, 20], Grid = [100],
            Color = [Height, [0, 0, 1], [0, 1, 1]]
       ] $ i = -10..10 );
```

At first a procedure named **kdv_two** *is implemented, which describes the so-called two-soliton solution of the well-known Korteweg-de Vries equation. This procedure then is used in the plot command in order to depict different t-slices of that solution. This can be achieved by using the sequence operator $ in the plot command, where each* **i** *is replaced by the corresponding value. To be more precise, the statement*

```
              [Mode = Curve, ...]  $ i = -10..10
```

creates 21 different objects.

Picture 3 *This example shows a two-dimensional scene containing four objects, each depicted with a different plot style. In addition each object has a title, whose position is specified by use of the option* **TitlePosition**. *Furthermore the axes are marked by tickmarks. The corresponding command is:*

```
>> plot2d(Axes = Origin, Labels = ["", ""],
         Labeling = TRUE, AxesOrigin = [0.000000, 0.000000],
         [Mode = Curve,
             [hold(u), hold(cos(u)*2)],
             u = [0.0, PI/2], Grid = [15],
             Title = "Points", TitlePosition = [2.97, 3.02],
             Color = [Height, [1, 1, 0], [1, 0, 0]],
             Style = [Points]],
         [Mode = Curve,
```

```
          [hold(u), hold(cos(u)*2)],
          u = [PI/2, PI], Grid = [15],
          Title = "Lines", TitlePosition = [4.15, 6.21],
          Color = [Height, [0, 1, 0], [1, 1, 0]]],
      [Mode = Curve,
          [hold(u), hold(cos(u)*2)],
          u = [PI, 3/2*PI], Grid = [15],
          Title = "LinesPoints",
          TitlePosition = [5.89, 6.81],
          Color = [Height, [0, 1, 1], [0, 0, 1]],
          Style = [LinesPoints]],
      [Mode = Curve,
          [hold(u), hold(cos(u)*2)],
          u = [3/2*PI, 2*PI], Grid = [15],
          Title = "Impulses", TitlePosition = [8.0, 4.0],
          Color = [Height, [0, 0, 1],[1, 0, 0]],
          Style = [Impulses]]);
```

Picture 4 *The commands*

```
>> loadlib("plotlib"):
   export(plotlib):

   xrotate(Axes = Box, Ticks = 0,
               [[u, 2*sin(u)], u = [0, 2*PI],
                angle = [0, 2*PI], Grid = [30, 30],
                Color = [Height, [0,0,1], [1,0,1]],
                Style = [ColorPatches, AndMesh]]);
```

are used to generate the fourth example. This shows the surface of revolution around the x-axis of the curve depicted in the following picture. The corresponding routine xrotate *is implemented in the library* plotlib, *which therefore has to be loaded by use of the command* loadlib("plotlib"). *The necessary arguments of this routine are:*

- *The parametrization of the curve to be rotated. In the example above the curve is parametrized through* [u, 2*sin(u)].

- *A range, in which the independent curve variable is evaluated,* u = [0, 2*PI] *in the example above.*

- *A range for the rotation angle,* angle = [0, 2*PI].

All other arguments listed in the plot command above like Axes, Ticks, Color *and* Style *are optional.*

Picture 5 *With the command*

```
>> plot2d(Axes = Corner, Ticks = 16, Labels = ["", ""],
          Labeling = TRUE, Arrows = TRUE,
          [Mode = Curve,
                [u, 2*sin(u)], u = [0, 2*PI],
                Grid = [100], Color = [Height]]);
```

the curve is depicted, which is rotated around the x- and the y-axis in pictures 4 and 6, respectively. In addition the axes are labeled by tickmarks and arrows, which can be achieved by use of the options Ticks = 16 *and* Arrows = TRUE.

Picture 6 *The commands*

```
>> loadlib("plotlib"):
   export(plotlib):

   yrotate(Axes = Box, Ticks = 0,
                [[u, 2*sin(u)], u = [0, 2*PI],
                 angle = [0, 3/2*PI], Grid = [30, 30],
                 Color = [Height, [1,0.3,0], [0,0,1]],
                 Style = [ColorPatches, AndMesh]]);
```

are used to rotate the curve depicted in picture 5 around the y-axis. For this purpose the routine yrotate *implemented in the library* plotlib *is used. Therefore this library has to be loaded by calling* loadlib("plotlib").

The following three examples are created by use of two other packages implemented as libraries in the MuPAD programming language: the libraries **Turtle** and **Lsys**, which stand for Turtle graphics and Lindenmayer systems (L-systems), respectively. L-systems are formal systems, which can be used to model the growth of plants. Furthermore a variety of so-called self-similar curves can be described by use of L-systems. In general L-systems are composed out of a finite alphabet, an axiom and a set of rules, which are used to replace the entries of the axiom. The words, which are produced by L-systems are often interpreted as curves. Therefore Turtle graphics is used. A turtle is a creature, which can turn to the left or to the right, which can move forward and change its color. The length of this movement as well as the angle of rotation is thereby fixed. The state of the turtle is described by its position, direction and color. In order to represent L-systems graphically, each entry of the alphabet is to be replaced by one the commands **move**, **left**, **right** and **color**.

Picture 7 *The commands*

```
>> loadlib("Lsys"):

    l1 := Lsys(90, "L",
                   "L" = "L+R+",
                   "R" = "-L-R",
                   "L" = line, "R" = line):

    Lsys::plot(l1, 11):
```

are used to generate the so-called dragon curve by use of Lindenmayer systems. By the call of `loadlib("Lsys")` *the above mentioned libraries* Lsys *and* Turtle *are loaded. The next command is used to create a L-system. The arguments are:*

- *The rotation angle for the turtle.*

- *The axiom, i.e. the word to start with.*

- *The list of rules, by which the entries of the axiom are replaced in each step.*

Afterwards the command `Lsys::plot(l1, 11)` *serves for the graphical representation of the eleventh generation of this L-system. In addition the colors were changed interactively.*

Picture 8 *The second example of L-systems is created with the call*

```
>> loadlib("Lsys"):

    l2 := Lsys(90, "F-F-F-F", "F" = "F-F+F+FF-F-F+F"):

    Lsys::plot(l2, 4):
```

It is to be noted, that in this example no rule of the form `"F" = line` *is given. Such rules can be left out, if these are unique. The command* `Lsys::plot(l2, 4)` *then serves for generating and plotting the fourth generation of the corresponding L-system. Again the colors were changed interactively.*

Picture 9 *These examples are created with the commands*

```
>> loadlib("Lsys"):

   p1 := Lsys(25.7, "L",
                   "L" = "R[+L][-L]RL",
                   "R" = "RR",
                   "L" = line, "R" = line):

   Lsys::plot(p1,7):
```

and

```
>> loadlib("Lsys"):

   p2 := Lsys(22.5, "L",
                   "L" = "LL-[-L+L+L]+[+L-L-L]",
                   "L" = line, "R" = line):

   Lsys::plot(p2, 5):
```

In contrast to the examples above, which consist only of a single path, these examples consist of multiple paths, which can be achieved by use of the commands push [and pop]. By use of these commands branching structures can be created, which are typical for plants. In addition the colors were changed interactively.

Picture 10 *A further library routine implemented in the library* plotlib *is* implicitplot, *which serves for generating implicit plots of functions depending on two variables.*

```
>> loadlib("plotlib"):
   export(plotlib):

   f := fun(args(1)^6+args(2)^6 + 3*args(1)^4*args(2)^2 +
            3*args(1)^2*args(2)^4 - args(1)^4-args(2)^4 +
            2*args(1)^2*args(2)^2):

   implicitplot(f, -1..1, -1..1, 7):
```

Again the corresponding library first is loaded. Then the function, which is to be plotted, is defined. This function has to be a pure function. Afterwards the routine `implicitplot` *is called. The arguments are the above pure function, the ranges for the independent variables, as well as an integer, which specifies the maximum number of interval bisections.*

Picture 11 *There are further routines implemented in the library* `plotlib`, *one of them is the routine* `polarplot`, *which serves for creating graphics in polar co-ordinates, as can be seen in this example. The corresponding commands are:*

```
>> loadlib("plotlib"):
   export(plotlib):

   polarplot(Axes = Origin, Ticks = 0,
                [[sin(3*u), cos(7*u)],
                 u = [-PI, PI], Grid = [250],
                 Color = [Height],
                 Style = [LinesPoints]]);
```

Picture 12 *The graphics shown in this example is created with the commands*

```
>> loadlib("plotlib"):
   export(plotlib):

   fieldplot(Axes = Origin,
                [[1, sin(y+cos(x))],
                 x = [-PI, PI], y = [-PI, PI],
                 Grid = [30, 30],
                 Color = [Height, [0, 0, 1], [1, 1, 0]]]);
```

In this example the routine `plotlib::fieldplot` *is used to depict a field described by the expression* `sin(y+cos(x))` *in the rectangular area defined by* $[-\pi, \pi] \times [-\pi, \pi]$.

The next two examples are used to demonstrate the possibilities of user-defined color functions.

Picture 13 *The first example for user-defined color functions models a simple light model. We define four light sources by their position and their color. For each sample point the distances from the different light sources are calculated.*

The color of the current sample point is computed, by taking into account, that the intensity of the light sources is diminished reciprocal to the square of the current distance.

```
>> #
    |       Definition of the light sources:
    |           A light source is defined by a list containing
    |           the position [x, y, z] and the color in RGB-
    |           specification.
    #
    LIGHT[1]  := [[ 1.0,-1.0,-1.0], [1, 0, 0]]:
    LIGHT[2]  := [[-1.0,-1.0,-1.0], [0, 0, 1]]:
    LIGHT[3]  := [[ 1.0, 1.0,-1.0], [1, 1, 0]]:
    LIGHT[4]  := [[-1.0, 1.0,-1.0], [0, 1, 0]]:


    #
    |       distance()
    |
    |           Procedure to calculate the distance between the
    |           sample point x, y, z and the light_coords.
    #
    distance := proc(x, y, z, light_coords)
    begin
        sqrt((x-op(light_coords, 1))^2 +
             (y-op(light_coords, 2))^2 +
             (z-op(light_coords, 3))^2):
    end_proc:


    #
    |       Color function
    #
    light_model := proc(x, y, z, u, v)
    local    i, j, dist;
    begin
        #
        | for every light source do
        #
        for i from 1 to nops(LIGHT) do
            #
            | Calculate the distance between the current point
            | and the light source i.
```

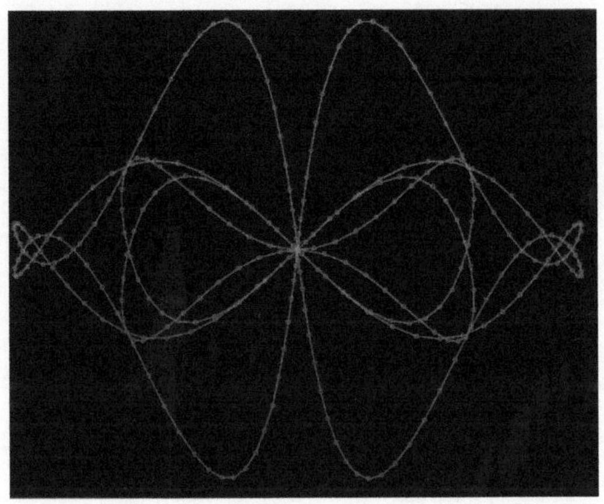

Picture 1

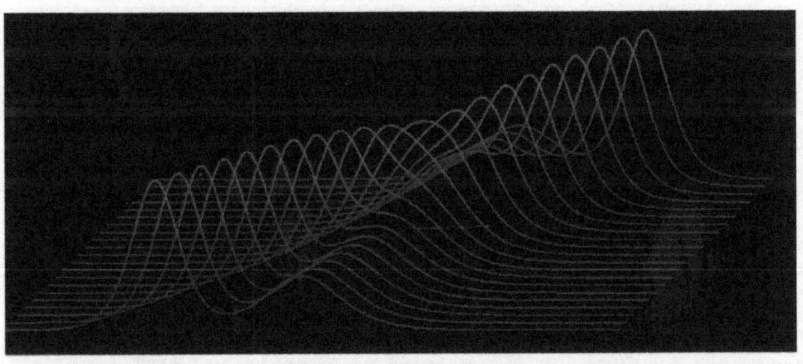

Picture 2

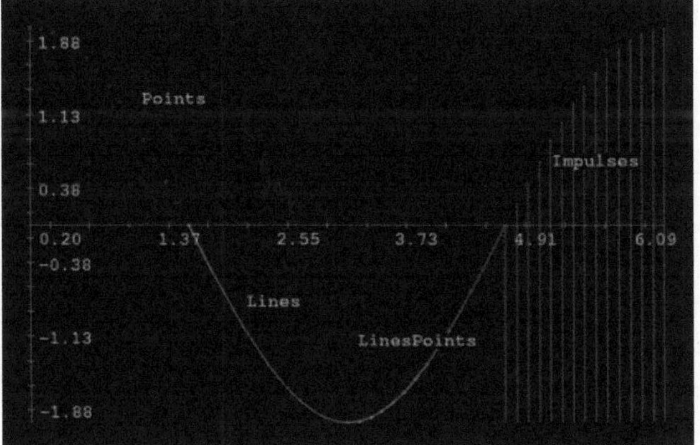

Picture 3

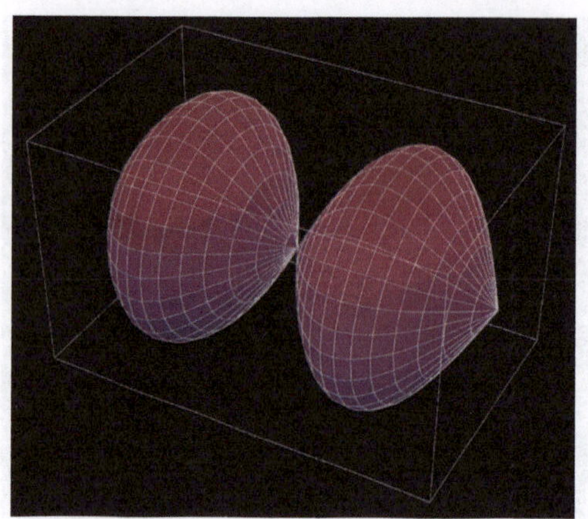

Picture 4

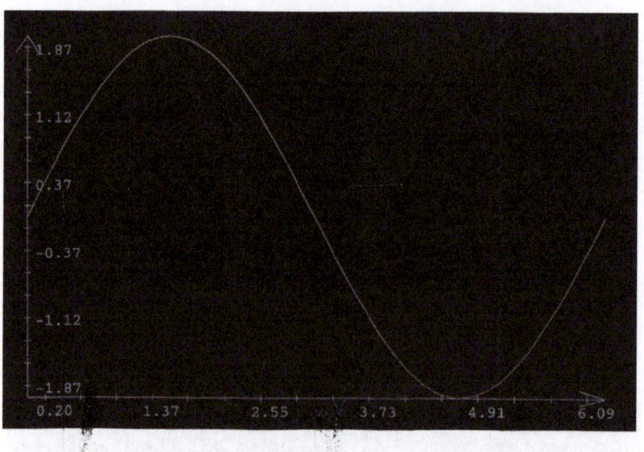

Picture 5

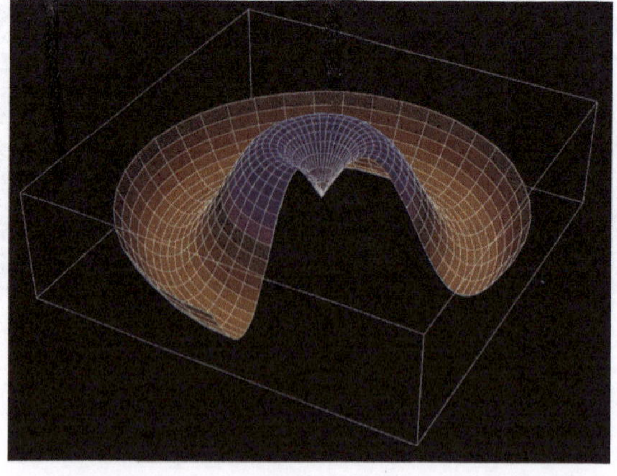

Picture 6

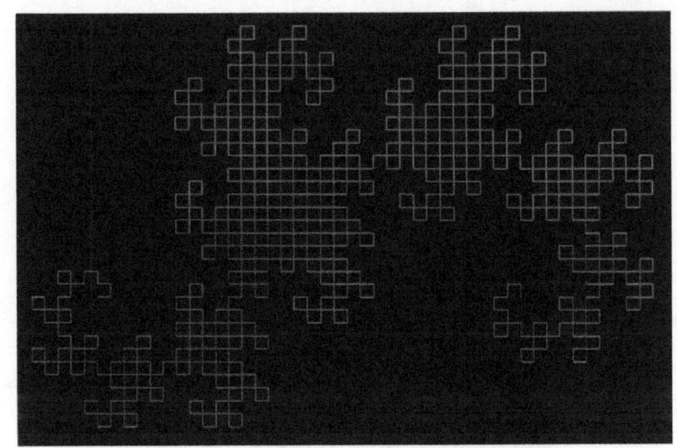

Picture 7

Picture 8

Picture 9

Picture 10

Picture 11

Picture 12

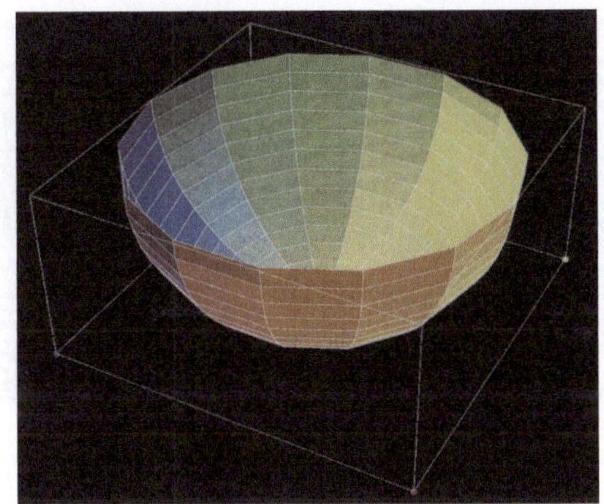

Picture 13

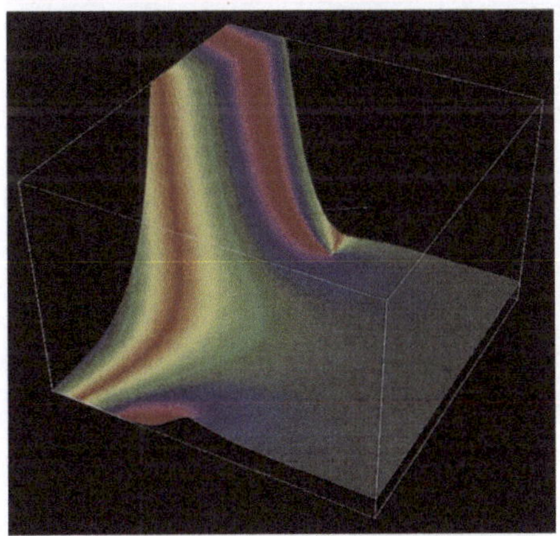

Picture 14

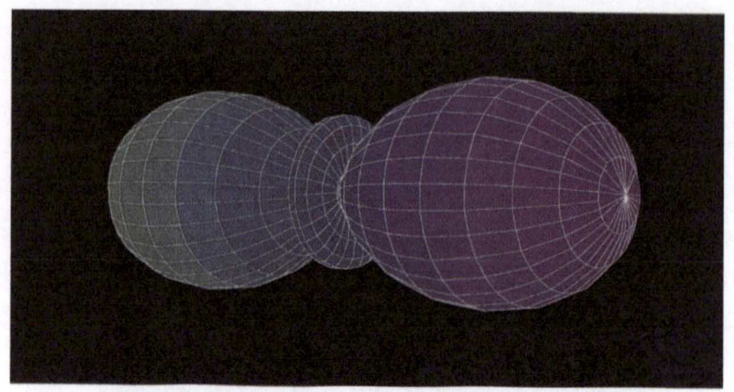

Picture 15

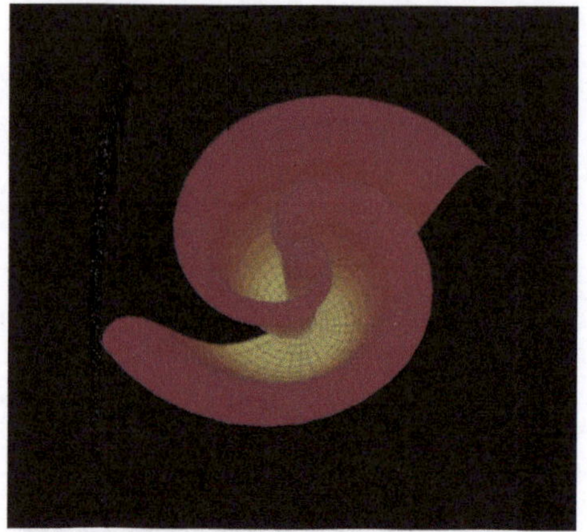

Picture 16

Picture 17

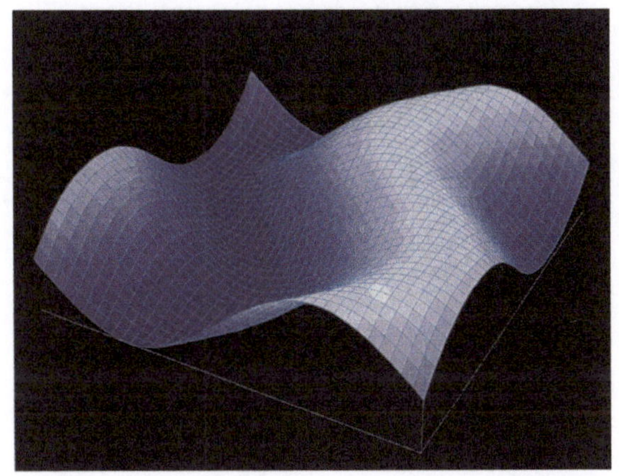

Picture 18

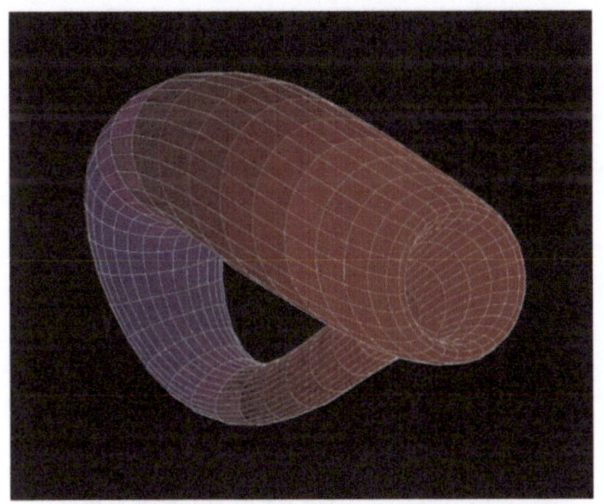

Picture 19

Picture 20

Picture 21

Picture 22

```
            #
            dist[i] := distance(x, y, z, op(LIGHT[i], 1)):
            if dist[i] <= 1 then
                dist[i] := 1.0:
            else
                dist[i] := 1.0/dist[i]^2:
            end_if:
        end_for:
        red   := _plus(dist[j]*op(LIGHT[j],[2,1])
                    $ j = 1..nops(LIGHT)):
        green := _plus(dist[j]*op(LIGHT[j],[2,2])
                    $ j = 1..nops(LIGHT)):
        blue  := _plus(dist[j]*op(LIGHT[j],[2,3])
                    $ j = 1..nops(LIGHT)):
        if red   > 1 then red   := 1.0 end_if:
        if green > 1 then green := 1.0 end_if:
        if blue  > 1 then blue  := 1.0 end_if:
        [red, green, blue]:
end_proc:


#
|    Plot Command
#
plot3d(Axes = Box, Ticks = 0,
        PointWidth = 10,
        PointStyle = FilledCircles,
        [Mode = Surface,
                [sin(u)*cos(v), sin(u)*sin(v), cos(u)],
                u = [PI/2, PI], v = [-PI, PI],
                Color = [Function, light_model],
                Style = [ColorPatches, AndMesh],
                Grid = [15, 15]],
        [Mode = List,
                [(point(op(op(LIGHT[i],1)),
                Color = op(LIGHT[i], 2))$ i=1..nops(LIGHT))]]);
```

In addition the different light sources are displayed in the picture as colored circles.

Picture 14 *The graphics depicted in this figure represents a surface plot of the modulus of the Riemann zeta-function in the complex plane. The colors*

*projected onto the surface indicate the complex phase of each sample point,
since every complex number z can be written as z = abs(z)*exp(I*phi).*

```
>> angle := proc(x, y, z, u, v)
   local    erg, real, imag, phi, value;
   begin
       erg  := zeta(u+I*v):
       real := op(erg,1):
       imag := op(erg,2):
       if abs(real) > EPS then
           phi := atan(imag/abs(real)):
       else
           phi := sign(imag)*float(PI/2):
       end_if:
       value := (phi-MINPHI)/(MAXPHI-MINPHI):
       if value > 4/5 then
          erg := [1, 5*(1-value), 0]:
       elif value > 3/5 then
          erg := [1-5*(4/5-value), 1, 0]:
       elif value > 2/5 then
          erg := [0, 1-5*(3/5-value), 5*(3/5-value)]:
       elif value > 1/5 then
          erg := [5*(2/5-value), 0, 1]:
       else
          erg := [1, 0, 1-5*(1/5-value)]
       end_if:
   end_proc:

   MAXPHI := float(PI/2):
   MINPHI := -MAXPHI:
   EPS    := 10^(-DIGITS):

   plot3d(Axes = None,
          [Mode = Surface,
              [u, v, min(abs(zeta(u+I*v)),10)],
              u = [-5, 9], v = [3/4, 15],
              Color = [Function, angle],
              Grid = [20, 20],
              Style = [ColorPatches, AndMesh]]);
```

The original figure shown in the color section was computed by use of the options Grid = [50, 50] *and* Smoothness = [1, 1], *whereas in the example command* Grid *is set to* Grid = [20, 20] *to shorten the time necessary for calculation.*

Picture 15 *The next example shows the application of the library routine* sphericalplot, *which serves for generating surface plots in spherical co-ordinates.*

```
>> loadlib("plotlib"):
   export(plotlib):

   sphericalplot(Axes = None,
                 CameraPoint = [6, -3, 6],
                    [[(5*cos(v)^2-1)/2, u, v],
                     u = [-PI, PI], v = [0, PI],
                     Grid = [30, 30],
                     Color = [Height,[0,1,0.5],[0.5,0,1]],
                     Style = [ColorPatches, AndMesh]]);
```

The necessary arguments of sphericalplot *are three expressions depending on two variables and the ranges for these variables. The three expressions are used to describe the radius as well as the horizontal and vertical angle of the sample points. As can be seen in picture 17 by use of* sphericalplot *multiple objects can be generated.*

Picture 16 *In addition to the already mentioned routines* sphericalplot *and* polarplot *there exists the routine* cylindricalplot *in the* plotlib *library, which can be used to depict surface plots in cylindrical co-ordinates.*

```
>> loadlib("plotlib"):
   export(plotlib):

   cylindricalplot(Axes = None,
                   CameraPoint = [16, -8, 41],
                   ForeGround = [0.5,0.5,0.5],
                       [[z*u, 2*u, -3*cos(z^2)],
                        u = [0, PI], z = [-2, 2],
                        Grid = [35, 40],
```

```
            Style = [ColorPatches, AndMesh],
            Color = [Height, [1,1,0], [1,0,1]]]);
```

Furthermore by use of the option ForeGround = [0.5, 0.5, 0.5] *the foreground color of the image is specified.*

Picture 17 *In order to show some of the different plot styles available for surfaces the next example has been chosen. Furthermore it shows, how multiple objects can be created in a call of* sphericalplot.

```
>> loadlib("plotlib"):
   export(plotlib):

   sphericalplot(Axes = None, Ticks = 0, LineWidth = 2,
                 [[1, u, v],
                   u = [-PI, PI], v = [-0, PI],
                   Grid = [15, 15],
                   Color = [Height, [1,1,0],[1,0,1]],
                   Style = [HiddenLine, Mesh]],
                 [[2, u, v],
                   u = [0, PI], v = [0, PI],
                   Grid = [15, 20],
                   Color = [Height, [1,0,1],[0,0,1]],
                   Style = [ColorPatches, AndMesh]],
                 [[3, u, v],
                   u = [-PI, PI], v = [PI/2, PI],
                   Grid = [25, 20],
                   Color = [Height, [0,0,1],[0,0.7,1]],
                   Style = [WireFrame, Mesh]]);
```

Picture 18 *This example is created by use of the following command.*

```
>> plot3d(Axes = Corner, Ticks = 0,
          ForeGround = [0, 0.82, 1],
          [Mode = Surface,
             [u, v, sin(u + sin(v))],
             u = [-PI, PI], v = [-PI, PI],
             Color = [Height, [0,0,1], [1,1,1]],
             Style = [ColorPatches, AndMesh]]);
```

The graphics shows the surface plot of `sin(u + sin(v))`. *In addition the axes style is chosen to* `Axes = Corner` *and the foreground color is set by* `ForeGround = [0, 0.82, 1]` *in the plot command above.*

Picture 19 *The next example depicts the so-called Klein bottle, a surface which like the Moebius strip does not have an outer and an inner side.*

```
>> klein := proc()
   local bx, by, pi, rad, u, v, x, y, z;
   begin

   bx := proc(u) begin 6*cos(u)*(1+sin(u)) end_proc;

   by := proc(u) begin 16*sin(u) end_proc;

   rad := proc(u) begin 4-2*cos(u) end_proc;

   x := proc(u, v)
     begin
       if pi < u and u <= 2*pi then
          bx(u)+rad(u)*cos(v+pi)
       else
          bx(u)+rad(u)*cos(u)*cos(v)
       end_if
   end_proc;

   z := proc(u, v)
     begin
       if pi < u and u <= 2*pi then
          by(u)
       else
          by(u)+rad(u)*sin(u)*cos(v)
       end_if
   end_proc;

   y := proc(u, v) begin rad(u)*sin(v) end_proc;

   pi := float(PI);
```

```
    plot3d(Axes = None,
           CameraPoint = [-16, -64, 100],
           [Mode = Surface,
                [hold(x(u, v)), hold(y(u, v)),
                 hold(-z(u, v))],
                u = [0, 2*PI], v = [0, 2*PI],
                Grid = [35, 30],
                Style = [ColorPatches, AndMesh],
                Color = [Height, [0,0,1],[1,0,0]]]);
end_proc():
```

Picture 20 *This example shows the so-called Sierpinski-tetra-hedron, drawn by use of graphical primitives. Starting from a tetra-hedron defined by four three-dimensional points, this tetra-hedron is recursively refined by calculating the midpoints of the different edges and combining these points with the corresponding original ones to four new tetra-hedrons.*

```
>> tetra_eder := proc(p1, p2, p3, p4)
   begin
       polygon(p1, p2, p4, Closed = TRUE, Filled=TRUE),
       polygon(p1, p3, p4, Closed = TRUE, Filled=TRUE),
       polygon(p2, p3, p4, Closed = TRUE, Filled=TRUE)
   end_proc:

   new_point := proc(p1, p2)
   local x, y, z, red, green, blue;
   begin
       #
       | Calculate the point between p1 and p2.
       #
       x      := op(p1,1) + (op(p2,1)-op(p1,1))/2.0:
       y      := op(p1,2) + (op(p2,2)-op(p1,2))/2.0:
       z      := op(p1,3) + (op(p2,3)-op(p1,3))/2.0:
       #
       | Calculate the color between the colors of
       | the points p1 and p2.
       #
       red   := (op(p1,[4,1])+op(p2,[4,1]))/2.0:
       green := (op(p1,[4,2])+op(p2,[4,2]))/2.0:
       blue  := (op(p1,[4,3])+op(p2,[4,3]))/2.0:
```

```
    #
    | Return the new point.
    #
    point(x, y, z, Color = [red, green, blue]):
end_proc:

tetra_rec := proc(p1, p2, p3, p4, n)
local np1, np2, np3, np4, np5, np6;
begin
    if n = 0 then
        tetra_eder(p1, p2, p3, p4):
    else
        np1 := new_point(p1, p2):
        np2 := new_point(p2, p3):
        np3 := new_point(p3, p1):
        np4 := new_point(p1, p4):
        np5 := new_point(p2, p4):
        np6 := new_point(p3, p4):
        tetra_rec(p1, np1, np3, np4, n-1),
        tetra_rec(np1, p2, np2, np5, n-1),
        tetra_rec(np2, p3, np3, np6, n-1),
        tetra_rec(np4, np5, np6, p4, n-1):
    end_if:
end_proc:

XMIN :=  -0.7:
XMAX :=   1.0:
YMIN :=  -0.5*sqrt(3.0):
YMAX :=   0.5*sqrt(3.0):
ZMIN :=   0.0:
ZMAX :=   2*YMAX:

a := point(XMAX, 0, 0,
           Color = [1, 0, 0]):
b := point(XMIN, YMAX, 0,
           Color = [0, 1, 0]):
c := point(XMIN, YMIN, 0,
           Color = [0, 0, 1]):
d := point(XMIN + (XMAX-XMIN)/2.0, 0, ZMAX,
           Color = [1, 1, 0]):
```

```
plot3d(Axes = None,
       CameraPoint = [0.010, -0.001, 10.0],
       BackGround = [0,0,0],
       ForeGround = [1,1,1],
       [Mode = List,
             [tetra_rec(a,b,c,d,3)]]);
```

Furthermore the points of the original tetra-hedron are colored with different colors, i.e. red, green, blue and yellow. For each calculated point the color is computed as the arithmetic-mean of the corresponding colors of the original points, in order to get a smoothly colored object.

Picture 21 *If the following commands*

```
>> sin_cos := proc(n)
   local    x, fsin, fcos;
   begin
     fsin := funcattr(sin, "float"):
     fcos := funcattr(cos, "float"):

     plot2d(Axes=None,
            Title = "Spiral n: ".n,
            [Mode = Curve,
                    [x*fcos(x), x*fsin(x)],
                     x = [0, 500], Grid = [n]
            ]):
   end_proc:

   spirals := proc()
   local    i;
   begin
       for i from 100 to 500 step 10 do
           sin_cos(i)
       end_for
   end_proc:
```

are evaluated, either a series of spirals can be generated by calling

```
>> spirals():
```

or the spiral depicted in the color section can be created by calling

539

```
>> sin_cos(190):
```

Picture 22 *The following commands serve for generating the graphics shown on the titlepage, which consist of two objects. The first object is a sphere, which is colored by use of a color function. The second object consist of some polygons, which are colored with respect to the height.*

```
>> #
   | Read the file, which contains the definition of
   | different polygons.
   #
   read("MuPAD_POLY.mu"):


   #
   | Color function used to color the sphere.
   #
   angle := proc(x, y, z, u, v)
   local    erg, real, imag, phi, value;
   begin
       erg  := 1/((x+y+2*I*z)^3+1):
       real := op(erg,1):
       imag := op(erg,2):
       if abs(real) > EPS then
           phi := atan(imag/abs(real)):
       else
           phi := sign(imag)*float(PI/2):
       end_if:
       value := (phi-MINPHI)/(MAXPHI-MINPHI):
       if value > 4/5 then
          erg := [1, 5*(1-value), 0]:
       elif value > 3/5 then
          erg := [1-5*(4/5-value), 1, 0]:
       elif value > 2/5 then
          erg := [0, 1-5*(3/5-value), 5*(3/5-value)]:
       elif value > 1/5 then
          erg := [5*(2/5-value), 0, 1]:
       else
          erg := [1, 0, 1-5*(1/5-value)]
       end_if:
   end_proc:
```

```
#
| Plot command
#
MAXPHI := float(PI/2):
MINPHI := -MAXPHI:
EPS    := 10^(-DIGITS):

plot3d(Axes = None, Scaling = Constrained,
       BackGround = [0.5, 0.5, 0.5],
       CameraPoint = [10.5, 0.35, 5.1],
       [Mode = Surface,
           [sin(x)*cos(y), sin(x)*sin(y), cos(x)],
           x = [0, PI], y = [-PI, PI],
           Color = [Function, angle],
           Grid = [40, 40], Smoothness = [3, 3],
           Style = [ColorPatches, AndMesh]
       ],
       [Mode = List, [outer_polygons],
           Color = [Height, [0.9, 0, 0], [0.9, 0.9, 0]]],
       [Mode = List, [inner_polygons],
           Color = [Height, [1, 0, 0], [1, 1, 0]]]);
```

By use of the first command the lists inner_polygons *and* outer_polygons *are defined. The procedure* angle *is used to project the complex phase of the map* 1/(z^3+1) *onto the sphere.*

Appendix F

Changes between MuPAD 1.2.1 and 1.2.2

This section describes the changes between versions 1.2.1 and 1.2.2 of MuPAD. Changes in the kernel:

- the operator `$` with one argument now only accepts numerical arguments. For instance `$x..y` now returns an error.

- expressions of the form `(a*b*c)^x` are now expanded only when `x` is an integer.

- lists, sets, tables and arrays, when used as operators, are mapped on the arguments, i.e. `([f,g])(x)` now returns `[f(x),g(x)]`.

- the function `args` now accepts a range too, for example `args(2..3)` returns the sequence of the second and third argument in a procedure.

- the function `divide` now accepts multivariate polynomials with non rational coefficients, for example `divide(a*x*y,b*x*y,[x,y],Exact)` returns `a/b`.

- the function `expr` now accepts also arguments that are not polynomials, and returns them unchanged (instead of an error).

- the function `fopen` now returns `FAIL` instead of an error when the file cannot be opened, and like `finput` and `ftextinput`, it does not consider any more `READ_PATH` and `LIB_PATH`.

- the variable `LIB_PATH` can now contain a sequence of paths.

- the functions `ceil`, `norm`, `round` and `trunc` now automatically do a floating-point evaluation. For example `trunc(PI)` returns 3 and the call `norm(PI*x+1)` returns `3.141592653`.

- `map` and `select` accept a sequence as first argument: `map((1,2,3),f,2)` returns `f(1,2),f(2,2),f(3,2)`, and `select((1,2,3),_less,3)` gives `1,2`.

- the function `read` is now a library function; the former kernel function is called `fread`.

- `select` and `zip` are now kernel functions, thus much faster.

- the function `sort` now evaluates its second argument in a boolean context: `sort([1,2,3],_less)` is now possible (`bool@_less` was previously required).

Changes in existing functions:

- non-evaluated mathematical functions now each have their own type. For example `type(bernoulli(n))` now returns `bernoulli` instead of `function`. Moreover all these functions can now be overloaded. For example if a domain `d` contains a method `exp`, and if `e` is an element of `d`, then `exp(e)` will automatically call this method. In addition, many of these functions have now an attribute `expand`; try for example `expand(sin(x+y))`.

- `atan` now accepts two parameters: `atan(a,b)` computes the complex argument of $a + ib$.

- the functions `conjugate`, `Re` and `Im` were extended to a larger class of expressions; try for example `Im(cos(I)^(2+3*I))`.

- the function `groebner::gbasis` now accepts expressions too, and not only polynomials.

- the function `loadproc` does not need any more the `..mu` suffix, and now first tries to load a binary file. Like `loadlib`, it now uses only the `LIB_PATH` variable (and no more `READ_PATH`).

- `powermod` now accepts a polynomial as third argument: the call of `powermod(f,n,g)` returns $f^n \bmod g$.

- the functions **abs** and **sign** have been extended to expressions containing symbolic variables, and some **abs** and **sign** attributes have been defined for many mathematical functions, for example **sign(exp(2))** now evaluates to 1, and **abs(I*PI)** evaluates to **PI**.

Changes in the **domains** package:

- the functions **_plus** and **_mult** are now overloaded like all other functions (terms of the same domain are no more grouped together like in version 1.2.1). For example, in a sum $a+b+c$, if the first operand whose domain contains a method **_plus** is b, then this method is called with a, b, c as arguments. The main advantage is that it is now possible to use scalars in addition or multiplication of domain elements, for example one can write **k*A** to multiply a matrix A by the scalar k. But on the other side the programmer now has to care about the domain type of the arguments while implementing a method like **_plus** or **_mult**.

- the method **domattr** can now be overloaded by **elemattr** for domain elements.

- the domain **ExpressionField** of the **domains** package, when called without arguments, now uses the system representation for its elements.

Changes in the **linalg** package:

- with the new overloading of **_plus** and **_mult**, the multiplication of a matrix by a scalar can now be written as **k * A**. If **k** cannot be converted into the coefficient domain of **A**, then the multiplication returns **FAIL** (instead of **k*A** in the previous version). This holds for all domains, for example **x*IntegerMod(6)(2)** returns **FAIL**.

- **ExpressionField(id,iszero)** is now the default coefficient domain for matrices, which means that the normalizer for expressions is **id**, and the test for zero is **iszero**.

- the number of rows and columns is no longer needed in the explicit definition of matrices, for example **Matrix()([[1,2],[3,4]])** is now valid.

- there is a new method **exp** to compute the exponential of a matrix (in fact this is not a function of the **linalg** package, but a method of the domains **Matrix** and **SquareMatrix**).

- similarly, the function `linalg::norm` is now a method of the domains **Matrix** and **SquareMatrix**.

- some function names were changed: now

  - **deleteColumn** is **delCol**,
  - **deleteRow** is **delRow**,
  - **gauss** is **gaussJordan**,
  - **getColumn** is **col**,
  - **getRow** is **row**,
  - **setColum** is **setCol**,
  - **orthogonalSystem** is **ogSystem**,
  - **orthonormalSystem** is **onSystem**,
  - **swapColum** is **swapCol**.

- the methods **adjoint** and **det** work now over any commutative ring (not only over integral domains). For example one can compute the determinant of a matrix over integers modulo 6.

Changes in efficiency:

- **factor** now uses Shoup's algorithm for the factorization of univariate polynomials over Z_p. Therefore **factor** is now much faster, especially when the degree of the polynomial and the logarithm of p are near.

Appendix G

New Functionality in MuPAD 1.2.2

This section describes the new functionality of the version 1.2.2 of MuPAD.

New packages:

- `RGB`: there is now a library `RGB`, that enables one to call colors by their name, instead of a list of three reals. For example after `loadlib("RGB")`, the expression `RGB::Red` can be used in `plot2d` as a color. A list of all color names is obtained with `RGB::ColorNames()`.

- `combinat`: combinatorial package. Contains currently only four functions: `permute` returns all permutations of a list, the function `powerset` returns all subsets of a set, `stirling1` and `stirling2` compute Stirling numbers of the first and second kind.

- `ode`: ordinary differential equation package. Currently it exports only the function `ode::solve`, which is in fact called directly by the overloading mechanism of the general solver `solve`. Therefore, to solve an ordinary differential equation, first define it using `ode`, then solve it using `solve`. See `?ode` for more details.

- `plotlib`: plot package. It contains currently `implicitplot` to plot two dimensional implicit functions, `cylindricalplot` for graphics in cylindrical coordinates, `fieldplot` to display vector fields, `polarplot` for graphics in polar coordinates, `sphericalplot` for graphics in spherical coordinates, `vector` to generate an arrow, `xrotate` and `yrotate` for surfaces of revolution around the x- and y-axis.

- **stats**: statistics package. Contains currently **mean** to compute the mean of a list of values, **ChiSquare** to compute the χ^2 distribution, **normal** to compute the normal distribution, **stdev** and **variance** to compute the standard deviation and the variance of a list of values.

- **Network**: Network package. Contains several functions for dealing with directed graphs.

New functions:

- **alias** and **unalias** enable to create and delete aliases.

- **asympt** for (generalized) asymptotic expansions.

- **besselJ** and **besselY** represent the Bessel functions of first and second kind respectively.

- **collect** collects coefficients of a polynomial.

- **combine** combines terms of the same structure (inverse of **expand**).

- **contfrac** computes a continued fraction representation.

- **decompose** computes a decomposition of a polynomial.

- **Factor** interactive interface to **factor**.

- **fread** reads a file without using **READ_PATH** and **LIB_PATH**.

- **gcdex** extended gcd algorithm.

- **int** definite and indefinite integration.

- **ldegree** computes the lowest degree of a polynomial.

- **length** returns the size of an expression.

- **limit** computes the limit of an expression.

- **linalg::cholesky** computes the Cholesky decomposition of a matrix.

- **linalg::curl** computes the "curl" of a vector.

- **linalg::divergence** computes the divergence of a vector.

- **linalg::eigenValues** computes the eigenvalues of a matrix.

- **linalg::eigenVectors** computes the eigenvectors of a matrix.

- `linalg::extractMatrix` extracts a submatrix.

- `linalg::grad` computes the gradient of a function.

- `linalg::isHermitian` tests for Hermitian matrices.

- `linalg::isOrthogonal` tests if a matrix is orthogonal.

- `linalg::isPosDef` tests for positive definite matrices.

- `linalg::jacobian` computes the Jacobian matrix.

- `linalg::jordanForm` computes the Jordan normal form.

- `linalg::normalize` normalizes a vector.

- `linalg::randomMatrix` generates a random matrix.

- `linalg::vectorPotential` computes the vector potential of a vector.

- `linsert` inserts an element in a list.

- `maprat` executes a function on general expressions.

- `numlib::decimal` infinite decimal representation of rationals.

- `numlib::pollard` finds factors of an integer with Pollard's rho method.

- `numlib::proveprime` primality proving using elliptic curves.

- `ode` defines an ordinary differential equation.

- `partfrac` computes a partial fraction decomposition.

- `pdioe` solves polynomial diophantine equations.

- `primpart` returns the primitive part of a polynomial.

- `radsimp` simplifies expressions with radicals.

- `randpoly` generates a random polynomial.

- `rectform` computes the rectangular form of a complex number.

- `series` computes series expansions.

- `simplify` simplifies expressions.

- `solve` solves all kinds of equations.

- **sum** definite and indefinite summation.

- **TeX** converts an expression to TeX-formatted output.

- **Type::ODE** recognizes ordinary differential equations.

- **Type::Series** recognizes series expansions.

- **userinfo** and **setuserinfo** enables the developer/user to print/see information while executing library procedures: **setuserinfo(All,3)** displays information for all library routines up to level 3.

New options to existing functions:

- there is a new term ordering **DegInvLexOrder** for polynomials, which corresponds to total degree ordering with ties broken by inverse lexicographic ordering. This ordering is mainly used for Gröbner bases.

New functionality in the **domains** package:

- in all domains of the domains package, a new method **random** has been added, that generates a random domain element.

- there is also a new method **pivotSize** that is used in Gaussian elimination to find the smallest pivot. This method is initialized to **length** for the domain **ExpressionField**, and for the domains **Polynomial** and **DistributedPolynomial** it is initialized to **degree**. For other domains, the first non zero element is taken as pivot.

- new category constructors: **HomogeneousFiniteProductCat**, **Module**, **RightModule**, **HomogeneousFiniteCollectionCat**.

- new category: **FiniteCollectionCat**.

- new domain constructors: **AlgebraicExtension**, **Product** for product of domains, and **TruncatedPowerSeries**.

- **infinity** represents a real positive infinite value, and **2*infinity-3** is automatically simplified.

- the domain **Interval** for interval arithmetic; after its initialization via **loadlib("domains")**, try **Interval(-3..2)^3**. It can be used as coefficient domain for matrices; try **MI:=SquareMatrix(2,Interval)**, then **A:=MI::random()**, and **A^3**.

Bibliography

[1] Borodin, Fagin, Hopcroft, and Tompa. Decreasing the Nesting Depth of Expressions Involving Square Roots. *Journal of Symbolic Computation*, 1:169–188, 1985.

[2] M. Bronstein. A Unification of Liouvillian Extension. *AAECC - Applicable Algebra in Engineering, Communication and Computing*, 1:5–25, 1990.

[3] M. Bronstein. The Transcendental Risch Differential Equation. *Journal of Symbolic Computation*, 9:49–60, 1990.

[4] Bruce W. Char, Keith O. Geddes, Gaston H. Gonnet, Benton L. Leong, Michael B. Monagan, and Stephen M. Watt. *Maple V: Language Reference Manual.* Springer-Verlag, 1991.

[5] H.I. Eppstein and B.F. Caviness. A Structure Theorem for the Elementary functions and Its Applications to the Identity Problem. *International Journal of Computer and Information Science*, 8:9–37, 1979.

[6] K.O. Geddes, S.R. Czapor, and G. Labahn. *Algorithms for Computer Algebra.* Kluwer Academic Publishers, Boston-Dordrecht-London, 1992.

[7] D. Gruntz. *On Computing Limits in a Symbolic Manipulation System.* Disseration, Swiss Federal Institute of Technology, Zürich, 1995.

[8] Richard D. Jenks and Robert S. Sutor. *AXIOM: The Scientific Computation System.* Springer-Verlag, 1992.

[9] MATHLAB Group. *MACSYMA Reference Manual.* Laboratory for Computer Science, MIT, Cambridge Mass., 1977.

[10] Stephen Wolfram. *Mathematica — A System for Doing Mathematics by Computer.* Addison-Wesley, 2nd edition, 1991.

Index